6th International Conference on Modern Technologies for Non-Destructive Testing 2017

IOP Conference Series: Materials Science and Engineering
Volume 289

Tomsk, Russia
9 – 14 October 2017

ISBN: 978-1-5108-5566-3
ISSN: 1757-8981

Printed from e-media with permission by:

Curran Associates, Inc.
57 Morehouse Lane
Red Hook, NY 12571

Some format issues inherent in the e-media version may also appear in this print version.

Printed by Curran Associates, Inc. (2018)

For permission requests, please contact the Institute of Physics
at the address below.

Institute of Physics
Dirac House, Temple Back
Bristol BS1 6BE UK

Phone: 44 1 17 929 7481
Fax: 44 1 17 920 0979

techtracking@iop.org

Additional copies of this publication are available from:

Curran Associates, Inc.
57 Morehouse Lane
Red Hook, NY 12571 USA
Phone: 845-758-0400
Fax: 845-758-2633
Email: curran@proceedings.com
Web: www.proceedings.com

Table of contents

Volume 289

6th International Conference: Modern Technologies for Non-Destructive Testing

9–14 October 2017, Tomsk, Russian Federation

Accepted papers received: 18 December 2017
Published online: 12 January 2018

Preface

Papers

IOP Conf. Series: Materials Science and Engineering **289** (2017) 011001 doi:10.1088/1757-899X/289/1/011001

1. Abstract

This issue of the journal is dedicated to the scientific and applied research presented at the VI International Conference "Modern Technologies for Non-Destructive Testing". The conference was held on October 9–14, 2017 in Tomsk (Russia).

The conference brought together experts from different countries and organizations who received a great opportunity to share knowledge in sessions, to initiate discussions on topics of interest to all participants.

The aim of the Conference was to support young talented people, to involve pupils, students, post-graduate students and young scientists in the solution of actual scientific problems of non-destructive testing, quality management and new educational technologies.

The main goals of the event were exchange of information and discussion of topics related to non-destructive testing, quality management, new educational technologies and widespread implementation of the obtained results.

The key topics of the conference were advanced technologies in non-destructive testing, effective quality management systems and educational technologies in the field of control and management.

The Conference Organizer is The Department of Physical Methods of Non-Destructive Testing of the Institute of Non-Destructive Testing of the National Research Tomsk Polytechnic University.

2. Committee:

Aleksandr S. Sigov - Dr., Prof., academician, president of Moscow State Institute of Radio-Engineering Electronics and Automation , general chair, Moscow, Russia
Eduard S. Gorkunov - Dr., Prof., academician, director of Institute of Engineering Science, UB RAN, Yekaterinburg, Russia
Anatoliy P. Surzikov - Dr., Prof., Head of Department of Physical methods of Non-destructive testing Tomsk Polytechnic University, vice-chairman, Tomsk Russia
Boris I. Kapranov - Dr., Prof., Sector leader Technical Tomography and Introscopy Tomsk Polytechnic university, vice-chairman, Tomsk, Russia
Eduard Siemens - Dr.-Ing., Prof., Anhalt University of Applied Sciences, Köthen, Germany.
V.A. Pomogaev - researcher of Radiation Lab, postdoc of Chemical and Biomedical engineering department, Notherdam University, Indiana, USA
Jao Ven - Prof., Northeastern University, Massachusetts, USA
Thomas Bohlen - Dr., Prof., Karlsruhe Institute of Technology, Karlsruhe, Germany
André Kurzmann - Institute of geophysics, Karlsruhe Institute of Technology, Karlsruhe, German
Klaus Eichhorn - researcher, Institute of learning photons and synchrotron emission, Karlsruhe Institute of Technology, Karlsruhe, Germany
Oldrih Stary - Faculty of Electrical Engineering, Czech Technical University in Prague, Prague, Czech Republic
S.B. Habibova - Chief of International Relations Department, The Ministry of Education of the Republic of Tajikistan, Tajikistan
Ilkay Yilmaz - Prof., vice-chairman, Department of Economics, Mersin University, Turke
B.T. Torobekov - PhD, professor, vice-rector for academic work Kyrgyz State Technical University after I. Razzakov , Kyrgyzstan
Aleksandr V. Rogachev - PDr., Prof., Rector of Francisk Skorina Gomel State University, Gomel, Republic of Belarus
Alexey Elyashevich - Dr., Prof., Peter the Great St.Petersburg Polytechnic University, St.Petersburg, Russia
Petr Ya. Kalita - DPhD, academician of Ukraine academy, president of Ukraine quality association, Academy of Quality Problems, Ukraine
Tatiana V. Fursa - Dr., leading researcher Electronics, Dielectrics and Semiconductors Lab Tomsk Polytechnic university, Tomsk, Russia
Inna V. Plotnikova - PhD, Associate Professor of Department of Physical methods of Non-destructive testing Tomsk Polytechnic university, Tomsk, Russia
Olga V. Galtseva - PhD, Associate Professor of Department of Physical methods of Non-destructive testing Tomsk Polytechnic university, Tomsk, Russia
Alexey N. Kalinichenko - PhD, Associate Professor of Department of Physical methods of Non-destructive testing Tomsk Polytechnic university, conference chair, Tomsk, Russia

3. The Organizing Committee of the VI International Conference "Modern Technologies for Non-Destructive Testing" wishes to thank the IOP Conference Series Materials Science and Engineering-Team for collaboration.

IOP Conf. Series: Materials Science and Engineering **289** (2017) 011002 doi:10.1088/1757-899X/289/1/011002

Peer review statement

All papers published in this volume of *IOP Conference Series: Materials Science and Engineering* have been peer reviewed through processes administered by the proceedings Editors. Reviews were conducted by expert referees to the professional and scientific standards expected of a proceedings journal published by IOP Publishing.

IOP Conf. Series: Materials Science and Engineering **289** (2017) 012001 doi:10.1088/1757-899X/289/1/012001

Influence of defects of nanostructured ZnO films on the photovoltaic characteristics of perovskite solar cells

D A Afanasyev[1], K Yu Mirzoev[2] and N Kh Ibrayev[3]
[1]Acting Director, Institute of Applied Mathematics, Karaganda, Republic of Kazakhstan
[2]Student, Karaganda State University named after E.A. Buketov, Karaganda, Republic of Kazakhstan
[3]Professor, Karaganda State University named after E.A. Buketov, Karaganda, Republic of Kazakhstan
E-mail: a_d_afanasyev@mail.ru

Abstract. The influence of the defects in ZnO films on the electrical and photovoltaic properties of perovskite solar cells was investigated in the work. According to the results of the research it was established that the defects in ZnO films affects the concentration of defects of perovskite films synthesized on the ZnO surface. However, the difference in the defect concentration in perovskite films is about 30%, while the concentration of defects in ZnO differs by 1000 times. A less significant influence is the concentration of ZnO defects on the electrical and photovoltaic properties of perovskite solar cells. The magnitude of the short-circuit photocurrent and the open voltage of the cells are affected by the concentration of perovskite defects and the quality of the perovskite-ZnO interface.

1. Introduction

At present, researchers are increasing attention to obtaining and studying the properties of nanostructured materials [1]. This interest is due to the great possibilities of practical application of such materials for the creation of elements of molecular electronics [2], optoelectronics [3], photovoltaic cells [4]. One of the promising materials on the basis of which it is possible to obtain various nanostructures is zinc oxide [5]. In a family of wide-band semiconductors, ZnO is a promising semiconductor in connection with the possibility of its use in the creation of various components of electronic equipment. ZnO is a recognized material for the creation of highly efficient light sources and receivers operating in the blue and ultraviolet range of the spectrum, solar cells, transparent contacts, ultrafast scintillators, gas sensors, etc. [6]. On the basis of ZnO films it is proposed to make perovskite solar cells (PSC). The basis of these cells is organo-inorganic compounds of the composition $OrPbX_3$, where Or - cation-organic compounds, such as, methylammonium, ethylammonium, etc., X – halogens Br, I, Cl. During the period from 2009 to 2014 PSC have increased their efficiency from 3.8% to more than 19% [7]. At the same time, the intensive growth of the efficiency of PSC continues [8]. Many properties of organo-inorganic perovskite materials have not been studied in detail and require further research. The degree of influence of the metal oxide film (TiO_2, ZnO, Al_2O_3), on the surface of which the perovskite films are synthesized, on the properties of solar cells remains little investigated. In this paper, the effect of defects of nanostructured ZnO films and perovskite films on the photovoltaic and electrical characteristics of perovskite solar cells with composition of ZnO-$CH_3NH_3PbI_3$-spiro-OMeTAD.

2. Experimental details

Initially, nanostructured ZnO films were fabricated by pulsed electrochemical synthesis and hydrothermal methods. The procedure of the synthesis of films is described in detail in [9].

Pulsed electrochemical deposition of ZnO nanorods was carried out in a three-electrode electrochemical cell with an aqueous electrolyte containing $NaNO_3$ and $Zn(NO_3)_2$ at a temperature of t = 70°C without mechanical stirring of the solution. Glass substrates coated with transparent electrically conductive layers of fluorine-doped tin oxide (FTO) from Sigma-Aldrich were used as a substrate. The opposite electrode was a graphite rod, and the reference electrode was a saturated Ag/AgCl electrode.

The rectangular potential pulses were applied to the substrate-cathode, the lower limit of which was U_{off} for the samples was – 0.8 V, the upper limit of U_{on} was – 1.4 V (the potentials are given relative to Ag/AgCl). The pulse rate was 2.5. The pulse frequency f was kept constant and equaled 2 Hz [10].

The synthesis of ZnO nanorods by the hydrothermal method consisted of two stages. The first stage is the deposition of a seed layer and the second stage is the synthesis of nanorods on a substrate with a seed layer of hydrothermal deposition. Zinc acetate ($Zn(CH_3COO)_2 \cdot 2H_2O$, Sigma Aldrich) and monoethanolamine (C_2H_7NO, Sigma Aldrich) used for synthesis of the seed layer ZnO The reagents were sequentially dissolved in isopropyl alcohol. The solution was applied to a pre-cleaned glass substrate coated with a FTO conductive layer by spin-coatingю This procedure was repeated 4 times. Further, to form the ZnO structure, the substrates were annealed in a muffle furnace at a temperature of 450 °C. Arrays of ZnO nanorods were synthesized on FTO substrates with a previously deposited seed layer of ZnO from an equimolar aqueous solution of zinc nitrate ($Zn(NO_3)_2$, Sigma Aldrich) and urotropine ($C_6H_{12}N_4$, Sigma Aldrich) for 10 hours. After every 2 hours samples were washed in deionized water and after the completion of the synthesis, the grown ZnO arrays were repeatedly washed in deionized water, dried and annealed at 400 °C [11].

For the synthesis of perovskite films, solutions of PbI_2 and CH_3NH_3I were used. The perovskite film $CH_3NH_3PbI_3$ ($MAPbI_3$) was deposited in two stages. In the first stage, PbI_2 was applied using the spin-coating method. The substrate was then dried at temperatures of 50 °C and 100 °C. In the second stage, a solution of ammonium methyl iodide CH_3NH_3I was applied to the sample. Later on, the perovskite film was annealed. The procedure for the synthesis of films is described in detail in [12]. All operations related to the synthesis of perovskite films were carried out in a box with an inert atmosphere (SPEKS GB 03-2M).

The morphology of the films was examined on an electron scanning electron microscope (SEM) with a field emission of Mira 3 (TESCAN). The absorption spectra of the films were recorded using a Cary 300 (Agilent) spectrophotometer. The fluorescence spectra were measured on a Cary Eclipse (Agilent) spectrofluorimeter. The kinetics of fast fluorescence of samples was measured with a pulsed spectrofluorometer with a picosecond resolution and recording in a time-correlated photon count (Becker & Hickl). Fluorescence excitation of the samples was carried out with a pulsed semiconductor laser with a λ_{gen} = 640 nm generation wavelength with – full width at half maximum (FWHM) τ = 80 ps.

Curent-voltage characteristics (CVC) were measured for PSC. The CVC of the cells was measured on Keithley 2400 instrument under standard simulated solar radiation AM 1.5, 100 mV/cm^2 (PET PHOTO Emission TECH., INC.) The electrical transport properties of PSC were studied using the method of electrochemical impedance spectroscopy. The impedance measurements were carried out under the standard simulated solar radiation AM 1.5, on the impedance meter Z-500PRO (Elins).

3. Discussion of results

The microstructure of synthesized ZnO films is shown in Figure 1. For the film obtained by the hydrothermal method, the average diameter of the rods was 150 nm. When using electrochemical synthesis, the average diameter of the rods was 660 nm. In the case of the hydrothermal method, the size of the rods is smaller than the size of the rods obtained by the electrochemical method. The rate of film synthesis is higher for the electrochemical method.

Spectra of absorption and luminescence were measured for ZnO films. The absorption spectra of ZnO are shown in Figure 2*a*. The optical density (*D*) of films obtained by different methods, and measured at 380 and 800 nm, has similar values. In the blue-green part of the spectrum, D of the films is significantly different. The luminescence spectra of the samples were also measured (Figure 2*b*). To determine the crystallinity of ZnO nanostructures, the ratio of the amplitude of the maximum intensity of the luminescence band in the visible region (I_{BO}) to the amplitude of the maximum of the edge luminescence band in the short wave region of the I_{UV} is often compared [13]. The I_{BO}/I_{UV} ratio for the electrochemical film was 2.6, and for the hydrothermal film 0.13. Therefore, the concentration of defects in ZnO nanorods obtained by the electrochemical method is much larger than the ZnO film obtained by the hydrothermal method. Indirect evidence of a high concentration of defects in the ZnO film obtained by the electrochemical method is the shape of the absorption curve 2 shown in Figure 2*a*. The film has a higher *D* value in the range of 400 to 800 nm.

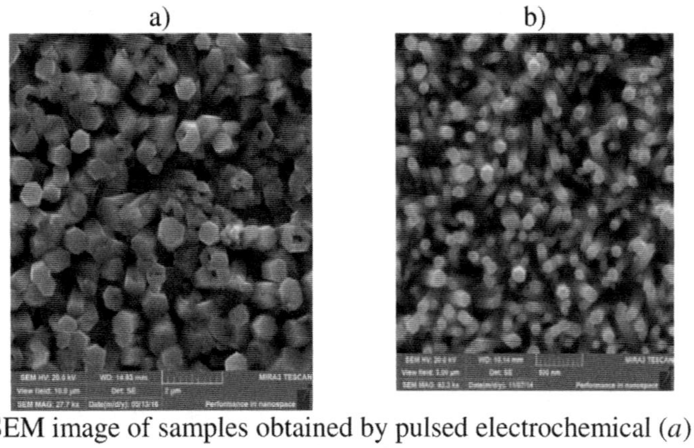

Figure 1. SEM image of samples obtained by pulsed electrochemical (*a*) and hydrothermal methods (*b*).

The concentration of defects in the films was estimated for ZnO samples. For the ZnO film obtained by the hydrothermal method on a non-conducting substrate, the resistivity of the sample and the Hall effect were measured. From the data obtained, the concentration of charge carriers in the sample was determined. Proceeding from the fact that the width of the forbidden ZnO band is high and at room temperature the contribution of the thermally activated intrinsic conductivity is small in comparison with the contribution of charge carriers from the defective levels of ZnO, the defect concentration in the film was estimated as ~ 3×10^{14} cm^{-3}. From the measurements of the resistivity of the ZnO film obtained by the electrochemical method, the concentration of defects in ZnO (electrochem. synthesis) was estimated by comparison with the specific resistance of the ZnO film (hydrothermal synthesis). This concentration was ~ 3×10^{17} cm^{-3}. Thus, the estimation of the defect concentration in ZnO films in electrical measurements agrees qualitatively with the data obtained from measurements of the luminescent properties of the samples.

The perovskite film had a granular structure with a granule size of the order of 100 nm. For the formation of films with hole conductivity, solutions based on Spiro-OMeTAD (N^2, N^2, $N^{2'}$, $N^{2'}$, N^7, N^7, $N^{7'}$, $N^{7'}$-octakis (4-methoxyphenyl) - 9, 9'-spirobi [9H-fluorene] - 2, 2', 7,7'-tetramine) in chlorobenzene were prepared. After the application of the film with hole conductivity, a silver electrode 60…100 nm thick was deposited in a vacuum (10^{-6} torr) using a thermal evaporation method, the deposition rate being 1.5 Å/s.

Figure 3 shows a section of a perovskite cell based on nanostructured ZnO (hydrothermal synthesis), including spiro-OMeTAD and a silver electrode. The total thickness of the sample with ZnO nanostructures (hydrothermal synthesis) was approximately ~ 1360 nm, of which ~ 1054 nm the thickness of the ZnO layer together with perovskite, the thickness of the spiro-OMeTAD film ~ 243 nm and Ag thickness ~ 60 nm. The thickness of the perovskite film without ZnO is 400 nm. For the

PSC with ZnO, obtained by the electrochemical method, the thickness was about ~ 1300 nm, of which ~ 600 nm the thickness of the ZnO layer, the thickness of the perovskite ~ 600 nm, the thickness of the spiro-OMeTAD film ~ 200 nm, and the thickness of Ag ~ 100 nm.

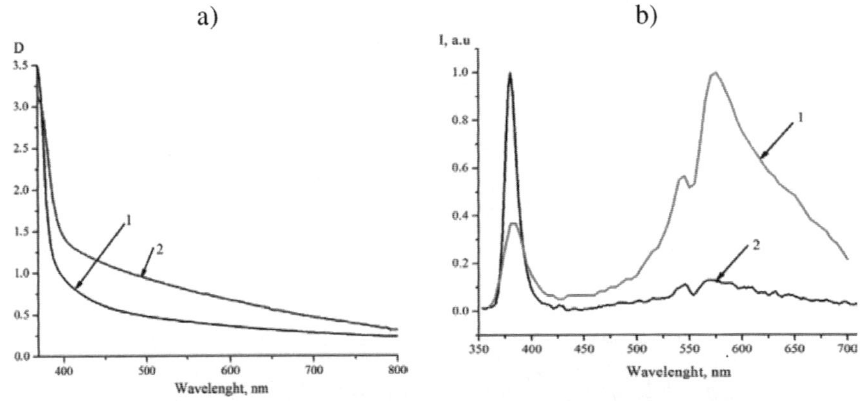

Figure 2. Absorption spectra (*a*) and luminescence (*b*) of ZnO nanorods obtained by pulsed electrochemical (1) and hydrothermal (2) methods.

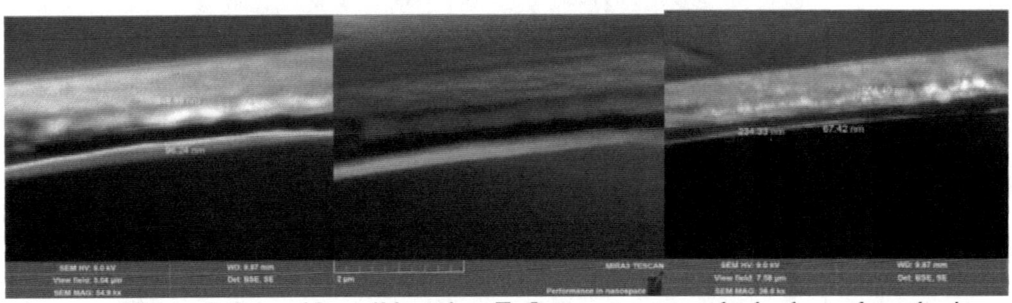

Figure 3. Perovskite cell based on ZnO nanostructures, hydrothermal synthesis.

The absorption spectra of $CH_3NH_3PbI_3$ ($MAPbI_3$) films are shown in Figure 4 (curves 1–3). The shape of the curves indicates the formation of a perovskite film. However, the form of the dependence of the absorption coefficient D on the wavelength indicates that the perovskite films have different qualities depending on which substrates for making $MAPbI_3$ were used. The absorption spectrum of a perovskite film based on hydrothermal ZnO rods differs significantly from the absorption spectra of perovskite films obtained on other ZnO coating (Figure 4*a*). Additional information can be obtained from the luminescence spectra.

The luminescence spectra of the $MAPbI_3$ films are shown in Figure 4*b*. For the samples obtained, the ratio of the luminescence intensity to the optical density of the film at a wavelength of 470 nm ($I_{MAX}/ D_{470\ nm}$) was estimated. These ratios are given in Table 1. The maximum ratio is observed for the perovskite film obtained on the ZnO surface (hydrothermal synthesis). The lowest value is observed for the ZnO film, (electrochem. synthesis). Thus, the perovskite film on the ZnO surface (the hydrothermal method) has less concentration of defects than the rest of the films. Large concentration of defects in ZnO is observed for perovskites on the surface of ZnO (electrochem. synthesis).

To study the effect of the properties of compact ZnO films on the efficiency of charge transfer in a perovskite film ($MAPbI_3$), the kinetics of the luminescence of perovskite films in the nanosecond time range was studied. The kinetics of the luminescence of films contains the stage of increase and quenching of the luminescence intensity. After the laser excitation is completed, the luminescence of the perovskite film is increased. The time after the end of laser excitation to the maximum of the luminescence intensity is in the range from 0.7 ns to 1.5 ns (Table 1). Then the luminescence intensity falls. In the case of the sample $CH_3NH_3PbI_3$ on the surface of the FTO film, the luminescence kinetics has an exponential form of damping, the lifetime is 22.3 ns. In the case of the $MAPbI_3$ film on the surface of the ZnO film, the luminescence lifetime decreases. For different ZnO films, the lifetime is different: $\tau_{hydrothermal.}$ = 17.2 ns; $\tau_{electrochym.}$ = 22.3 ns.

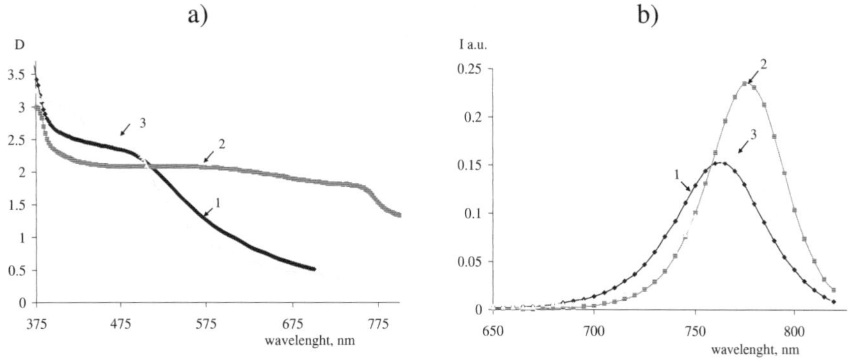

Figure 4. Absorption *(a)* and luminescence spectra *(b)* of $MAPbI_3$ films obtained on the ZnO surface: 1) FTO; 2) ZnO nanorods, hydrotherm. synthesis; 3) ZnO nanorods, pulsed electrochem. method.

Table 1. Luminescent characteristics and effect of n-type and p-type layers on suppression of luminescence of perovskites.

Sample	$I_{MAX}/D_{470\ nm}$	k (I_{MAPbI3} /$I_{MAPbI3\ -spiro-OMeTAD}$)	k (τ_{MAPbI3} /$\tau_{MAPbI3\ -spiro-OMeTAD}$)	Time of increase of luminescence intensity Δt, (ns)
FTO–perovskite	0.065	15.2	3.19	1.53
ZnO (electrochem.) - perovskite	0.064	2.35	3.46	1.34
ZnO (hydrothermal) perovskite	0.113	2.07	1.69	0.7

As is known, the efficiency of charge transfer can be estimated by quenching the intensity or lifetime of luminescence [14]. Analysis of the data shows that the best charge transfer is observed for the perovskite–ZnO sample (hydrothermal synthesis) (Table 2). For the perovskite-ZnO sample (electrochem. synthesis), charge transfer proceeds less intensively. The effect of the spiro-OMeTAD film on the intensity and lifetime of the luminescence of perovskite films. The luminescence intensity of $MAPbI_3$ is quenched less than the decrease in the lifetime of the perovskite luminescence. This is due to a more efficient electron transfer in the samples of the perovskite-nanorods ZnO. The short rise time of the luminescence in the perovskite-ZnO film (hydrothermal synthesis) indicates a high carrier transport velocity in this sample. In the case of the ZnO sample (electrochem. synthesis)–$MAPbI_3$–Spiro-OMeTAD, the maximum rate of hole transport from perovskite to the Spiro-OMeTAD layer is observed. A higher rate of hole transport is an indirect indication of a lower rate of electron transfer from perovskite layer to ZnO.

For perovskite films, it is possible to estimate the concentration of defects in accordance with the method described in detail in [15, 16] when photoexcitation of a sample by a short laser pulse with low energy (where Auger recombination can be neglected) and assuming that the recombination of trap states is much slower, than edge radiative recombination. The initial concentration of charge carriers after photoexcitation $n_c(0)$ can be defined as:

$$n_C\,(0) = \frac{\sum_i n_{TR}^i(0)\left(1-e^{-a_i\tau_0\frac{I_{PL}}{k}}\right)}{k} \quad (1)$$

where $n_{TR}^i(t)$ is the trap state density and a_i is the parameter related to the capture cross section and the velocity of the charge carriers. Assuming that the ratio between the integrated photoluminescence intensity (I_{PL}) of the band ($I_{PL} = k\left(\int_0^\infty {n_C(t)}/{\tau_o dt}\right)$, where k is the constant), the concentration of defects in the perovskite film (n_{TR}^i) can be determined from equation 1.

Table 2. Luminescence lifetime of perovskite films.

Interface (sample, type of film, synthesis method)	Constants of electron transfer (k, ns^{-1})
$MAPbI_3 \rightarrow$ FTO	0.045
$MAPbI_3 \rightarrow$ ZnO (hydrothermal synthesis)	0.058
$MAPbI_3 \rightarrow$ ZnO (electrochem. synthesis)	0.045
$MAPbI_3 \rightarrow$ Spiro-OMeTAD (FTO), ($k_{FTO-MAPbI3-Spiro-OMeTAD} - k_{FTO-MAPbI3}$)	0.098
$MAPbI_3 \rightarrow$ Spiro-OMeTAD, (hydrothermal synthesis)	0.040
$MAPbI_3 \rightarrow$ Spiro-OMeTAD, (electrochem. synthesis)	0.110

For $MAPbI_3$-ZnO samples, the dependence of the luminescence intensity of the perovskite film on the laser excitation energy was measured. To determine the trap concentration in perovskite films, plots of the luminescence intensity (I_{PL}) were plotted against the concentration of free charge carriers (n_c) (Figure 5*a*). The concentration of charge carriers was estimated on the basis of the main characteristics of laser excitation and the optical density of the perovskite film. The form of the approximating curve obtained using Equation 1 is shown in Figure 5*b*. The data obtained from the approximation of the curves for perovskite films are given in Table 3. It can be seen from the table data that a perovskite film ($MAPbI_3$) with a higher defect concentration is formed on the ZnO film with a higher defect concentration.

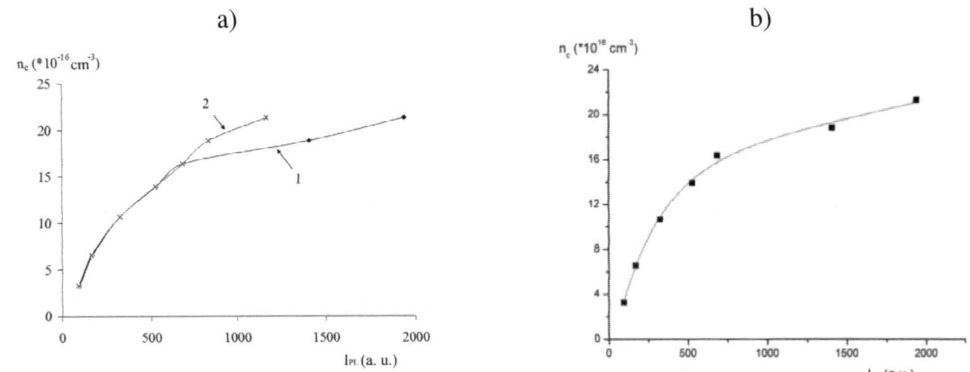

Figure 5. Determination of the trap state densities: *a)* Dependence of n_c on luminescence intensity: 1) perovskite-ZnO (electrochem. synthesis); 2) perovskite-ZnO (hydrothermal synthesis). *b)* Approximation of the experimental data using equation 1 for the perovskite–ZnO film (electrochem. synthesis).

The effect of the morphology and defect structure of ZnO on the photovoltaic properties of PSC is studied. The light and dark CVC of the perovskite cells are shown in Figure 6. The data obtained are also given in Table 4. It can be seen from the light CVC that for a solar cell with a ZnO film

IOP Conf. Series: Materials Science and Engineering **289** (2017) 012001 doi:10.1088/1757-899X/289/1/012001

(electrochem. synthesis) a higher value of the idling voltage (V_{oc}) the value of the short-circuit current (J_{sc}) as compared to a cell based on ZnO (hydrothermal synthesis). A smaller value of V_{oc} for the perovskite-ZnO sample (hydrothermal synthesis) is associated with a high recombination rate at voltages *U*> 0.4 V compared to cells based on ZnO (electrochem. synthesis), Figure 6*b*.

Table 3. Concentration of defects in perovskite films.

Sample (substrate)	Concentration of defects n_{tr}
Perovskite-ZnO, (hydrothermal synthesis)	1.2×10^{16} см$^{-3}$
Perovskite-ZnO, (electrochem. synthesis)	1.7×10^{16} см$^{-3}$

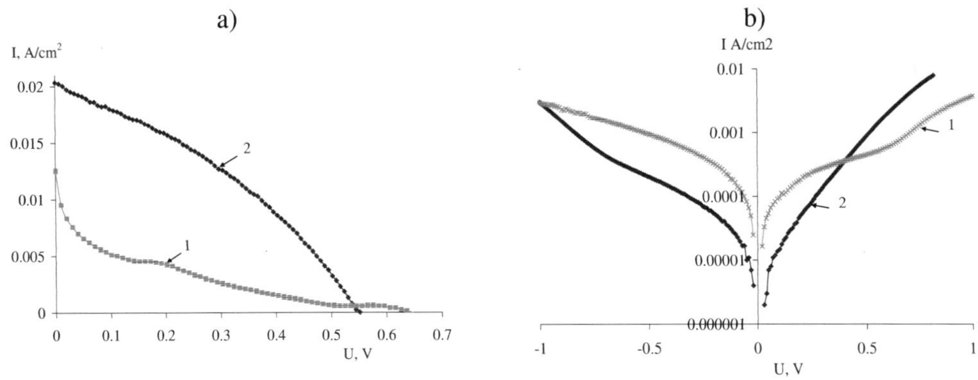

Figure 6. Light (*a*) and dark (*b*) CVC of solar cells based on nanostructured ZnO films: 1) ZnO, electrochem. synthesis; 2) ZnO, hydrotherm. synthesis.

The dark CVC of the cells are shown in Figure 6*b*. It can be seen from Figure 6*b* that for a cell with a ZnO film (electrochem. synthesis) large leakage currents are observed on the straight part of the CVC at a low voltage value of up to 0.4 V. At high voltages, the leakage currents are higher for a cell with a ZnO film (hydrotherm. synthesis). The results obtained make it possible to analyze the properties of the perovskite film and the perovskite-ZnO interface. The smaller concentration of defects in the perovskite film and the ZnO film (hydrothermal synthesis) leads to lower dark currents in the cell at low operating voltages. High leakage currents at voltages greater than 0.4 V indicate the emergence of additional channels for the recombination of charge carriers in a cell at high voltages (*U*> 0.4 V). In this case, no recombination losses are observed in the perovskite–ZnO film (electrochem. synthesis).

Table 4. Photovoltaic and electrical properties of perovskite solar cells.

The method of manufacturing ZnO films	J_{sc}, A/cm^2	U_{oc} B	FF	η %	R_{SER} Ом/cm^2	R_{SH} Ом/cm^2	*S*, cm^2
Electrochem. synthesis	0.012	0.636	0.12	1.02	327.1	51.24	0.24
Hydrothermal synthesis	0.021	0.551	0.30	3.85	35.95	82.56	0.21

The impedance of the PSC was measured. Figure 7 shows the impedance spectrum for the perovskite-ZnO cell (hydrothermal synthesis). For a more accurate determination of the recombination time characteristics in perovskite cells, the simplest model of the solar cell R1–(R2, C1) was used. The electrical schematic of the model is shown in the Figure 7. In the solar cell model, R1 is associated with the series resistance of the contacts and layers in the solar cell, R2 is related to the electrical resistance of the perovskite film, and C1 is the cell capacitance. The values of the indices of the solar cell chain were estimated. The effective lifetime was determined as (τ = *R2*×*C2*). The results are shown in Table 5.

The data obtained for the cells by the method of impedance spectroscopy (Table 5) correlate with the data on the measurement of the light CVC (Table 4). The high value of the series resistance of the R_{SER} cell corresponds to the high value of R1 obtained when measuring the impedance of the cells and vice versa.

Comparison of the recombination constants shows that when the cells are illuminated, the recombination rate in a cell with ZnO (electrochem. synthesis) is 25% higher than the recombination rate in a cell with ZnO (hydrothermal synthesis). This indicates that the concentration of defects of nanostructured ZnO films does not significantly affect the rate of recombination of electron-hole pairs at the ZnO–perovskite interface. A comparison of the changes in the rate of recombination in PSC (25%) with the data of Table 3 indicate that defects in perovskite have a greater significance for recombination in PSC and that defects in ZnO films are practically unaffected.

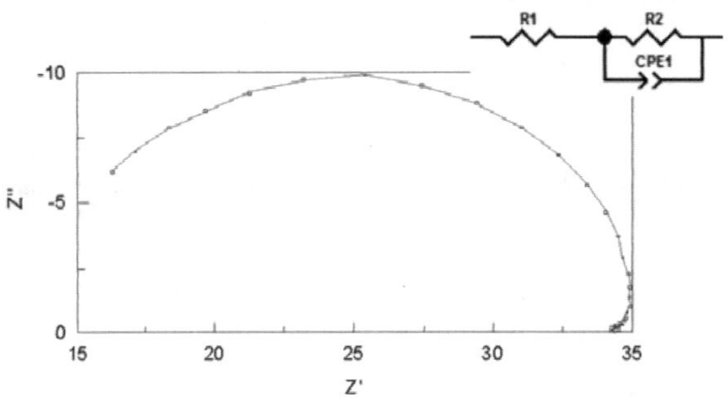

Figure 7. Nyquist plot for a solar cell based on ZnO (hydrothermal synthesis). In the inset, the model of the solar cell electrical circuit.

Table 5. The values of the elements of the chain of perovskite solar cells under illumination, in accordance with the equivalent circuit.

Solar cell with ZnO (method of synthesis)	R_1, Ω	R_2, Ω	C_1, F	n_1	τ (μs)
Hydrotherm. method	12.5	23.6	6.7×10^{-7}	0.863	15.7
Electrochem. method	178	294	4.0×10^{-8}	0.96	11.8

It is important that for ZnO films obtained by different methods, the values of the series resistance of cells significantly differ (Table 4, 5). The specific resistance of the ZnO film obtained by the electrochemical method is significantly lower than the resistance of the ZnO film obtained by the hydrothermal method. This resistance of PSC's on the basis of ZnO films (electrochem. synthesis) is an order of magnitude higher than the resistance of cells based on ZnO films (hydrothermal synthesis). This is possible if there is a high resistance at the ZnO–perovskite interface. Thus, these results indicate a significant effect of the ZnO-perovskite interface on the photovoltaic characteristics of the perovskite solar cells obtained.

4. Summary

The paper presents the results of a study of the influense of defect concentrations of nanostructured ZnO films on the electrical and photovoltaic properties of PSC. The different methods determine the concentration of defects in zinc oxide films and perovskite films. On ZnO samples with a higher defect concentration, a perovskite with a higher defect concentration is formed. However, if the concentration of defects in ZnO differs by a factor of 1000, then in perovskite films the defect concentration's differs only by 30%.

For a solar cell with a ZnO film (electrochem. synthesis), a higher value of the open circuit voltage (V_{oc}) but a smaller value of the short-circuit current (J_{sc}) is observed compared to a cell based on ZnO (hydrothermal synthesis). The smaller value of J_{sc} is associated with a lower charge transfer rate in the $MAPbI_3$→ZnO film (electrochem. synthesis). A smaller value of V_{oc} for $MAPbI_3$-ZnO sample (hydrothermal synthesis) is associated with a high recombination rate at voltages U>0.4 V compared to cells based on ZnO (electrochem. synthesis).

A smaller concentration of defects in the perovskite-ZnO film (hydrothermal synthesis) leads to lesser dark currents in the cell. High leakage currents at voltages exceeding 0.4 V indicate the occurrence of additional recombination processes at high voltages. In this case, no perovskite-ZnO (electrochem. synthesis) of these recombination processes is observed.

Acknowledgments
This work was carried out with the financial support of the Ministry of Education and Science of the Republic of Kazakhstan.

References

[1] Ryzhonkov D I, Levina V V, Dzidziguri E L 2010 *Nanomaterials* (Moscow: Binom, Laboratoriy znanii) (In Russian)
[2] Ghosh A W 2011 *Semicond. Sci. Tech.* **5** 383–479
[3] Iwasaki Y, Osasa T, Asahi M, Matsumura M, Sakaguchi Y, Suzuki T 2006 *Phys. Rev. B.* **74** 195209 doi: 1103/PhysRevB.74.195209
[4] Ito S, Murakami T N, Comte P et al. 2008 *Thin Solid Films* **516** 4613–4619 doi: 10.1016/j.tsf.2007.05.090
[5] Djurisic A B, Leung Y H 2006 *Small* **2** 944–961 doi: 10.1002/smll.200600134
[6] Djurisic A B, Chen X Y 2010 *Progress in quantum electronics* **34** 191–259 doi: 10.1016/j.pquantelec.2010.04.001
[7] Snaith H J 2013 *J. Phys. Chem. Lett.* **4** 3623–3630 doi:10.1021/jz4020162
[8] Best Research-Cell Efficiencies [Electronic resource] URL: https://www.nrel.gov/pv/assets/images/efficiency-chart.png
[9] Afanasyev D A, Aitymov Zh K, Ilyassov B R et al. 2017 *Bulletin of the Karaganda-physics* № 1(85) 22–27
[10] Klochko N P, Khrypunov G S 2012 *Semiconductor* **46** 825–831 doi: 10.1134/S1063782612060127
[11] Ilyasov B R, Ibrayev N Kh, Nuraje N 2015 *Materials Science in Semiconductor Processing* **40** 885–889 doi: j.mssp.2015.07.087
[12] Ibrayev N Kh, Afanasyev D A, Mirzoev K Yu, Smagulov G K 2016 *Eurasian physical technical journal* **13** 45–51
[13] Li D, Leung Y H, Djurisic A B et. al. 2004 *Applied Physics Letters* **85** 1601–1603 doi: 10.1063/1.1786375
[14] Pydzin´ska K, Karolczak J, Kosta I et. al. 2016 *ChemSusChem* **9** 1647–1659 doi: 10.1002/cssc.201600210
[15] Xing G, Mathews N, Lim S S et. al. 2014 *Nature materials* **13** 476 – 480 doi: 10.1038/nmat3911
[16] Peng W, Anand B, Liu L et. al. 2016 *Nanoscale* 8 1627–1634 doi: 10.1039/C5NR06222E

Steady operation of the electric drive of pipeline armature in the emergency situation at low ambient temperatures

S V Bordunov[1], V V Kuznetsov[2], K R Muratov[3] and O Yu Zavyalova[4]
[1]Director, Public Corporation Scientific – Innovation Enterprise "Echteh", Tomsk, Russia
[2]Associate Professor, Omsk State Technic University, Omsk, Russia
[3]Associate Professor, Tyumen Industrial University, Tyumen, Russia
[4]Associate Professor, Tomsk Polytechnic University, Tomsk, Russia
E-mail: bordunov@live.ru

Abstract. This scientific work is devoted to the study of the electric drive operation of pipeline armature at low ambient temperatures. Hit of moisture into reducer and rare inclusions in operation of locking regulator are led to curdling lubricant that causes the increased wear of mechanical knots. There is a probability of freezing mechanical components; it leads to emergency situations. The problem of improving working efficiency of the electric drive of shut-off regulating armature at low ambient temperatures of the environment is solved in this work. A simulation model of the GUSAR electric drive was developed to solve this problem. Studies of the simulation model show the need to limit the torque increase rate on a drive motor shaft. The algorithm of setting of PI speed controller to obtain acceptable transient processes is suggested. Recommendations for the use of the algorithm in the microprocessor control system of electric drive are proposed. It is shown that the electric drive operation algorithm with torque increasing limitation on the motor shaft will be smoothly working off the perturbing actions that occur in pipeline armature.

1. Introduction

Nowadays an intelligent electric drive of any production mechanism, in particular shut-off regulating element, is a complex system; its mechanical and electrical parts are in constant interaction [1–5]. In general, the intellectual part of the electric drive consists of a series of printed circuit boards containing internal memory, an electric power converter, a microprocessor part, drivers and a power unit, jointed by magnetic and electrical connections together. The mechanical part of the drive device contains an inertial multimass structure with elastic mechanical bonds [6].

At present, automated control systems for electric drives (ED) provide reliable and uninterrupted operation of mechanisms in various fields of industry and technology. The capabilities of modern intelligent electric drives are determined by the parameters of the power part and the functional characteristics of the control systems [7, 8].

In the age of high technologies, the most common is an asynchronous ED with a microprocessor-based control device, through which it is possible to organize the regulation of output variables in a wide range with high speed and accuracy [9, 10].

Rare inclusions at operation of the shut-off regulating body lead to emergency operating conditions, especially at low ambient temperatures. The determining influence on the working capacity and the ED service life has not only the power inverter and asynchronous motor, but also the whole reducer of shut-off regulating armature [11]. The freezing ED mechanical part is due to of moisture hit into reducer at low ambient temperatures and also due to curdling lubricant.

2. Research Method

At present the actual goal is to improve the stability of the asynchronous electric drive of shut-off regulating pipeline armature at low ambient temperatures [12]. The most vulnerable type of structural failure of the shut-off regulating pipeline element is the wear of the ED mechanical part at ambient temperatures below - 40^0C. Therefore it is expedient to look for increasing ways of the durability of working parts under conditions of increased wear. It is necessary to limit the torque on the shaft of the electric drive to achieve optimum operating characteristics at low ambient temperatures.

Studies of the GUSAR oil pipeline armatures by SPA "Sibirskiy Mashinostroitel" (Russia) are presented in this work. GUSAR electric drive with electronic control system is designed for multi-function control by pipeline armatures and is used in oil, gas, petrochemical and other industries; it is designed for a temperature range from +20°C to –60°C. Figure 1 shows the exterior of the GUSAR electric drive. The electric drive consists of electronic control unit, asynchronous motor and wave-type reducer.

Three-phase asynchronous motor with squirrel-cage rotor of low-power explosion-proof execution of DAT-156M-02 type is used in this ED. Nominal power is 0.25 kW.

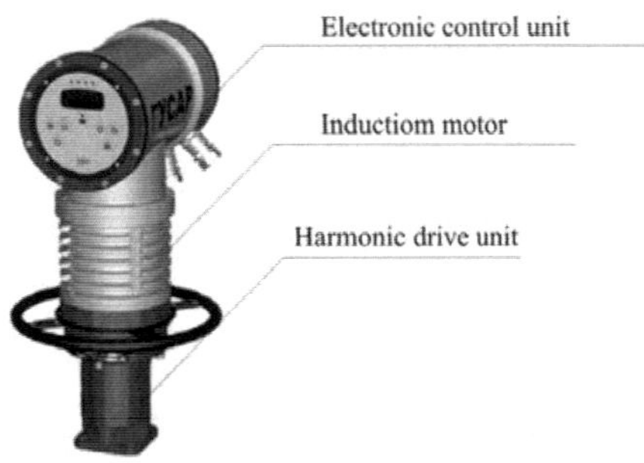

Figure 1. The exterior of the GUSAR electric drive.

Wave transmission with intermediate rolling elements is used in the reducer of the GUSAR electric drive (Figure 2); currently this transfer is in a stage of rapid development [13].

The main advantages of this transmission are a large gear ratio (i = 80 ÷ 320) at a small number of parts, improved mass-dimensional characteristics in comparison with conventional gears, high kinematic accuracy and smooth running, high output capability [14].

Figure 2*a* shows the profile of wave gear with intermediate rolling elements.

a)

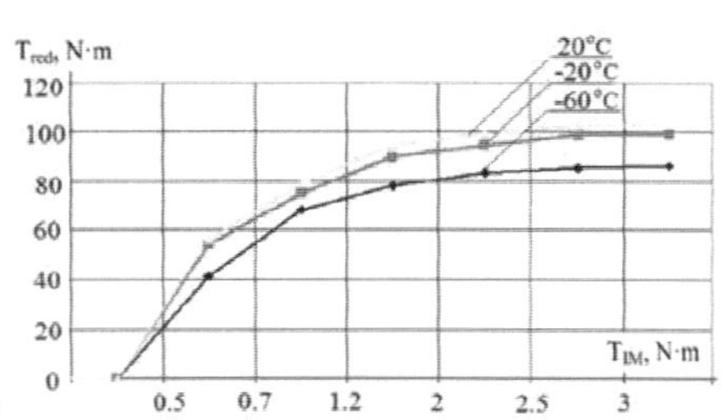

b)

Figure 2. Wave reducer, where *a)* is profile of the wave gear with intermediate rolling elements; *b)* are wave gear characteristics.

The wave reducer has significant drawbacks. Figure 2*b* shows of the motor torque dependencies, namely, the dependencies of motor torque T_{IM} on the torque on wave reducer output T_{red}. These dependencies were obtained experimentally and have a nonlinear character. It means that the moment of losses in the gear will depend on the moment on the reducer input. This nonlinearity is more manifested at values less than half the nominal torque: T_{IMrate} = 1.7 N·m. Also, the torque of losses increases at decreasing ambient temperature. The torque on the output link of the electric drive of pipeline armature will decrease at decreasing temperature at other equal conditions. The dependence of the moment of losses of the wave reducer on the ambient temperature and on the torque of the electric motor violates the classical principles of control systems construction for electric drives. Obligatory identification of torque at the output link of ED occurs with an error in such mechanisms; in turn, it influences on the control system. It is necessary to take into account the experimentally obtained characteristics of the wave reducer for correct operation of the electronic control unit by drive and elimination of emergency situations in the control system.

Figure 3 shows simplified functional diagram of ED. A predetermined speed signal ω_{ref} is supplied into the direction of "OPEN" or "CLOSED" to the electronic control unit. The signal p_{ref} of specified position (usually in percent) is set to put the intermediate position of the output link of the pipeline armature.

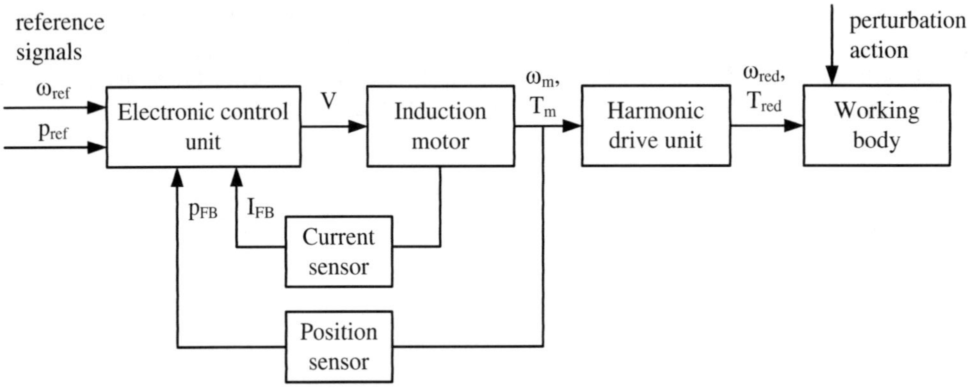

Figure 3. Functional diagram of the GUSAR electric drive.

The frequency converter of the electronic control unit generates a three-phase voltage of given frequency and amplitude V for an induction motor [15]. The electric drive control system is constructed according to the principles of vector control law [16, 17]. Feedback signals (the motor current and motor shaft position) are fed into the control unit; the motor shaft position is read in by means of absolute encoder installed on the motor shaft.

The speed required for the control system is calculated from the position signal. The motor torque T_m is converted into a torque T_{red} by means of the wave reducer, which rotates the working element. Perturbing impact is the load torque on output link of the ED. The electric drive must ensure operation at given range of load torque and stop the motor at exceeding threshold torque.

The simulation model of the GUSAR electric drive was developed by using MATLAB software (Figure 4) for electric drive control system creation of pipeline armature and study of various operating modes. This model includes an asynchronous motor with squirrel-cage rotor in two-phase coordinate system dq, a frequency converter with flow support channel and speed control channel, mechanical part, including a nonlinear dependence of losses torque of the wave reducer (Figure 2*b*). This simulation model allows to study of electric drive of pipeline armature in different dynamic operation modes.

3. Results and Discussion

Figure 5 shows the transient processes graphs of the electric drive at full operating cycle. The graph of the moment T is shown at the top of Figure 5, and the velocity graph ω is at the bottom of Figure 5. The operation cycle includes soft start, load surge and braking [7]. Soft start is provided with the help of power-up sensor, which forms the temp of increasing the electric drive speed. The use of power-up sensor allows deleting beats in the mechanical part of the electric drive, which occur in the gear, couplings and other nodes containing clearances.

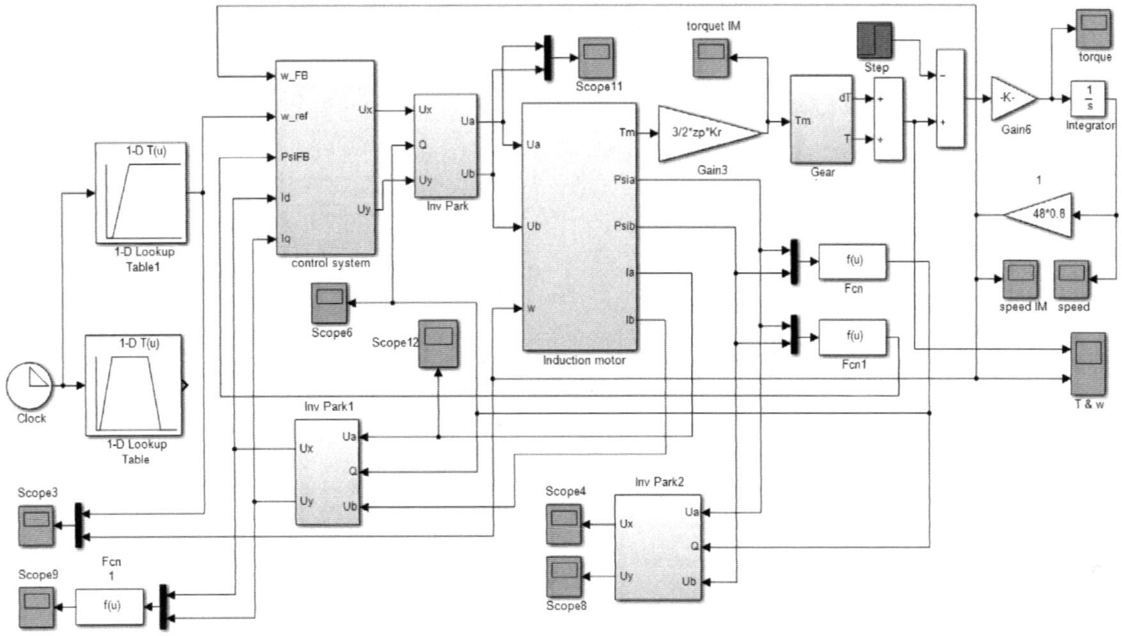

Figure 4. Simulated model of the GUSAR electric drive.

The transient processes graphs of the electric drive at load surge are shown in Figure 6. Enhanced scale of image of load surge allows making a conclusion about control system high operating speed. This is very important at the operation of pipeline armature in such modes as maintaining pressure or flow of gaseous or liquid media. However, there are operating modes and application areas of such electric drives, in which this speed can lead to emergency modes.

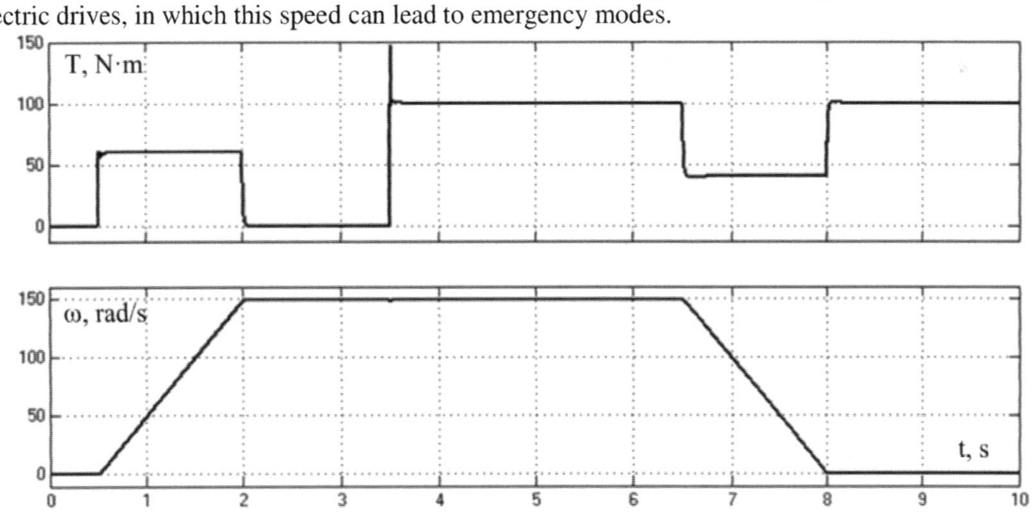

Figure 5. Transient curves of full operating cycle of the electric drive.

Such modes are connected with low ambient temperatures very often. Rare use of ED, curdling lubricant, freezing of mechanical parts of the drive lead to the fact that there is a danger of breakage of mechanical parts at sharp increasing torque on the electric motor shaft. It is necessary to increase gradually the torque in such modes; it is realized using the speed controller. A softer setting of the speed controller will allow limiting the rate of increasing the given parameters on the torque, which is the speed controller output. In this case the decreasing drive operating speed is expected.

Figure 7 shows transient processes graphs of the electric drive at the load surge with a smooth increasing torque. It is sufficient to reduce the proportional part of the PI speed controller for this. Speed sagging at the level of 20% of the nominal value occurs in this case. The operating speed loss avoids possible emergency conditions in this situation.

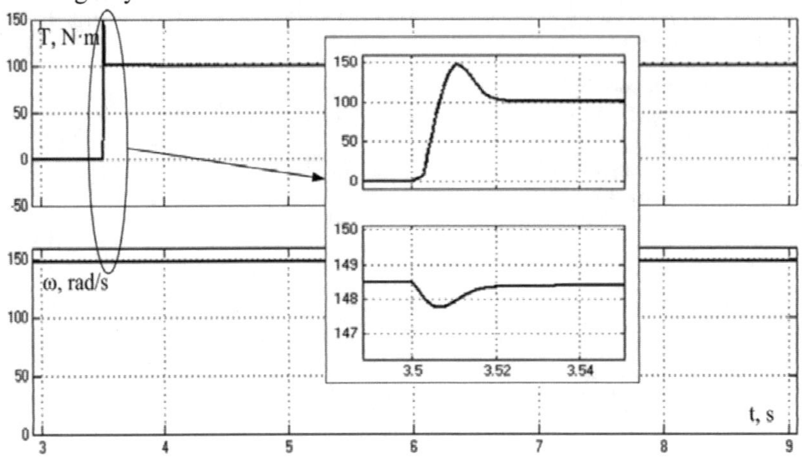

Figure 6. Transient curves of the electric drive at the load surge.

Smooth increasing the moment will allow estimating the torque on the motor shaft and making an appropriate decision by the microprocessor system. This situation was realized using a simulation model. This situation was realized using a simulation model. Figure 7 shows the results of this simulation. The maximum torque of the ED is set at level of 125 N·m. The control system monitors the torque. When this value is exceeded, the drive brakes by decreasing the given speed to zero at the normal speed controller settings. The torque is negative at the process of braking; after this it is equal to the load torque at zero speed.

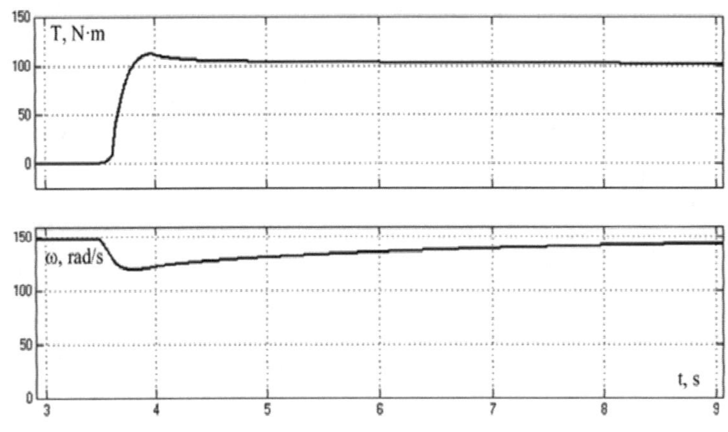

Figure 7. Transient curves of the electric drive at the load surge with a smooth increasing torque.

A milder setting of the speed controller is needed in two cases: to simulate "accident" at excess of the given torque by the user and smooth tearing off of the frozen mechanical parts at low ambient temperatures. For the second case it is necessary to base on the data of the ambient temperature and also to keep the event log (for calculating the time from the last successful start-up of the motor) at setting the PI speed controller (Figure 8).

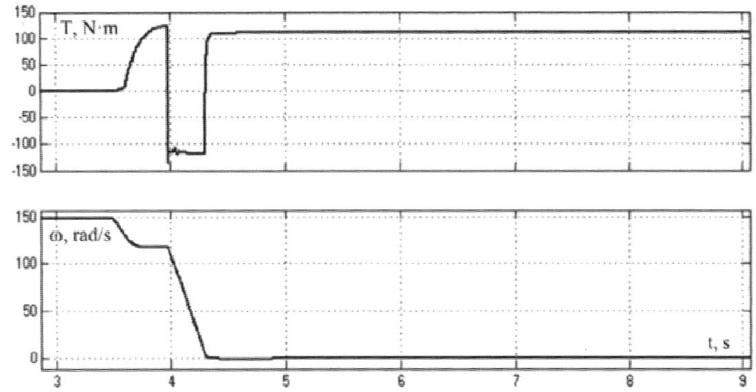

Figure 8. Transient curves of the electric drive at the load surge at disaster shutdown.

4. Conclusion

The freezing of the electric drive mechanical part due to curdling lubricant takes place at low ambient temperatures. The probability of an emergency at the dynamic modes of the electric drive is increased. Increasing sustainable operation at low temperatures is a priority task.

Experimental data show that it is necessary to limit the rate of increasing torque of the motor at dynamic conditions of operation at low ambient temperature. This limitation is proposed to be made by means of the corresponding setting of PI speed controller. The setting of the speed controller can be adapted during the operation of the device depending on the ambient temperature and on how often the drive operation is used.

References

[1] Payuk L *et al* 2016 *MATEC Web of Conferences* **79** 01060 doi: 10.1051/matecconf/20167901060
[2] Aristov A *et al* 2016 *J. of Physics: Conf. Ser.* **671** 012002 doi: 10.1088/1742-6596/671/1/012002
[3] Karakulov A S 2002 *Proc. of MTT'2002* 1213755 doi: 10.1109/SPCMTT.2002.1213755
[4] Odnokopylov I G *et al* 2017 *IOP Conf. Ser.: Mater. Sci. Eng.* **189** 012010 doi: 10.1088/1757-899X/189/1/012010
[5] Langraf S *et al* 2017 *MATEC Web of Conferences* **102** 01026 doi: 10.1051/matecconf/201710201026
[6] DeWall K *et al 1997 Motor-operated valve (MOV) actuator motor and gearbox testing* (United States) doi: 10.2172/515583
[7] Bing D , JunMin P 2001 *Proc. of IECON'2001* 7282671 doi: 10.1109/IECON.2001.976478
[8] Wang S *et al* 2018 *Appl Surf Sci* **428** 1070–1078 doi: 10.1016/j.apsusc.2017.09.225
[9] Granjon P 2011 *Proc. of 8th. International Conference on Condition Monitoring and Machinery Failure Prevention Technologies* 109–119
[10] Tosun G *et al* 2016 *Proc. of PEMC'2016* 7752055 doi: 10.1109/EPEPEMC.2016.7752055
[11] Xu L *et al* 2012 *Proc. of ISCID'2012* **2** 6406021 doi: 10.1109/ISCID.2012.248
[12] Klueber P 1983 *IEEE T Ind Appl* **IA-19** 962–967 doi: 10.1109/TIA.1983.4504321
[13] Tjahjowidodo T *et al* 2013 *Mechatronics* **23** 497–504 doi: 10.1016/j.mechatronics.2013.04.002
[14] Abdelraheem S *et al* 2017 *Proc. of MEPCON'2016* 7836869 doi: 10.1109/MEPCON.2016.7836869
[15] Pankratov V V and Kotin D A 2009 *Russ Electr Engin* **80** 651–656 doi: 10.3103/S1068371209120037
[16] Takahashi A *et al* 2008 *Proc. of ICCAS'2008* 4694213 doi: 10.1109/ICCAS.2008.4694213
[17] Vasendina E *et al* 2016 *IOP Conf. Ser.: Mater. Sci. Eng.* **110** 012070 doi: 10.1088/1757-899X/110/1/012070

IOP Conf. Series: Materials Science and Engineering **289** (2017) 012003 doi:10.1088/1757-899X/289/1/012003

Integrated microelectromechanical gyroscope under shock loads

T G Nesterenko[1], A N Koleda[2] and E S Barbin[3]
[1] Associate Professor, National Research Tomsk Polytechnic University, Tomsk, Russia
[2] Assistant, National Research Tomsk Polytechnic University, Tomsk, Russia
[3] Engineer, National Research Tomsk Polytechnic University, Tomsk, Russia
E-mail: ntg@tpu.ru

Abstract. The paper presents a new design of a shock-proof two-axis microelectromechanical gyroscope. Without stoppers, the shock load enables the interaction between the silicon sensor elements. Stoppers were installed in the gyroscope to prevent the contact interaction between electrodes and spring elements with fixed part of the sensor. The contact of stoppers occurs along the plane, thereby preventing the system from serious contact stresses. The shock resistance of the gyroscope is improved by the increase in its eigenfrequency at which the contact interaction does not occur. It is shown that the shock load directed along one axis does not virtually cause the movement of sensing elements along the crosswise axes. Maximum stresses observed in the proposed gyroscope at any loading direction do not exceed the value allowable for silicon.

1. Introduction

The recent years have seen scientific and technical and technological advances in microsystems engineering which are widely used in different fields of information technology including microsensors. Microelectromechanical systems (MEMS) are one of the most prospective trends in modern electronics. MEMS technology utilizes methods similar to those applied in chipping technology allowing to create systems during the integration process, which possess mechanical electrochemical, optical and electronic properties and sizes comparable with ordinary integrated circuits.

Gyroscopes and accelerometers are the most popular MEMS devices suitable for numerous applications of consumer and engineering electronics, namely: car's safety systems, navigation and orientation of automotive vehicles, monitoring facilities, non-destructive testing monitoring systems for buildings, industrial facility platform stabilization, medical monitoring devices, oil and gas extraction devices [1–4].

The dissemination of MEMS sensor in all areas of the life is achieved largely by MEMS technology which has become popular and is not expensive. The process of MEMS development and production requires the extensive infrastructure starting from specialized tools and ending with production capabilities which provide specific procedures such as deep silicon etching.

A rapid development of MEMS sensors is conditioned by the following advantages [4–8]:

Microminiature circuits. Microcircuit technology allows producing micromechanical and optical assemblies with considerably less size than it is possible by using traditional technologies.

High operability. The incorporation of microcircuits and sensors, processing circuits and actuating device in one structure allows creating complex comprehensible systems in a miniature housing.

High production efficiency and repeatability. MEMS manufacturing mostly utilizes tried out and controlled operating procedures, which produce products with desired parameters.

Narrow spread of parameters within the product. The production of components in the single technological cycle allows obtaining virtually indistinguishable parameters for the same components.

High-performance operability. Electronics as well as electric communication channels made by the integrated technology and having small sizes allow improving such parameters as operating frequency, signal-to-noise ratio, and others.

Low cost mass production.

At the same time, there is a number of problems and weak points connected with the sensor operation. Operating conditions for MEMS include strong vibrations and impacts. In order for MEMS sensors to be used on mobile objects under ultrahigh inertial loads, the strength analysis is required for the gyroscope structure and conservation of its metrological characteristics [9–15].

The main goal of this paper is to analyse the shock load on the two-axis MEMS gyroscope.

2. Configuration and operation principle

Microelectromechanical (MEMS) gyroscopes are devices that measure angular velocity of various objects. The proposed two-axis MEMS gyroscope comprises the electronic module and the silicon sensor, general configuration of which is illustrated in Figure 1. The silicon sensor converts the input signal (angular velocity) into the electric signal which is then processed by the electronic module to obtain output information.

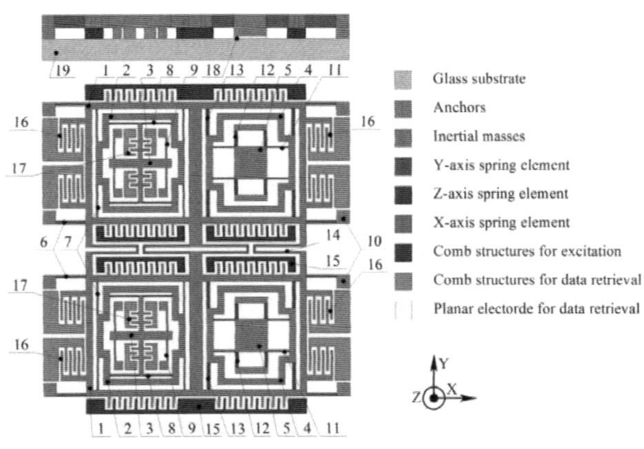

Figure 1. Configuration of MEMS gyroscope silicon sensor: *1* – frames; *2, 4* – decoupling frames; *3, 5* – inertial mass; *6, 8, 11, 14* – *Y*-axis spring elements; *7, 9* – *X*-axis spring elements; *10* – anchors; *12, 13* – *Z*-axis spring elements; *15* – comb drive actuator; *16* – comb structure for data retrieval along *Y*-axis; *17* – comb structure for data retrieval along *X*-axis; *18* – planar electrodes for data retrieval along *Z*-axis; *19* – glass substrate.

This gyroscope measures angular velocities Ω_x and Ω_z around *X*- and *Z*-axes, respectively. Its operation principle is based on the Coriolis effect. The gyroscope induces antiphase primary oscillations of frames (Figure 1) due to electrostatic forces created by the comb drive actuators. The displacement amplitudes of appeared secondary oscillations of inertial masses are proportional to Ω_x and Ω_z angular velocities.

3. Shock loads

The shock load analysis aims at the determination of deformations and mechanical stresses arising in micromechanical elements. The shock load causes the oscillations of the mechanical structure of the silicon sensor. Shock loads are characterized by the acceleration rate, shock duration, and shock pulse waveform.

During the operation, MEMS gyroscope under the shock load should maintain the shock resistance both in on and off conditions and also restore their normal operation after this load. The requirements for operability and conservation of metrological characteristics are not usually regulated.

The structural design of the silicon sensor should be anticipated by the analysis of occurring stresses and the identification of more vulnerable stress locations. The spring element of the silicon sensor is the most vulnerable to the shock load because its rigidity is considerably lower than that of the entire structure. Shock-induced stresses in the spring element should not exceed those allowable for the sensor material.

According to the fourth failure hypothesis [16] of material, the amount of specific potential energy of distortion stored by the deformed element is accepted as the strength criterion. The dangerous condition is experienced when the specific potential energy of distortion achieves its limiting value which is determined by simple tension and compression tests. Hence, the stress condition is defined by

$$\sigma_{eq} \le [\sigma_{lim}] \quad (1)$$

where σ_{lim} is the limit stress of material. The equivalent stress σ_{eq} shown in Figure 2 is calculated as

$$\sigma_{eq} = \sqrt{\frac{(\sigma_1-\sigma_2)^2+(\sigma_2-\sigma_3)^2+(\sigma_1-\sigma_3)^2}{2}} \quad (2)$$

where $\sigma_1, \sigma_2, \sigma_3$ are primary stresses in the structure.

Figure 2 illustrates the schematic layout of the specific potential energy of distortion.

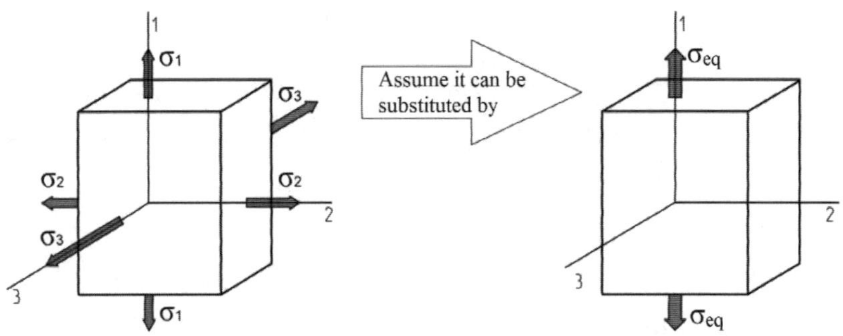

Figure 2. Schematic layout of specific potential energy of distortion.

Let us assume that the shock pulse waveform is a sinusoidal half wave with 3500 Hz amplitude and 500 μs length. The shock load can be either long- or short-term. Its behavior depends on the eigenfrequency ω_0 of the system oscillations and is driven by the shock load.

The relative movement of sensing elements during and after the shock load is the product of the eigenfrequency and duration of the shock pulse $\omega_0 \cdot \tau$. For the proposed gyroscope, $\omega_0 \cdot \tau \approx 33.4$ on each axis, *i.e.* less than unity. So, the shock is long-term, and the movement of sensing elements induced by shock accelerations can be considered as a quasi-static.

In addition to the deformation of spring elements, the contact velocity of electrodes represents a threat to the sensor material. The material fracture due to shock can occur if the shock speed increases the value of [16, 17]

$$V_{im} = \frac{\sigma_{lim}}{\sqrt{E \cdot \rho}}, \quad (3)$$

where ρ is the silicon density; *E* is the Young's modulus.

The allowable shock load is 25 m/s in <110> direction of the silicon single crystal. The shock load is $V_0 = W/\omega_0$ for the considered long-term shock, where W is the shock acceleration at the time of contact. The first three eigenfrequencies of the gyroscope are observed in the vicinity of 10 kHz. As the shock speed is 0.51 m/s, the fracture of material does not occur when contacting.

However, the movement of sensing elements at certain shock parameters exceeds the values of allowable gaps. Thus, the configuration of the silicon sensor implies the contact interaction between its elements. Figures 3–5 contain the simulation plots of the shock load applied to the gyroscope while in operation, however, without the contact interaction.

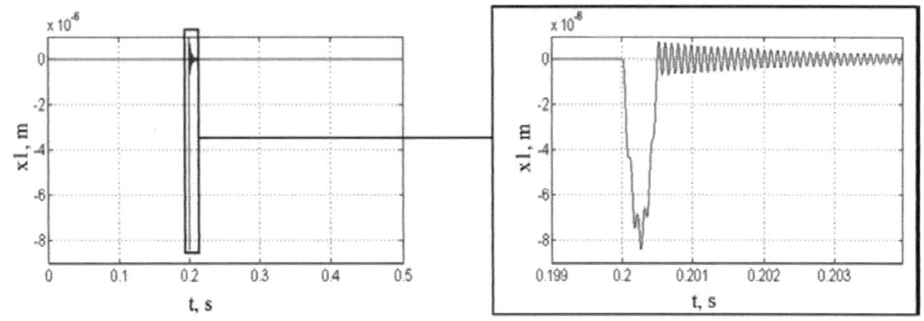

Figure 3. Computer simulation of shock-induced movement of sensing elements along *X*-axis.

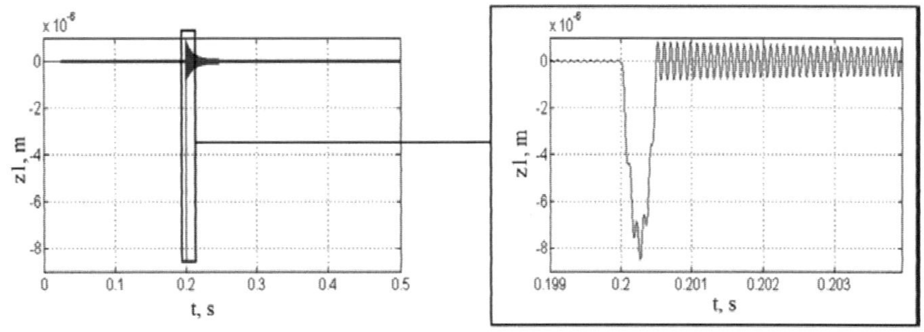

Figure 4. Computer simulation of shock-induced movement of sensing elements along *Z*-axis.

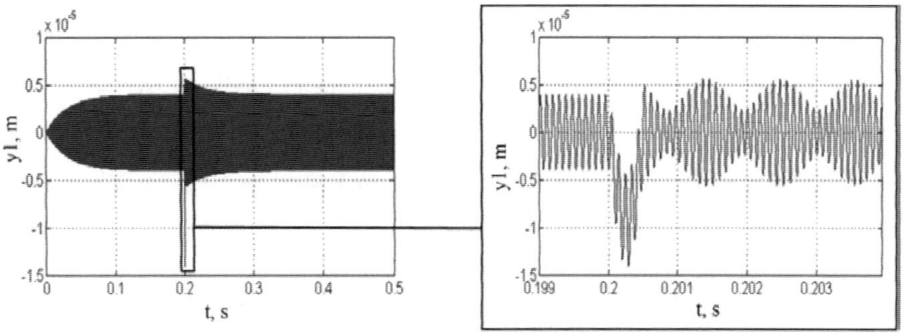

Figure 5. Computer simulation of shock-induced movement of sensing elements along *Y*-axis.

The sensing elements can move to a following distances during the shock loading if no limit is placed on their movement, namely: 9.5×10^{-6} m along *X*-axis; 9.2×10^{-6} m along *Y*-axis and 9.2×10^{-6} m along *Z*-axis. Minimum gaps between moving and fixed parts of the structure are several times less than these distances. So, the contact interaction is observed in both spring element and electrode structures, thereby leading to their fracture and the fracture of the electronic module due to the short-circuit failure.

In order to prevent the silicon sensor from the shock load, the gyroscope configuration is provided with stoppers depicted in Figure 6. These stoppers limit the movement of moving parts within 4 μm. Given that the gap width in comb electrodes is 5 μm, there is no any contact between them, whereas it is observed in places where the stoppers are located.

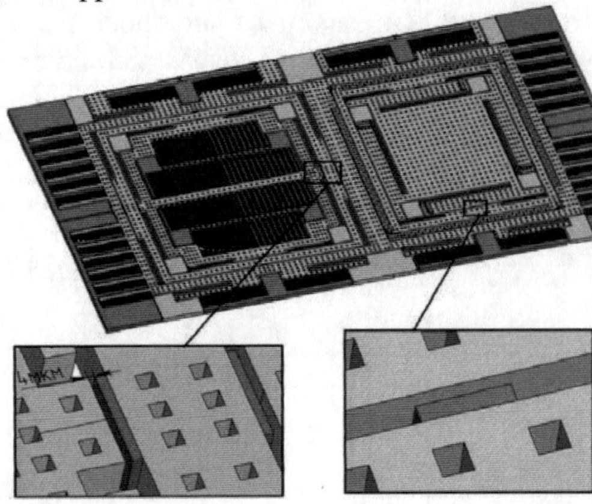

Figure 6. MEMS sensor configuration with stoppers.

The finite element method is used to analyze the contact interactions between the gyroscope structure and the stoppers. The possible directions of sensing element movement, movement of sensing elements before the contact, and occurred stresses are measured successively. This is illustrated in Figures 7–9.

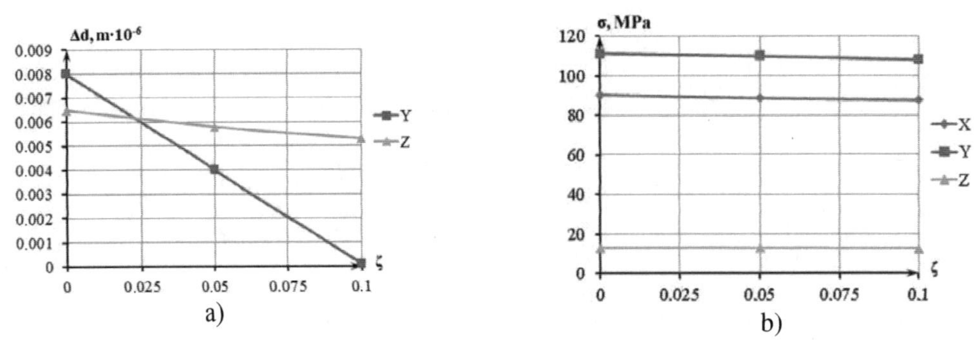

Figure 7. Movements (a) and stresses (b) due to shock on *X*-axis.

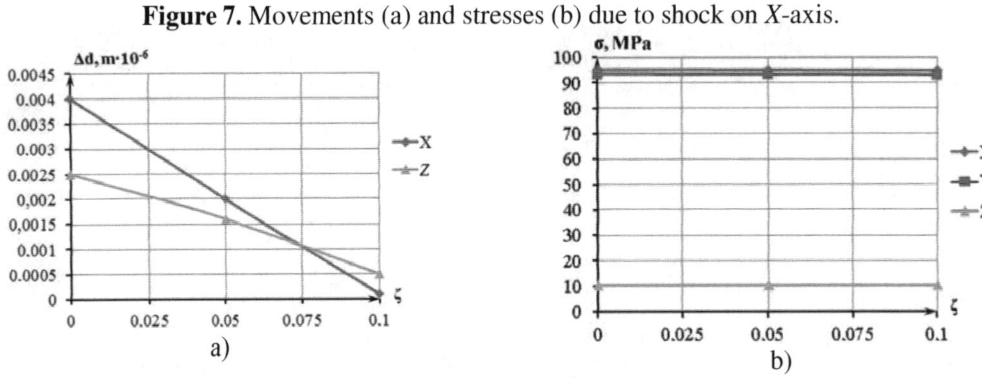

Figure 8. Movements (*a*) and stresses (*b*) due to shock on *Y*-axis.

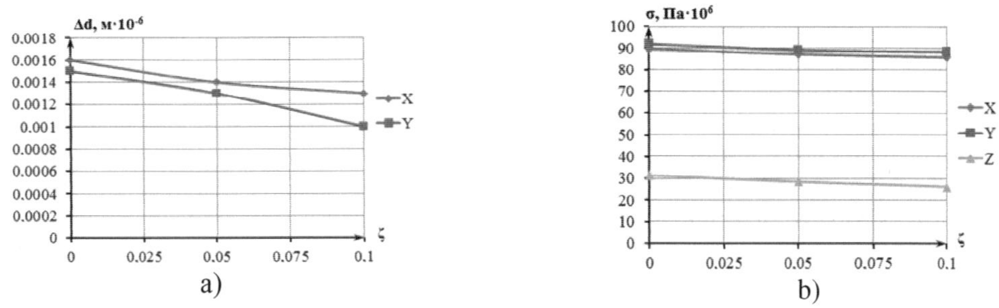

Figure 9. Movements (a) and stresses (b) due to shock on *Z*-axis.

As can be seen from these figures, the movement of sensing elements along the crosswise axes are less, and the occurred stresses are also less than their limit value.

4. Conclusions

In this paper, the authors proposed the shock-proof two-axis microelectromechanical gyroscope. It was shown that without stoppers, the shock load enabled the contact interaction between the different parts of the silicon sensor. Stoppers were installed in the gyroscope to prevent the contact interaction between electrodes and spring elements with fixed part of the sensor. The contact of stoppers occurred along the plane enabling it to prevent serious contact stresses. At the increased shock resistance caused by the increase in gyroscope eigenfrequency up to 13500 Hz the contact interaction did not occur.

It was shown that the shock load directed along one axis did not virtually cause the movement of sensing elements along the crosswise axes. Maximum stresses observed in the proposed gyroscope at any loading direction did not exceed the value allowable for silicon.

Acknowledgment

This work was performed in National Research Tomsk Polytechnic University. The authors like to acknowledge the financial support from the Ministry of Education and Science of the Russian Federation (Agreement № 14.578.21.0232, unique identifier RFMEFI57817X0232).

References

[1] Butakov N A 2004 *International Journal of Open Information Technologies* **2** 2307–8162 (in Russian)
[2] Georgy J, Noureldin A, Korenberg M J, Bayoumi M M 2010 *IEEE Transactions on Intelligent Transportation Systems* **11** 856–872 doi: 10.1109/TITS.2010.2052805
[3] Lane N D, Miluzzo E, Lu H, Peebles D, Choudhury T, and Campbell A T 2010 *IEEE Communications Magazine* **48** 140–150 doi: 10.1109/MCOM.2010.5560598
[4] Khan W Z, Xiang Y, Aalsalem M Y and Arshad Q 2013 *IEEE Communications Surveys and Tutorials* **15** 402–427 doi: 10.1109/SURV.2012.031412.00077
[5] Furfie B 2010 *Engineering and Technology* **5** 34–35 doi: 10.1049/et.2010.0406
[6] Neul R, Kehr K, Döring C, Götz S, Hauer J and Veith M 2007 *IEEE Sensors Journal* **7** 302–309 doi: 10.1109/JSEN.2006.888610
[7] Or E M and Pundik D 2007 *IEEE Transactions on Consumer Electronics* **53** 1508–1512 doi: 10.1109/TCE.2007.4429245
[8] Murari B 2003 *IEEE Micro* **23** 36–44 doi: 10.1109/MM.2003.1209465
[9] Evstifeev M I, Eliseev D P, Rozentsvein D V and Chelpanov I B 2012 *Gyroscopy and Navigation* **3** 51–55 doi: 10.1134/S2075108712010063
[10] Sundaram S et al 2011 *Journal of Micromechanics and Microengineering* **21** 045022 doi: 10.1088/0960-1317/21/4/045022
[11] Naumann M, Dietze O, McNeil A, Mehner J and Daniel S 2013 *TRANSDUCERS and*

EUROSENSORS 2013 570–573 doi: 10.1109/Transducers.2013.6626830

[12] Nekrasov Y A, Moiseev N V, Belyaev Y V, Pavlova S V, Lyukshonkov R G 2017 *Gyroscopy and Navigation* **8** 31–37 doi: 10.1134/S207510871604009X

[13] Younis M I, Jordy D and Pitarresi J M 2007 *Journal of Microelectromechanical Systems* **16** 628–638 doi: 10.1109/JMEMS.2007.896701

[14] Zhou J, Jiang T, Jiao J W and Wu M 2014 *Microsystem Technologies* **20** 137–144 doi: 10.1007/s00542-013-1833-9

[15] Nesterenko T, Barbin E, Koleda A, Zorina E 2016 *International Siberian Conference on Control and Communications* 7491845 doi: 10.1109/SIBCON.2016.7491845

[16] Fedosev V I 1979 *Strength of Materials* (Moskva: Nauka Press)

[17] Starovoitov E I 2008 *Strength of Materials* (Moskva: Fizmatlit Press)

IOP Conf. Series: Materials Science and Engineering **289** (2017) 012004 doi:10.1088/1757-899X/289/1/012004

Development of a software-hardware complex for studying the process of grinding by a pendulum deformer

A P Borisov
Associate professor, Altai State Technical University, Barnaul, Russia
E-mail: boralp@mail.ru

Abstract. The article is devoted to the development of a software and hardware complex for investigating the grinding process on a pendulum deformer. The hardware part of this complex is the Raspberry Pi model 2B platform, to which a contactless angle sensor is connected, which allows to obtain data on the angle of deviation of the pendulum surface, usb-cameras, which allow to obtain grain images before and after grinding, and stepping motors allowing lifting of the pendulum surface and adjust the clearance between the pendulum and the supporting surfaces. The program part of the complex is written in C # and allows receiving data from the sensor and usb-cameras, processing the received data, and also controlling the synchronous-step motors in manual and automatic mode. The conducted studies show that the rational mode is the deviation of the pendulum surface by an angle of 40^0, and the location of the grain in the central zone of the support surface, regardless of the orientation of the grain in space. Also, due to the non-contact angle sensor, energy consumption for grinding, speed and acceleration of the pendulum surface, as well as vitreousness of grain and the energy consumption are calculated. With the help of photographs obtained from usb cameras, the work of a pendulum deformer based on the Rebinder formula and calculation of the grain area before and after grinding is determined.

1. Introduction

Having conducted a study of the state of the food industry in Russia, it can be concluded that the level of development of technical production to the world level does not correspond. Based on the results of the research, 19% of the active part of production assets correspond to the declared level, 20% need modernization, and 42% should be replaced [1].

At present, flour production is developing in two directions:

1. Shorter work schedules are inherent in small agricultural mills and small farms.
2. Full technology, which includes advanced modes of screen separation, grinding, and hydrothermal processing.

In both cases, the fundamental process is the grinding of grain materials, which determines the quality of flour, issuance and energy consumption.

2. Relevance

The technology of multiple grinding of grain operates and the production of intermediate products with their further delivery to flour of various grades, and bran at present time.

As the main equipment of grain processing enterprises for the production of flour, roller machines are used, the main working part of which is a pair of rolls, making rotational movement with different speeds towards each other. This process is partially automated and lasts continuously, but also such production has its drawbacks, namely the high power consumption (up to tens of millions of kWh per year). Also, roller machines have other disadvantages, such as: complex technology; large costs for the production of a rolling mill bed, re-grinding grain, and so on.

Despite the sufficient complexity of the process for obtaining flour in mill production, the production of endosperm from the grains is not more than 63...68%, although the endosperm in it is up to 85%. At present, the processes of additional obtaining of endosperm are perfected only by increasing the number of machines or creating multi-roll machines, which only increases energy consumption.

The most energy-consuming parts of the grinding process are the first and second torn systems, where engines up to 17.5 kW are installed, and on subsequent grinding and grinding systems it reaches 7.5 kW. In large mills the machine stock is up to 15 machines and more, which indicates the complexity and energy intensity of this process.

3. Principle of operation

An alternative to rolling machines is a pendulum deformer, developed by the professor at the Altay state technical university V.L. Zlochevsky [2]. The principle of operation of this installation is as follows. To destroy the grain, two cylindrical surfaces are used, one of which is the outer surface of the cylinder and the other is the inner surface.

After grabbing the grains (Figure 1), the process of elastic deformation is performed. When the ultimate strength of the shell is reached, the grain breaks down to maximize the opening of the endosperm. Such destruction occurs at the weakened place of the grain - groove. Comparing such a process with grain cutting, which occurs when using corrugated roll systems, we find that using a pendulum deformer, most of the grain will be used to produce flour.

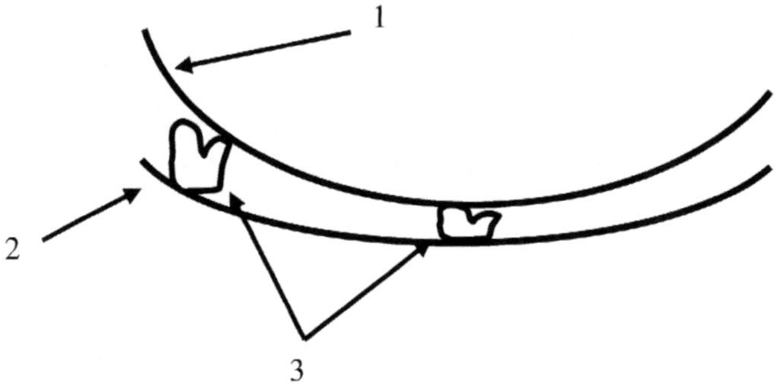

Figure 1. Interaction of grain with surfaces: 1 – pendulum surface; 2 – bearing surface; 3 – grain.

The pendulum surface has amplitude and frequency of oscillations, which allows controlling the speed, acceleration of the pendulum and energy costs for grinding. In order to make it possible to change the methods of grain deformation, change the size of the gaps, etc., the supporting (lower) surface of the pendulum is regulated. Using such control it becomes possible to change the geometry of the working area, as well as the working surfaces [2].

The general view of the pendulum deformer is shown in Figure 2.

The pendulum deformer consists of a base on which a rack, guides and a lead screw are installed. A pendulum is mounted on the stand with a shaft. A stepper motor is connected to it. The curved deck is fixed on the pendulum. Based on the substrate, located under the pendulum. The substrate has a concave structure. To the substrate, as well as to the pendulum, a stepper motor is connected. Its necessity consists in that it can be used to adjust the height of the substrate, thereby changing the angle of the meeting of the grain with the surface of the pendulum and the degree of grain refinement [3].

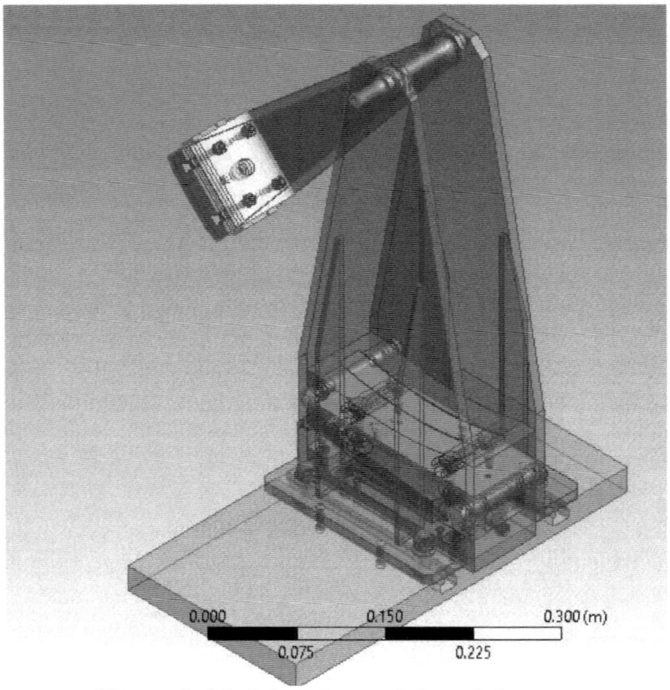

Figure 2. Model of the pendulum deformer.

A distinctive feature of the pendulum deformer from roller is that it does not require constant rotation of the working cylinder. Also, in this installation, the problem with overheating of the grains and working surfaces was solved, since the whole grinding process proceeds very quickly, and the working surfaces of the pendulum do not touch each other.

The pendulum deformer can be used in appropriate dimensions, such as a pre-system in front of roller machines, and in laboratory grinds to adjust the operation of rolling machines, as well as in determining the properties of the grain. To control the grinding process and to study the properties of grain, a software and hardware complex was developed.

4. Software and hardware

The automated system consists of the following elements:

- Pendulum deformer, with an angle sensor connected to it, USB-cameras and stepper motors.
- Software on a personal computer.
- Microcomputer Raspberry Pi B +.

The connection diagram of all the elements of the system is shown in Figure 3 [4].

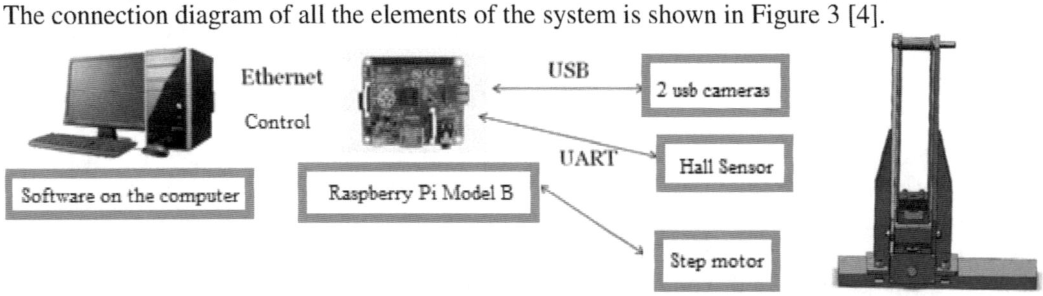

Figure 3. The device of the milling products formation system.

Determination of the energy of grain destruction is based on the identification of unambiguous values of the stored potential energy that can be obtained at the extreme points of the pendulum surface with a damped oscillatory process. It is also necessary to take into account the energy losses due to friction. Thus, the sequence of actions of the method:

- retraction of the pendulum surface by a certain angle α_1 by means of a stepper motor controlled from a personal computer;
- placing the test grain on the support surface;
- release of the pendulum surface;
- fixing the angle α_2 on the opposite side;
- calculation of energy costs of the pendulum deformer, glassiness and strength of grain.

It should be noted that the definition of losses is possible in two ways:

- making an analytical calculation using known theoretical information;
- calculating the losses for the standard oscillation of the pendulum, which is taken to a given angle.

In this paper, the first method is used, although it is possible that in the future the second method will be used.

It is known that for a pendulum:

$$E = m \cdot g \cdot l_m \cdot (1 - \cos\alpha),$$

where m – mass of the pendulum, E – pendulum-stored potential energy, α – angle of deviation of the pendulum, l_m – reduced length of the pendulum. Proceeding from the fact that the damping decrement of the pendulum is equal to:

$$d = ln\left(\frac{\alpha_1}{\alpha_2}\right) = \beta T,$$

where T – period of fluctuation,

it is possible to calculate the energy dissipated in one period:

$$E_{\text{p}} = m \cdot g \cdot l_m \cdot (cos\alpha_2 - cos(e^{-\beta T}\alpha_1)) .$$

Therefore, in order to determine the energy value for the destruction of the grain, it is necessary to calculate the damping decrement and the period of natural oscillations of the pendulum for a given angle of initial deviation, and also fixing the extreme positions of the surface of the pendulum.

To determine the angular position of the pendulum surface, an industrial encoder was first used to convert the value of the angle into a sequence of pulses. However, due to the low resolving power (0.25 °), as well as the high cost of the copies with a higher resolution, it was decided to abandon this method of angle measurement. An alternative to the encoder was the non-contact magnetic proximity sensor KMA200 from NXP, which indicated a resolving power of at least 0.04 ° in digital mode. As it turned out, in the analog mode it is possible to obtain higher values, which was the reason for choosing the analog mode of the sensor operation. The analog interface of the sensor is a differential analog output, varying within 5...95% of the supply voltage, depending on the angular position of the magnet. By the experimental method it was possible to obtain the resolution of the sensor at the analog output 0.01 ° [5].

To control the grinding process using a pendulum deformer and control its characteristics, the sensors use a Raspberry Pi model B + microcomputer with the control of the developed program. The Raspberry Pi model B + microcomputer controls the lifting of the pendulum surface and the adjustment of the reference surface by means of synchronous stepper motors and the transfer of all collected information from a contactless angle sensor to a computer. The software installed on the computer allows you to view the incoming information in an accessible format, and also allows you to control the pendulum deformer using the command system for the microcontroller and, accordingly, control the stepping motors. Also, the speed and acceleration of the pendulum surface are calculated based on the data obtained from the angular position sensor KMA200.

USB-cameras are used to determine the perfect work of the pendulum deformer. For this we use the formula proposed by P.A. Rebinder:

$$A_p = k_v \cdot k_n \cdot V_m + \alpha_{пов} \cdot \Delta S$$

where A_p – energy consumption for destruction; k_v – coefficient considering what part of the volume of the particle deforms; k_n – coefficient characterizing the physical and mechanical properties of the body being destroyed; V_m – volume of the body being destroyed; $\alpha_{пов.}$ – specific surface energy of the body being destroyed; ΔS – formed when a new surface is destroyed.

Using the openCV library, the grain area is calculated before grinding and after grinding, which allows the software to perform calculations using the Rebinder formula. Images are obtained with USB cameras, by constantly comparing the photos until they become the same, that is, the grinding process stops. On the basis of these photographs, namely the photograph of the grain after grinding, information is taken on the area of the crushed grain, which in turn is used in calculating the work of the pendulum deformer.

The software (Figure 4) for controlling the hardware of the complex is written in the C # programming language. To connect the microcomputer with the software and with the pendulum deformer, a wired connection is used via the Ethernet port.

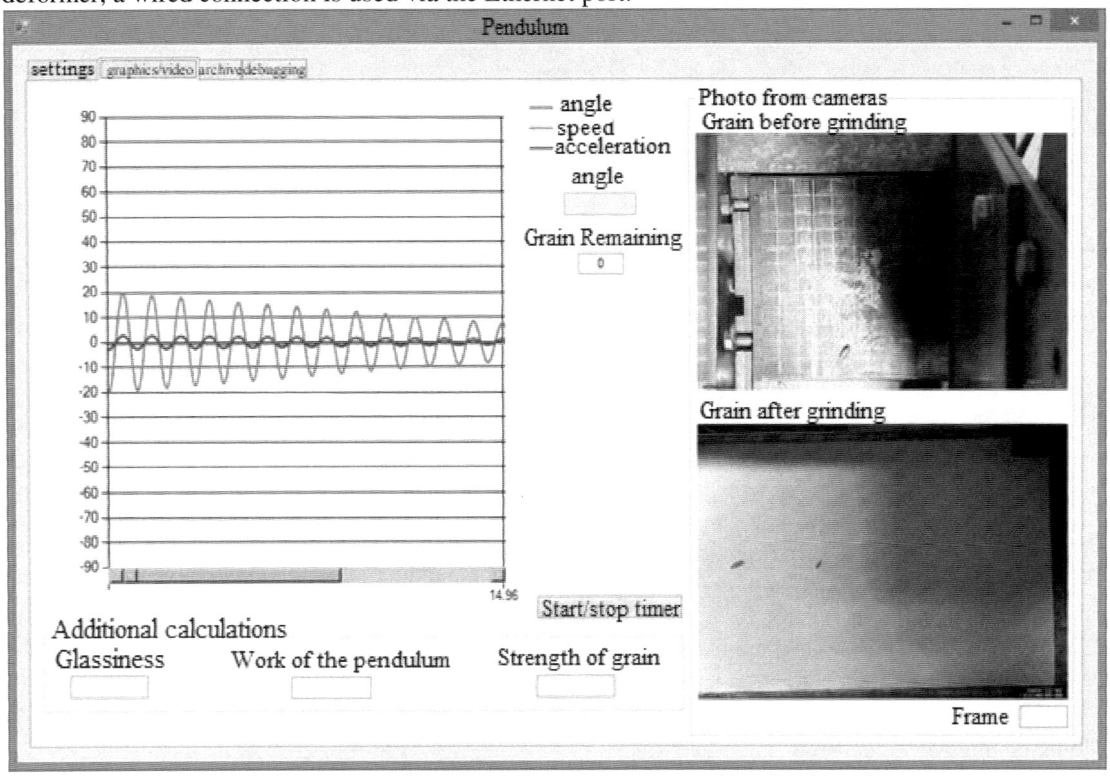

Figure 4. Main program screen, graphics / video tab.

A series of experiments revealed the regime of maximum efficiency of the pendulum deformer, shown in Figure 5.

At the current time, there are the following grain characteristics and tuning of the pendulum deformer, which were experimentally verified and confirmed:

1) Position of the grains on the deck (in the beginning, in the middle, in the end).

2) Angle of deviation of the pendulum (deviates by 30^0, 40^0, 45^0);

3) The angle of rotation of the grain in the square of the deck (can lie - vertically, horizontally, diagonally).

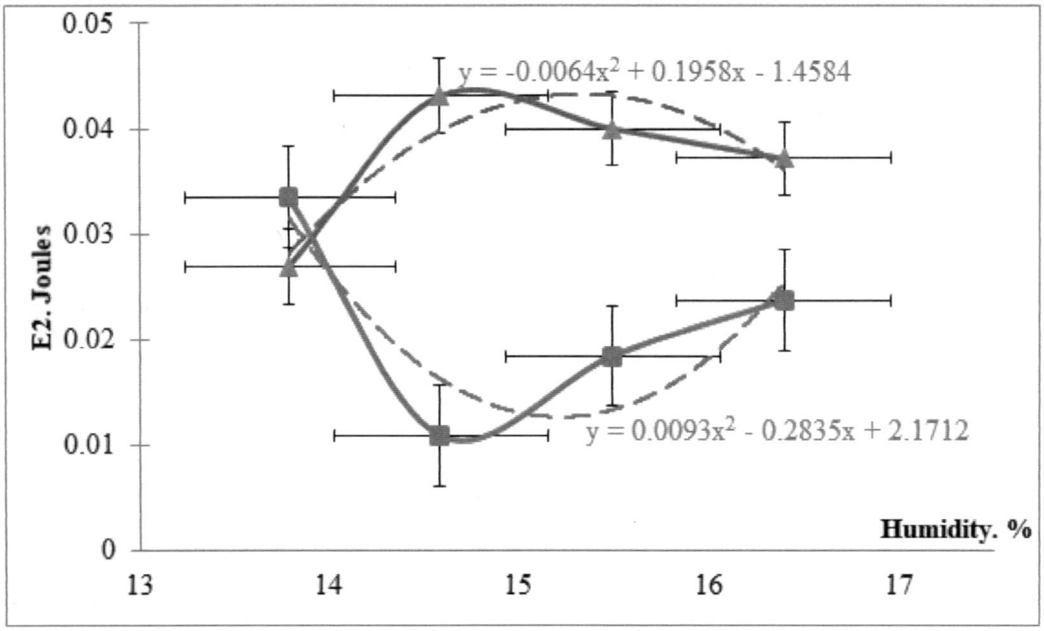

—■— The vertical arrangement of the grain, the deviation of the surface from the vertical by 45 degrees, the location of the grain on the supporting surface, shifted from the plumb zone by 40 mm down.

—▲— The horizontal arrangement of the grain, the deviation of the surface from the vertical by 30 degrees, the location of the grain on the supporting surface, shifted from the plumb zone by 40 mm down.

Figure 5. Range of rational use of the pendulum deformer.

Analysis of the experimental data shows:

1. Depending on the selected regime for one grain, energy costs for the destruction of grain materials are from 0.01 to 0.05 Joules. When grinding on a roller machine, the energy costs are of the order of 0.114–3.24 Joules per grains.
2. Since in grains 1, 6, 8 the grain lies at the beginning of the deck, the kinetic energy reserve is minimal and there is no grain breakage. Therefore, these regimes are not rational.
3. Mode No. 9 has the greatest deviation of the pendulum surface and the grain is located at the beginning of the support surface, then the greatest energy consumption (on average 0.042 Joules) and the maximum speed of the pendulum movement from all modes are observed.
4. Since the grain is located in the middle of the support surface, where the largest values of speed and kinetic energy, due to the location of this zone at the point of the plumb and the angle of the deviation of the pendulum surface 40^0, the most rational is mode No. 5, having throughout the experiments average values of energy consumption 0.026–0.028 Joules) (Figure 6).

Thus, the characteristics of the rational operating mode of the pendulum deformer are the deviation of the pendulum deformer by an angle of 40°, the horizontal arrangement of the grain material, and its location along the center of the support surface.

The software in the beginning is being taught on different batches of grain with known parameters to study the glassiness. It can be concluded that when the glassiness decreases, the energy consumption for grain refinement decreases, proceeding from the studies on grain refinement (Figure 7) [6].

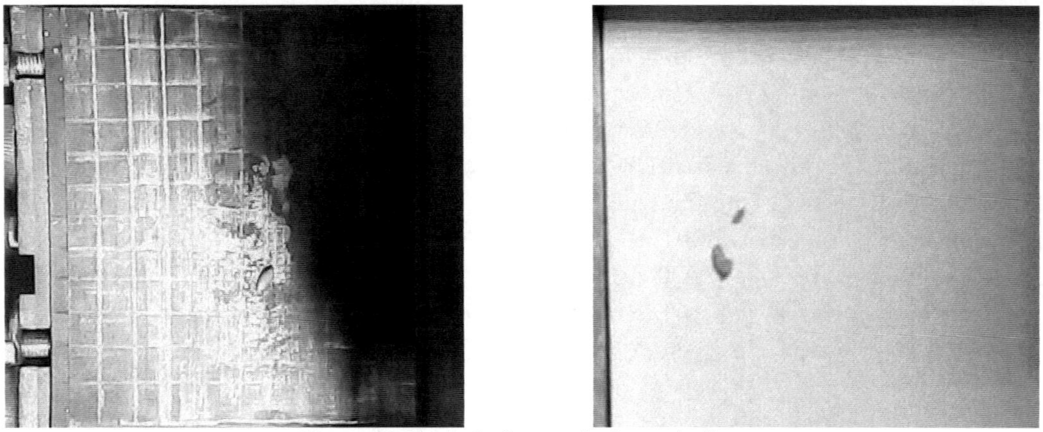

Figure 6. Grain before and after grinding.

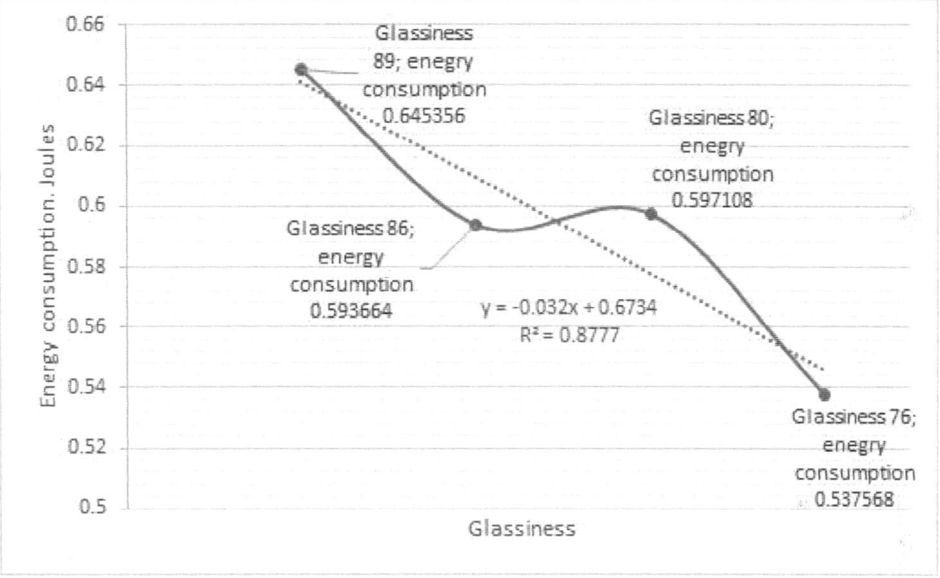

Figure 7. Dependence of energy consumption on grinding grain from glassiness.

5. Conclusion

The result of the work was the development of a controller for the non-contact determination of the angle of the pendulum surface position in order to control the stepper motors and determine the energy consumption for grinding grain. The conducted laboratory tests using the software and hardware complex have shown that the energy consumption when grinding grain on a pendulum deformer is less than when grinding on a roller machine. Also the use of USB-cameras allows to measure the work of the pendulum deformer, and thanks to the "learning" of the program – to measure the glassiness of the grain without violating GOST.

Acknowledgement

The study was carried out with the financial support of the Russian Foundation for Basic Research in the framework of the scientific project No. 17-48-220056.

References

[1] Lyukshin V S, Barsuk A V and Fazleev R R 2015 *IOP Conf. Ser.: Mater. Sci. Eng.* **91** 012047 doi: 10.1088/1757-899X/91/1/012047

[2] Zlochevskiy V L, Zlochevskiy A V 2005 *Sposob formirovaniya zernovykh produktov razmola* [A method of forming a grinding cereals] Patent RF № 2263544

[3] Edakin N V, Borisov A P, Zlochevskiy V L 2017 *Proceeding of Technologies for the food and processing industry of aic - healthy food* 84–92 (in Russian)

[4] Edakin N V, Borisov A P 2016 *Avtomaticheskoe upravlenie protsessom izmel'cheniya zerna* [Automatic control of the grain milling process] Patent RF № 2016610021

[5] Solopov V S, Borisov A P 2013 *The Polzunovsky Herald* 161–164 (in Russian)

[6] Zlochevskiy V L, Borisov A P 2015 *Proceeding of Mezhdunarodnoe nauchnoe periodicheskoe izdanie po itogam Mezhdunarodnoy nauchno-prakticheskoy konferentsii* 20-23 (in Russian)

IOP Conf. Series: Materials Science and Engineering **289** (2017) 012005 doi:10.1088/1757-899X/289/1/012005

Hybrid membranes for fuel cells

S S Bochkareva[1], S S Shashkina[2]
[1]Associate professor, Irkutsk National Research Technical University, Irkutsk, Russia
[2]Student, Irkutsk National Research Technical University, Irkutsk, Russia
E-mail: sv.b58.00@mail.ru

Abstract. Fuel cells are a very efficient, reliable, durable, and environmentally friendly energy source. Membranes for fuel cells were developed based on nitrogen-containing high-molecular compounds and organic–inorganic composites. Their electrical conductivities were measured. The influence of a silicon block of composites on the proton exchange properties of membranes was proved.The comparative characterization of the studied materials was performed.

1. Introduction

The recent studies on the development of fuel cells show the clear trend to the transition from the classic fuel cells with a liquid electrolyte to fuel cells based on a proton exchange polymer membrane (Figure 1).

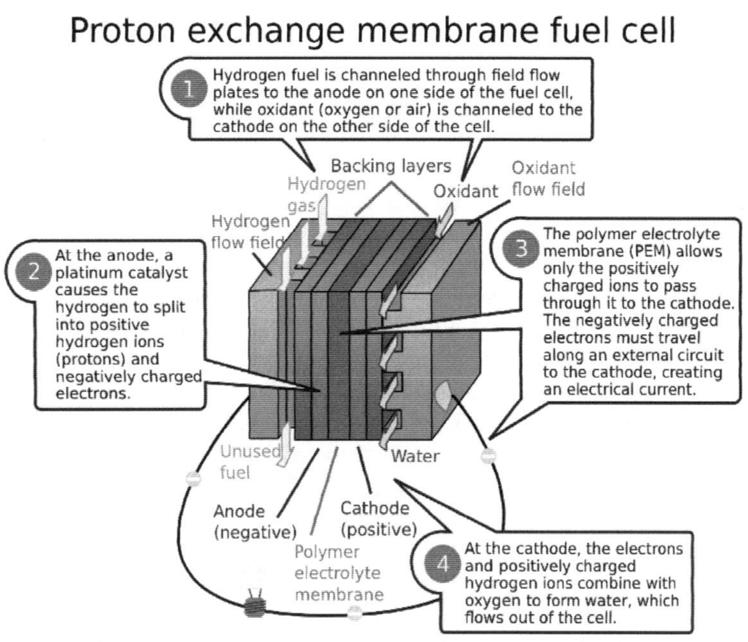

Figure 1. Proton exchange membrane (PEM) fuel cell.

The main function of a proton exchange membrane is the transfer of a proton formed as a result of the hydrogen ionization on an anode into the cathode region. A number of requirements are imposed on the proton-conducting membrane. For the efficient work of a fuel cell, the ohmic loss on a membrane must be minimized, which is achieved in case of high proton conductivity of a membrane. The specific conductivity at the level of 10^{-1}–10^{-3} S/cm is considered sufficient for the use in fuel cells. A membrane must possess significant mechanical strength so that electrodes could be pressed to a membrane. Besides water, highly chemically active hydrogen peroxide can be formed on a cathode as a result the side reaction, thus, a membrane must possess high chemical resistance to both the oxidation and the hydrolysis. During the operation of a fuel cell, water released on a cathode is partially absorbed by a membrane, which leads to its swelling and, thus, the change in its geometric dimensions. Thus, it is desirable that the degree of swelling of a membrane were as low as possible [1].

Most of these requirements are met by the membranes based on hybrid organic–inorganic polymers [2, 3]. Similar characteristics that determine the prospects of the use in diverse electrochemical devices are typical for membranes based on spatially cross-linked interpolymer complexes synthesized, e.g., by the reaction of epichlorohydrinhomopolymers or copolymers of epichlorohydrin and ethylene oxide with poly-4-vinylpyridine. The ionic conductivity of such ammonium interpolymer complexes activated by orthophosphoric acid and additionally moistened in water is $5.5 \cdot 10^{-3}$ S/cm at room temperature. They are thermally stable (at up to 250°C) and possess the conductivity of 10^{-2}–10^{-4} S/cm at the temperature of 150°C [4, 5].

Compared to the commercial "Nafion" perfluorinated proton exchange membranes [1], products of the sol-gel synthesis with the participation of alkoxysilanes possess a number of advantages [6–10]. The function of the silicon block of such systems is mainly reduced to improving the transport of protons generated by a polymer electrolyte owing to the surface phenomena. These membranes show high proton conductivity both at room temperature and at the temperature higher than the boiling point of water (up to 120–140°C), which is unavailable for polymer membranes. It was shown that the proton conductivity of such membranes in a wide temperature range is provided by silanol groups and functional groups of the organic block doped with mineral acids. The structure of sol-gel synthesis products is determined by the combination of the hydrophobic part and side chains containing ionogenic groups (the hydrophilic part) in their composition. In the presence of water in a polymer, it is localized near ionogenic groups forming the nanosized hydrated region. In this region, besides water, there are different hydrated forms of a proton, which move freely. Thus, the hydrophilic part of the polymer provides the efficient proton transport while the hydrophobic part stabilizes the morphology of membranes providing their mechanical strength.

The described prospects of products of the sol-gel synthesis with the participation of alkoxysilanes and ionogenic organic polymers for creating proton exchange membranes brought us to the study of the transport properties of products based on the synthesized composites.

The aim of the study is to reveal the effect of the silicon block of the composites on the proton exchange properties of the formed membranes. For more correct evaluation of this effect, first, the specific electrical conductivity of the studied polymer systems not containing the silicon block, namely, poly-1-vinylpyrazole (PVP), poly-2-methyl-5-vinylpyridine (PMVP), copolymers of 2-methyl-5-vinylpyridine with vinylchloride (MVP–VC), 2-methyl-5-vinylpyridine with vinylacetate (MVP–VA) and copolymers of 1-vinylpyrazole with vinyltetrahydroindole (VP–VTHI), and 1-vinylpyrazole with vinylacetate (VP–VA), was determined.

2. Experimental

The membrane was formed from solutions of polymers and copolymer in dimethylformamide (DMFA). For increasing the membrane elasticity, a film former (polyvinylbutyral) was added to the initial (co)polymer solution. The films were created by pouring, i.e., by the uniform distribution of homogenized mixtures of (co)polymer solutions with a film former on the glass surface. The surface activation, after drying and heat treatment of films at 110^0C, was performed by doping the formed

IOP Conf. Series: Materials Science and Engineering **289** (2017) 012005 doi:10.1088/1757-899X/289/1/012005

membranes with orthophosphoric acid solutions. The electrical conductivity of films was measured by complex impedance under the isothermal mode.

3. Results and discussion

Under doping films with a 9 M H_3PO_4 solution, for homopolymers of PVP and PMVP and for MVP–VA copolymer, the destruction of their integrity is observed. Films created using solutions of MVP–VC copolymer under doping are not destroyed. The same films are characterized by the highest specific electrical conductivity values (Table 1). Keeping the integrity under doping is also typical for VP–VTHI–VA copolymers. The results obtained for VP–VTHI copolymers indicate the increase in the electrical conductivity under the increase in the fraction of the heterocyclic fragment in the copolymer composition (Table 1). A noticeable increase in the elasticity and electrical conductivity of films is caused by the introduction of polyvinylacetate (PVA) in the composition of the initial mixture (the VP–VA copolymer case) as the second film former (Table 1).

Table 1. Proton conductivity of polymer membranes based on homopolymers and copolymers.

No	Membrane basis (polymer, copolymer composition)	Specific conductivity, S/cm
1.	PVP	2.8×10^{-5}
2.	PMVP	6.7×10^{-5}
3.	MVP–VC (80:20 mol %)	9.1×10^{-3}
4.	MVP–VC (90:10 mol %)	6.7×10^{-3}
5.	MVP–VA (67:33 mol %)	5.5×10^{-5}
6.	VP–VTHI (91: 9 mol %)	2.17×10^{-4}
7.	VP–VTHI (76:24 mol %)	3.5×10^{-5}
8.	VP–VTHI (14:86 mol %)	3.0×10^{-5}
9.	VP–VA (80:20 mol %)	1.8×10^{-5}
10.	VP–VA (80:20 mol %), PVA	7.65×10^{-3}

The specific electrical conductivity was measured by the van der Pauw Method four-point probe technique at the frequency of 500 Hz and the temperature of 25 °C.

The results (Table 1) show that the electrical conductivity of polymer membranes based on homopolymers and copolymers is insufficiently high. It was assumed that the electrical conductivity of membranes can be increased by the formation of films based on organic–inorganic composites (Figure 2).

With consideration of the high hydrolysis rate of alkoxy- and chlorosilanes used for the synthesis of composites, the main goal of the film formation was slowing down the hydrolytic polycondensation process with the purpose of synthesizing soluble products. The full hydrolysis of these silanes leads to insoluble products, based on which the formation of elastic films is impossible.

Among the studied polymer systems, only composites based on alkoxysilanes proved to be applicable for it. Regulating the rate of the autocatalytic process of the hydrolytic polycondensation of trichlorosilanes is a highly complex and, in some cases, insoluble problem.

The initial solutions of composites for the formation of polymer films were synthesized by the hydrolysis of tetraethoxysilane (TEOS), chloromethyltriethoxysilane (CMTES), and 3-aminopropyltriethoxysilane (APTES) in the presence of (co)polymers from water–alcohol solutions in the absence of a catalyst (alkali or acid one). The product of such a process was a soluble hydrolysate, which was a mixture of polyorganylsiloxanes, mainly linear-structured ones, containing a significant number of free silanol groups (scheme 1, the case of homopolymers) [11].

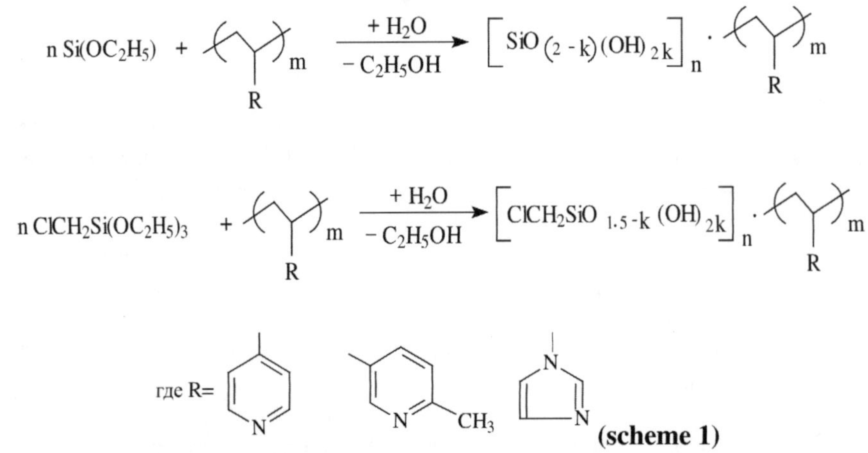

(scheme 1)

Figure 2. Electron microscopic image of surface composite $C1CH_2SiO_{1.5}$:PVP.

Thus, the scheme of the synthesis of polymer membranes based on composites is described by the following phases:

1. The hydrolysis of alkoxysilanes from water–alcohol solutions in the presence of (co)polymers and in the absence of a catalyst under different silane–organic (co)polymer ratios. Thefractionofthesiliconblockwasvariedwithconsiderationofthenecessaryelasticityofthefinalsamplesanddidnotexceed 10 mol %.
2. The introduction of a film former (polyvinylbutyral or PVB) in the resulting hydrolysate in the ratio composite: PVB = 1 : 0.04, the homogenization of the mixture.
3. The application of the mixture of a hydrolysate and a film former on the glass surface.
4. Curing films in the process of drying in air.
5. The heat treatment at 1100C.
6. The separation from the substrate (glass) and the membrane activation with a 9 M orthophosphoric acid solution.

In accordance with the described scheme, a series of films were created. However, not all the samples were applicable for further operation.

Films based on 3-aminopropyltriethoxysilane were brittle, and the attempts to dope them with a H_3PO_4 solution led to their full destruction. Also, films based on composites including 1-vinylimidazole were of low quality.

All films based on composites including TEOS and two samples based on CMTES were applicable for the proton conductivity study.

The data shown in Table 2 clearly show the total increase in the electrical conductivity of films based on composites compared to films based on homopolymers and copolymers. Higher specific conductivity values are shown by polymer films including MVP fragments. The increase in the content of the silicon block in the film composition shown for the case of a SiO_2: MVP–VC copolymer system (nos. 5–7, Table 2) indicate the increase in the electrical conductivity under the increase in the SiO_2 content in a composite. The influence of the number of silanol groups on the transport properties is shown by the comparison of the conductivity of films based on TEOS and CMTES. The CMTES hydrolysis product, because of the presence of chloromethyl substitute, other things being equal, has less silanol groups compared to the analogous TEOS hydrolysis product (scheme 1). This results in a significant decrease in the specific electrical conductivity value (nos. 1, 2 and 3, 4, Table 2).

An indirect proof of the crucial importance of the silicon block for the achievement of the high proton conductivity are the data obtained for a membrane based on TEOS and copolymers of 1-vinylpyrazole and VTHI. Rather high transport activity of these samples (nos. 9, 10, Table 2), in the absence of a clearly marked predominant center, is mainly caused by the presence of the silicon block in its composition. Similar results were shown for samples based on TEOS and copolymers of 1-vinylpyrazole and methyl methacrylate (MMA) (nos. 11, 12, Table 2). However, in this case, the manifestations of the proton exchange activity owing to the hydrolysis of ester group and, thus, the formation of an additional proton generation center cannot be excluded.

The studied membranes are films of the thickness of 40–50 μm characterized by high heat resistance (up to 300°C) and good mechanical properties: the tensile strength is 51.1 ± 2.3 MPa, the relative elongation is 25 ± 5 %.

Table 2. Proton conductivity of polymer membranes based on composites.

No.	Composite composition (molar ratio of components)	Specific conductivity, S/cm
1.	SiO_2 : PVP (1 : 19)	5.6×10^{-3}
2.	SiO_2 : PMVP (1 : 19)	2.0×10^{-2}
3.	$ClCH_2SiO_{1.5}$: PVP (1 : 19)	7.6×10^{-4}
4.	$ClCH_2SiO_{1.5}$: PMVP (1 : 19)	3.0×10^{-4}
5.	SiO_2 : copolymer MVP–VC (1 : 19)	1.2×10^{-3}
6.	SiO_2 : copolymer MVP–VC (1 : 13)	8.5×10^{-3}
7.	SiO_2 : copolymer MVP–VC (1 : 9)	4.0×10^{-2}
8.	SiO_2 : copolymer MVP–VA (1 : 19)	1.0×10^{-2}
9.	SiO_2 : copolymer VP–VTHI (1 : 19)	1.1×10^{-3}
10.	SiO_2 : copolymer VP–VTHI (1 : 13)	1.7×10^{-3}
11.	SiO_2 : copolymer VP–MMA (1 : 19)	7.8×10^{-3}
12.	SiO_2 : copolymer VP–MMA (1 : 13)	1.4×10^{-3}

4. Summary

Thus, copolymers and composites based on nitrogen-containing heterocyclic compounds are materials that are of interest for the development of proton exchange membranes for hydrogen–air fuel cells, which can compete with the best Russian and foreign analogs in terms of performance characteristics.

References

[1] Dobrovolsky YA, Chikin AI, Sanginov EA et al. 2015*Alternative Energy and Ecology (ISJAEE)* 22–45 doi:10.15518/isjaee.2015.04.02(In Russian)
[2] Bonnet B, Jones D J, Roziere J et al. 2000*Electrochem. Syst* **3** 87–92

[3] Pichonat T, Gauthier-Manuel B, Hauden D 2004 *Chemical Engineering Journal* **101** 107–111 doi:10.1016/j.cej.2004.01.019
[4] Ivanchev S S, Mjakin S V 2010 *Russian Chem. Reviews* **79** 101–117 doi:10.1070/RC2010v079n02ABEH004070
[5] Lebedeva O V, Sipkina E I, Pozhidaev Yu 2016 *Membranes and membrane technology* **6** 138–143 doi:10.1134/S2218117216020097
[6] Kuznetsova O V, Kuznetsova A N, Begunova L A 2017 *IOP Conf. Series: Materials Science and Engineering***189** 012007 doi:10.1088/1757-899X/189/1/012007
[7] Orhokova E A 2016 *Izvestiya vuzov. Applied chemistry and biotechnology* **6** 15–22 doi: 10.21285/2227-2925-2016-6-2-15-22
[8] Lebedeva O V, Chesnokova A N, Badlueva T V et al. 2015 *Petroleum Chemistry* **55** 333–338 doi:10.1134/S096.5544115040040
[9] PozhidaevYu N, Lebedeva O V, Bochkareva S S et al. 2008 *Russian Journal of Applied Chemistry* **81** 1837–1841 doi: 10.1134/S1070427208100248
[10] Pozhidaev Yu N, Lebedeva O V 2012 *Russian Journal of Applied* **85** 244–247 doi:10.1134/S1070427212020140
[11] Bochkareva S S 2016 *Izvestiya vuzov. Applied chemistry and biotechnology* **6** 81–93 doi: 10.21285/2227-2925-2016-6-3-81-93

IOP Conf. Series: Materials Science and Engineering **289** (2017) 012006 doi:10.1088/1757-899X/289/1/012006

The in-process control of PVC sheath of a double core cable

N S Galeeva[1], V V Redko[2] and L A Redko[2]
[1] Assistant Professor, National Research Tomsk Polytechnic University, Tomsk, Russia
[2] Associate Professor, National Research Tomsk Polytechnic University, Tomsk, Russia
E-mail: nadushasns@sibmail.com

Abstract. In this work the possibility of the sheath hermiticity testing by measuring of the cable capacity per unit length variation during spark testing is considered. The research object is 2×0.75 HO3VVH2-F cable. According to the physical modelling it is proved that such defect of sheath as pinhole through the whole thickness of sheath can be registered for the test length 10 cm with test voltage frequencies 1kHz and 10kHz.

1. Introduction

Cable products are applied for transmission of electrical energy and information at a distance. Cable products are long-length objects, so its production is a resource-intensive process. Thus, to decrease mass marriages and production costs a regulation of the process should be conducted at time. It is possible to achieve with quality in-process control [1–5].

One of the most typical cable construction parts is nonmetallic cable sheath [1–12]. This construction part is applied for overall mechanical, chemical, weather and electrical protection.

According to the regulatory documentation [13, 14] the in-process control of nonmetallic cable jacket is carried out by spark testing. According to this testing method the cable core is grounded and a high test voltage is applied to the insulation surface with specialized electrode. When an insulation flaw is passing through the controlled area, a breakdown occurs and is registered by the automatics.

The regulatory documentations for each type of cable products determine the requirements for the nonmetallic cable sheaths. There are two main requirements. First of all, decreasing of a sheath thickness must not be excess 15…25 % from the defect-free sheath for each type of cable products. Secondly the cable sheath must be hermetical [15–18].

Nowadays, the in-process control of sheath hermiticity is carried out only for sheaths applied over a metallic shield or an armour. Otherwise the test voltage is not possible to apply to the cable sheath. But the hermiticity of cable sheath must be tested for any cable construction according to the regulatory documentation. Thus, the current problem is to develop a method that is able to test the nonmetallic cable sheath which is applied over the nonmetallic parts of cable.

In the previous article [19–24] the method was offered for cable capacitance recording during spark testing. In this article the possibility of the hermiticity test of cable sheath is explored. This research is conducted for cable type 2×0.75 HO3VVH2-F (it is European analog of Russian cable SHVVP 2×0.75).

2. Object of testing

2×0.75 HO3VVH2-F cable is a flexible double core cable with a double PVC coating (Figure 1).

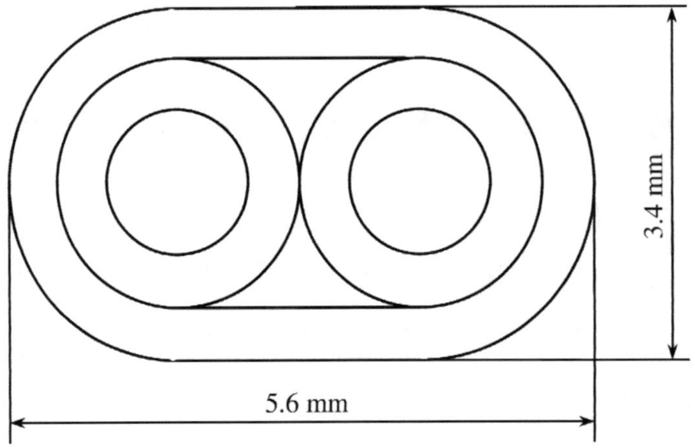

Figure 1. Construction of 2×0.75 HO3VVH2-F.

Processing of this cable has 2 stages. On the first stage melted plastic pellets are applied over conductor during the extrusion process. On the second stage melted plastic pellets are applied over pair of insulated core. According to the regulatory documentation the sheath and the insulation of its cable must be able to be separated freely [15–18]. Thus, there is no adhesion between these elements. Because the cable work voltage is 380 V (for Russian analog), gaps between the insulated cores and the sheath are not filled with insulating material.

3. Hypothesis

In the absence of adhesion it is suggested that there is a thin air layer in the place of insulation and sheath contact. The value of a high voltage for the spark testing is chosen in accordance with the standards [13, 14]. While the high testing voltage is applied to the cable surface, the initial value of electric field strength for the air gaps is much more the dielectric strength of air. Thus, the air gaps between the insulation and sheath in a strong field may be considered as conducting areas. Based on this hypothesis the mathematical model of the cable has been developed in the program of finite element analysis. Using this model the electric field and potential distributions for cable cross section during the spark testing have been obtained. In the Figure 2 the sheath is defect-free.

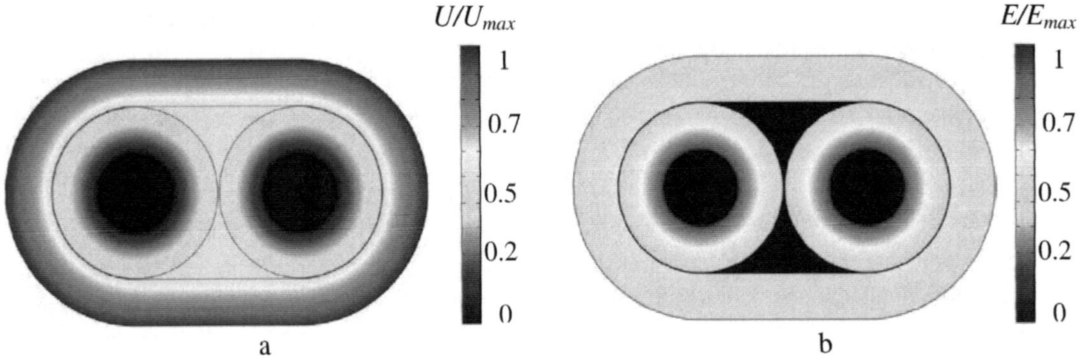

Figure 2. The potential (*a*) and electric field (*b*) distributions for cable 2×0.75 HO3VVH2-F cross section during the spark testing in case of defect-free sheath.

The cable hermiticity is broken in case of the pinhole through the whole thickness of sheath. The pinhole can be represented as thin cylindrical air gap from external to internal sheath surface.

The electric field and potential distributions for cross section of cable with defect sheath during the spark testing significantly differ from the defect-free one (Figure 3).

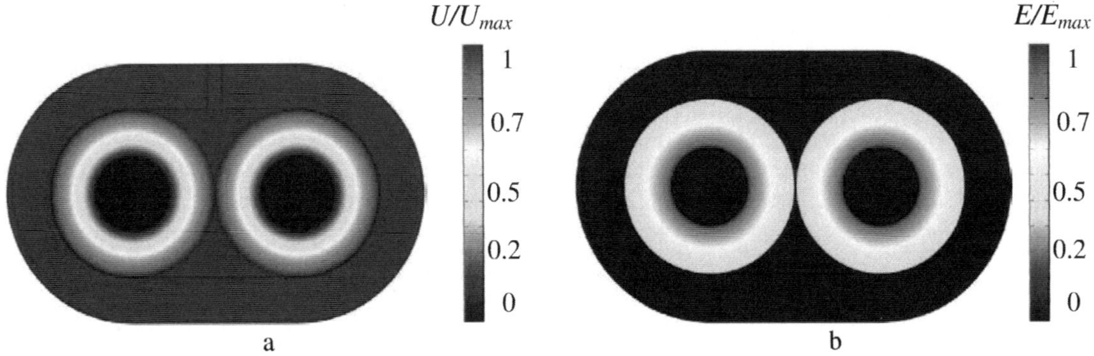

Figure 3. The potential (*a*) and electric field (*b*) distributions for cable 2x0.75 HO3VVH2-F cross section during the spark testing in case of defect sheath.

It can be noted that such potential and electric field distributions is not only in cross section of pinhole defect, but in whole test length of cable, according to the hypothesis.

The electrical equivalent circuit of a cable with defect-free sheath is presented in the Figure 4*a*. The capacity per unit length (C_Σ) is

$$C_\Sigma = \frac{2 \cdot C_s \cdot C_{ins}}{C_s + 2 \cdot C_{ins}},\text{ where}$$

C_s is the equivalent capacity per unit length of cable sheath, C_{ins} is the equivalent capacity per unit length of single insulated core.

In case of sheath with pinhole the equivalent capacity C_s is shunted (Figure 4*b*).

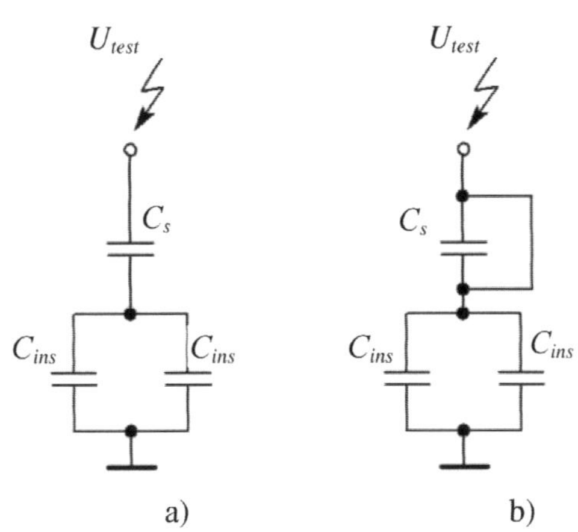

Figure 4. The equivalent electric circuit of the cable HO3VVH2-F with a defect-free sheath (*a*) and defect sheath (*b*).

Then, the capacity per unit length (C'_Σ) is

$$C'_\Sigma = 2 \cdot C_{ins}.$$

Thus, according to the hypothesis the capacity per unit length is significantly increase in case of pinhole through the whole thickness of sheath. Due to this effect the defect can be registered during the in-process control.

To verify this state the physical modeling of the cable with defect sheath in a strong electric field is provided.

4. Physical modelling

Initial experiments were carried out in weak electric fields, i.e. the electric field in the cable was much weaker than air dielectric strength. Measurement of the cable capacity was carried out according to the instructions of the state standard for measuring the electrical capacity of the wire [18].

To match this physical model with a real cable in strong fields, the air gap between the insulations and the sheath of the cable should have good conductive properties. To fulfill this condition, the air gap was filled with a concentrated 20 % NaCl salt solution.

5. Experiment

To provide the experiment two test samples of the cable is prepared. The first test sample is the cable with defect-free sheath, the second one is with defect sheath. To prevent an end effect the length of the test samples is 700 mm. The sheath defect of second test sample is pinhole from the external to the internal sheath surface. The pinhole diameter is 0.75 mm. The test samples are immersed partly in a metal tank with water, according to the measurement procedure specified in the standard [25–30]. Measurements are provided with the digital LCR meter AM-3001, which basic accuracy does not exceed 0.05 %. One of the LCR meter contacts is connected to the conductive parts of the cable, the other contact is connected to the metal tank with water, which has been grounded. The drive voltage of the LCR meter is 1 V, a parallel equivalent circuit of measurements is chosen.

The empirical dependence of the relative variation of cable capacitance δ from the test length l is presented for the several frequency values (Figure 5). The formula for the calculation of the parameter δ is

$$\delta = \frac{C_{\mathrm{d}} - C}{C} \cdot 100\% ,$$

where C_d is the electric capacity of the cable with defect sheath, C is the electric capacity of the cable with defect-free sheath. The test length is the length of immersed part of the test sample.

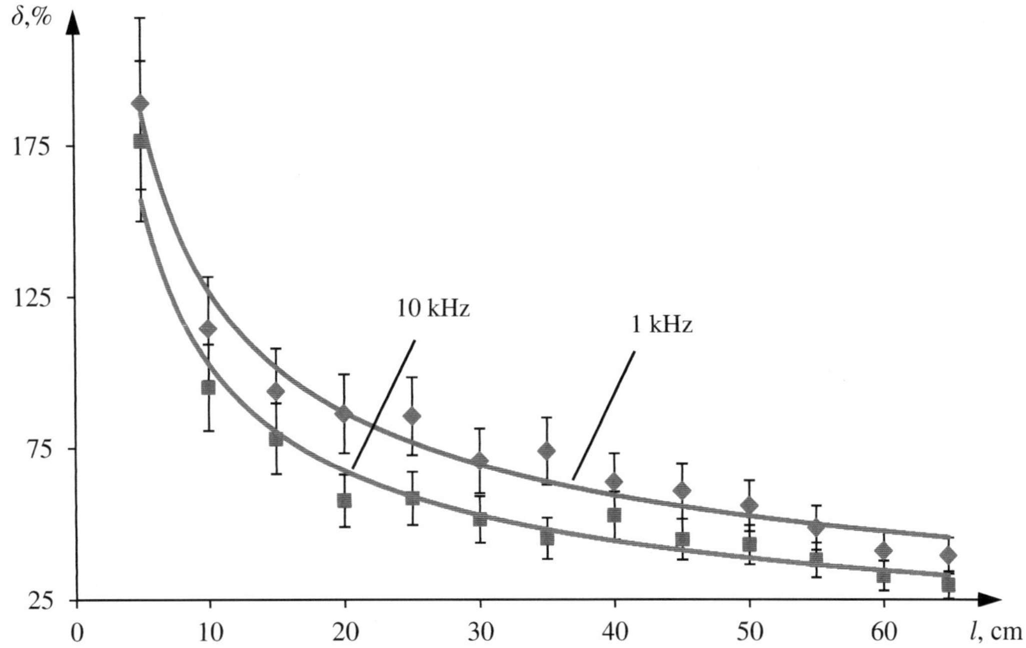

Figure 5. The empirical dependence of the relative variation of cable capacitance from the test length.

Analysis of the obtained data has shown the exponential behavior of the dependences. For the maximal test length l = 65 cm the sheath pinhole leads to capacity variation about 9 % with test voltage frequency 100 kHz, 25 % variation with frequency 10 kHz and 38 % variation with 1 kHz frequency. The most important results for this experiment are obtained for the test length l = 10 cm, because this is the most commonly used electrode length for the in-process spark testing. The variation for this length is about 120 %, 100% and 22 % for the 1 kHz, 10 kHz and 100 kHz test voltage frequencies respectively.

6. Summary

This work describes the possibility of the sheath hermiticity testing by measuring of the cable capacity per unit length variation during spark testing. The cable 2×0.75 HO3VVH2-F is proposed as a model. It is important, that spark testing is in-process method of testing. For this reason dead zone is 20 % from nominal value of measuring cable capacity. This value is defined to reduce the number of the false alarm. In case the test length is 10 cm, the physical modeling results show that variation of the capacity is higher than the insensitivity level. Thus, it was found that the defect through the whole thickness of sheath can be registered by given method with frequencies 1 kHz and 10 kHz.

The aim of the further research is verification of the obtained results by the measurements of the cable capacity per unit length variation performed with the spark tester with additional options of cable capacity measuring.

References

[1] Cadick J 2004 *Cables and Wiring* (New York: Delmar Publishers) p 244
[2] 1956 *Electrical wire and cable handbook* (Pittsburgh: United States Steel Corporation) p 237
[3] Gorodetsky S S and Lackernik R M 1971 *Testing of cables and wire* (Moscow: Energy) p 272
[4] Sclater N 1991 Wire and cable for electronics:a user's handbook (New York: McGraw Hill) p 237
[5] Goldshtein A E and Fedorov E M 2010 *Russ. J. Nondestr. Test.* **46** 424–430 doi: 10.1134/S1061830910060069

[6] 1969 *Engineering Design Handbook: Electrical Wire and Cable* (Alexandria, VA, National Technical Information Service) p 297
[7] Cousins K 2000 *Polymers for Wire and Cable – Changes Within an Industry* (Shawbury, UK Smithers Rapra Technology) p 122
[8] Privezentsev V 1973 *Fundamentals of Cable Engineering* (Moscow: Mir) p 407
[9] 1971 *Cable Engineering* (Fort Belvoir, Defense Technical Information Center) p 54
[10] *Term and Definition* (Fontana, CA, Custom Wire Harnesses & Specialty Connectors) [Electronic resource] URL: http://www.dsmt.com/pdf/resources/glossary.pdf
[11] Horn F W and Bleinberger W E 1966 *Design and manufacture of plastic insulated aluminum conductor telephone cable* (Fort Belvoir, Defense Technical Information Center) p 31
[12] 2007 *Advances in high voltage engineering* (Stevenage: The Institution of Engineering and Technology) p 669
[13] 2004 BS 5099:2004 *Electric cables. Voltage levels for spark testing* (UK: BSI)
[14] 2000 UL 1581 *Reference Standard for Electrical Wires, Cables, and Flexible Cords* (Underwriters Laboratories Inc., Northbrook)
[15] 2003 GOST 7399-97 *Wires and cords of voltage up to and including 450/750 V. Specifications.* (Minsk:IPK Izatel'stvo standartov) (in Russian)
[16] 2003 GOST 2990-78 *Cables, wires and cords. Methods of voltage test.* (Moscow: IPK Izatel'stvo standartov)
[17] 2011 EN 50525-2-71:2011 *Electric cables - Low voltage energy cables of rated voltages up to and including 450/750 V (U0/U) – Part 2-71: Cables for general applications - Flat tinsel cables (cords) with thermoplastic PVC insulation* (UK, BSI)
[18] 2011 EN 50525-2-11:2011 *Electric cables - Low voltage energy cables of rated voltages up to and including 450/750 V (Uo/U) – Part 2-11: Cables for general applications - Flexible cables with thermoplastic PVC insulation* (UK, BSI)
[19] Redko V V, Starikova N S, Redko L A and Vavilova G V 2015 *IOP Conf. Ser.: Mater. Sci. Eng.* **81** 012083 doi: 10.1088/1757-899X/81/1/012083
[20] Drobny J G 2012 *Polymers for Electricity and Electronics: Materials, Properties, and Applications* (New York: Wiley) p 352
[21] Tsaliovich A 1995 *Cable Shielding for Electromagnetic Compatibility* (US: Springer) p 469
[22] Leonov A P and Charkov D I 2017 *MATEC Web Conf.* **91** 01030 doi: 10.1051/matecconf/20179101030
[23] 2012 *GOST R 54813-2011 Electric cables, wires and cords. Spark test method* (Moscow: Standartinform) (in Russian)
[24] 2008 *GOST R 23286-78 Cables, wires and cords. Standards for insulation and sheath thickness and voltage tests* (Moscow: Standartinform) (in Russian)
[25] Blohm W 1999 *Proc. Intl. IWMA Conf. Economical Processing of Rod to Wire & Cable* 67–75
[26] Gavrilin A, Moyzes B, Cherkasov A, Mel'nov K and Xiaoliang Zhang 2016 *MATEC Web of Conf.* **79** 01078 doi: 10.1051/matecconf/20167901078
[27] Goldshtein A E, Vavilova G V and Belyankov V Yu 2015 *Russ. J. Nondestr. Test.* **51** 86–93 doi: 10.1134/S1061830915020047
[28] 2016 *CAPAC® /FFT Capacitance measurement systems for cable production lines* (Zumbach Electronic AG, SCHWEIZ: Hauptsitz) [Electronic resource] URL: http://www.zumbach.com/pdf/Literature_DE/ Catalogs/CAPAC/CAPAC_Familie_CAPA. 002.0002.D.pdf
[29] 2016 *CAPACITANCE 2000: Capacitance measuring devices for cable production lines* (Sikora AG, Germany: Headquarters) [Electronic resource] URL: http://sikora.net/wp-content/uploads/2016/03/ CAPACITANCE_ 2000_ GB.pdf
[30] Yakimov E V and Ustyugov D A 2016 *MATEC Web Conf.* **79** 01069 doi: 10.1051/matecconf/ 20167901069

IOP Conf. Series: Materials Science and Engineering **289** (2017) 012007 doi:10.1088/1757-899X/289/1/012007

Non-destructive method of surface mapping to improve accuracy of mechanical stresses measurements

N A Djuzhev[1], E E Gusev[2], A A Dedkova[3], M A Makhiboroda[4]
[1]Director R&D center "MST and BEC", National Research University of Electronic Technology "MIET", Moscow, Russia
[2]Engineer R&D center "MST and BEC", National Research University of Electronic Technology "MIET", Moscow, Russia
[3]Lead Engineer R&D center "MST and BEC", National Research University of Electronic Technology "MIET", Moscow, Russia
[4]Deputy Director NMST, National Research University of Electronic Technology "MIET", Moscow, Russia
E-mail:bubbledouble@mail.ru

Abstract. In this paper, we describe a non-destructive method for determining the mechanical stresses based on the mapping of the sample surface. The method shortens the measurement time. The loading of PC's RAM decreases during measurement process. Determine the area of local surface irregularities. The magnitude of the curvature of the structure estimate over the entire surface of the substrate. This makes it possible to increase the accuracy of the calculation of the magnitude of mechanical stresses using the Stoney formula. The use of the proposed technique increases probability of success of substrates bonding operations and contact lithography.

1. Introduction

It is important to monitor and change mechanical stresses in thin films. This allows you to create structures with controlled stresses [1]. The mechanical stresses can be controlled in various ways: change the parameters of the technological process of formation of structures, perform heat treatment operations [2], form an additional layer on the sample [3], apply external influences. Using the above methods can reduce the number of operations to prepare the surface with the required relief.

Various methods of monitoring mechanical stress exist. One of the non-destructive methods is the control of the bending of the plate with the subsequent calculation of the stress value by the Stoney formula. Forming a surface map allows you to obtain an array of values of the measured parameter in the local area. When calculating stresses using the Stoney formula, following variables are used: substrate thickness, film thickness, and curvature of surface. Thus, by measuring each parameter in the local region of substrate, it is possible to reduce the error in measuring the value of mechanical stresses [4].

This method is applied to products of microelectronics, especially to devices based on microelectromechanical systems (MEMS). In a number of cases, MEMS is a thin micro- or nanosized film on a silicon wafer. Membrane is the area of the film in the air. Contact of the measuring device with MEMS membrane can lead to deformation of the device. Consequently, the development and application of non-contact methods for monitoring mechanical stresses allow to improve the yield of suitable ones.

2. Goal

Previous authors have been developed a non-destructive method for determining the mechanical stresses in the thin films on the wafer [5]. The surface profile passing through the center of the plate perpendicularly or parallel to the base cut was analyzed. In comparison with the method described in [4], the calculation of the local curvature of the surface was carried out by analyzing the geometric arrangement of points without the use of derivatives. This allowed to increase the accuracy of measurement of mechanical stresses.

The purpose of this work is to develop and test a method of mapping the surface. This approach will allow us to localize regions with maximum curvature values over the whole surface of the sample. Reduce the load on the PC's RAM. Also reduce the time to complete this task. More accurately calculate the Stoney formula mechanical stresses on the substrate.

3. Description of the method

Surface of the sample can be planar or relief. A damping table is used in the implementation of the described method. This allows the sample to be placed on a flat surface. The polished surface of the sample also require to detect the reflected signal.

On the example of the non-contact optical profilometer Veeco Wyko NT9300 follows the description of the method. The standard measurement mode allows to obtain a three-dimensional image of a surface of 0.84×1.1 mm. A longer length of the region is needed to analyze the curvature of the plate surface (100 mm for substrate Ø100, 150 mm for substrate Ø150 etc.). Thus, a series of sequential measurements is carried out (Figure 1). The software of the optical profilometer makes it possible to carry out a series of sequential measurements in an automatic mode. The «Stitching» mode is used, which includes setting the start and end points for stitching images, the degree of overlap of the images (about 30%), the scanning range for height, etc.

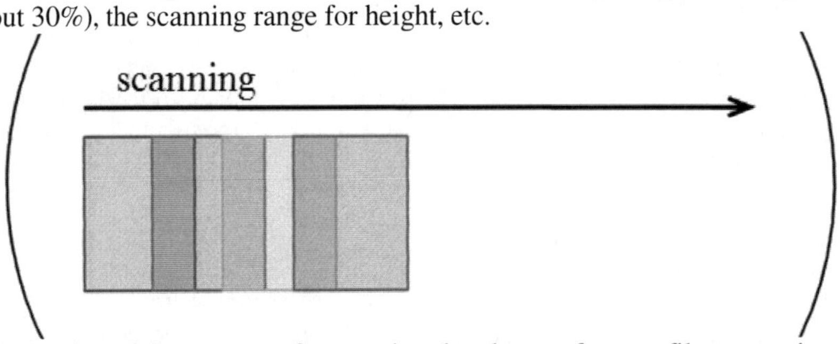

Figure 1. Illustration of the process of measuring the plate surface profile as a series of successive measurements with an overlap of about 30%.

The principle of the interferometer is as follows. The beam of electromagnetic radiation is spatially divided into two coherent beams. Each beam passes a different optical path to the screen. Thus, an interference pattern is created, along which the phase shift of the beams is established.

An important advantage of optical profilometers is that they create a three-dimensional image without contact with the surface. Thus, the method is non-destructive. This makes it possible to work with samples regardless of the mechanical strength of the structure under study. The method allows recording the features of the relief, starting from a roughness of nanometer scale to steps of millimeter height. For example, after vertical etching of silicon (Bosch process), distance between minimum and maximum point of the surface relief can be hundreds of micrometers, as shown in Figure 2. The ability to perform a high-precision analysis of the surface and record the amount of deflection of the sample, which amounts to millimeters, determined the use of an optical profilometer to solve this problem. Detection of the optical response is performed in the mode of vertical scanning interferometry.

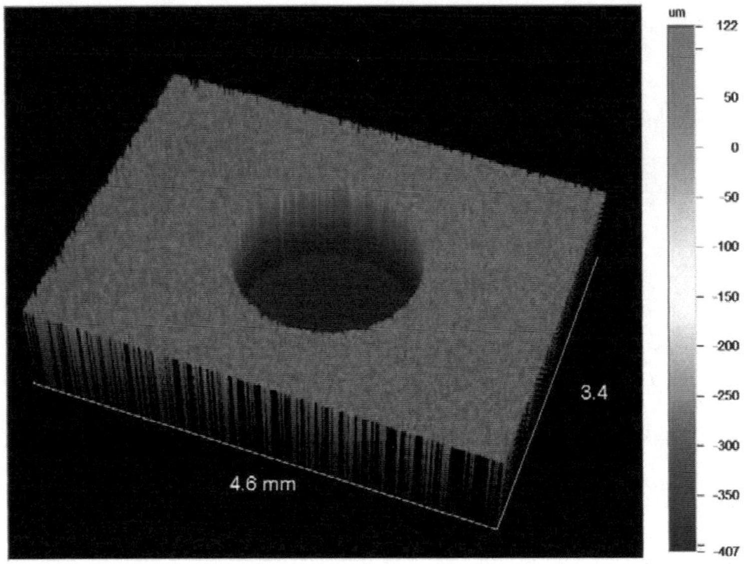

Figure 2. Measurements on the optical profilometer of the surface relief.

In the case of presence of local irregularities (depressions, elevations) on the sample, scanning along two different lines lagging behind by several millimeters can lead to different results (Figure 3). Therefore, tags are set for repeated measurements of samples. In Figure 3, dark areas show local surface irregularities (depressions or elevations).

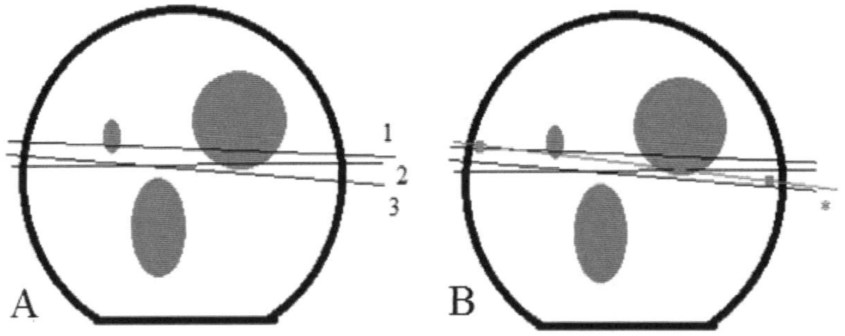

Figure 3. Effect of selecting measurement direction on final profile.

It should be noted that when scanning along two lines there is a high probability to overlook the presence of significant local irregularities. This is due to the fact that the scan lines will not cross them. The presence of such local defects and irregularities can have a significant negative impact on the conduct of subsequent technological processes (for example, bonding substrates, contact lithography) and cause the end product to be inoperative.

Thus, one can see the prospects and the need to develop a method for mapping surface, which will allow us to evaluate entire surface of sample and localize surface irregularities.

One way to carry out measurements using a technique similar to the one described above is to use the "Stitching" mode and overlap along the horizontal and vertical axis. However, in most cases this is not possible due to the large amount of data that exceeds the storage and processing capabilities of standard personal computers. Therefore, the "Stitching" method can be useful in the process of studying small-diameter substrates or using supercomputers. During the measurement on an optical profilometer, use of reference points (Figure 4) instead of overlapping data will reduce the loading of the RAM of the personal computer.

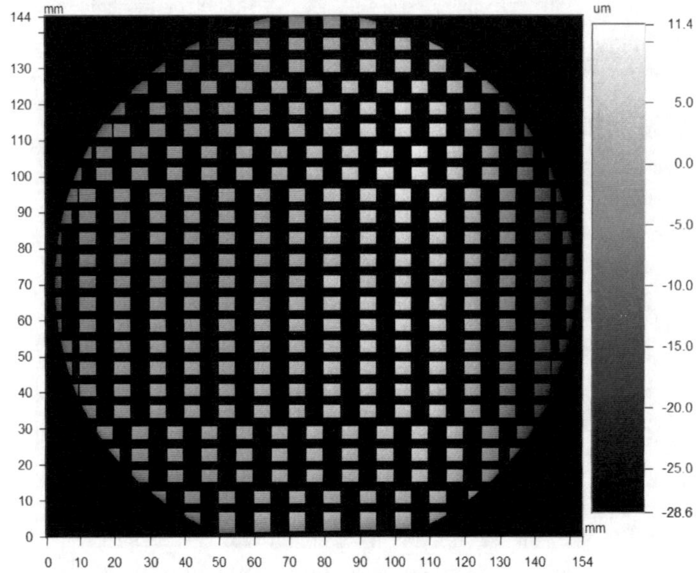

Figure 4. Result of scanning the surface using the reference points.

A series of sequential measurements is performed in automatic mode. In this case, a mode is used in which the coordinates of each of the scanning areas are set (300 regions for Ø150 substrates). Due to this approach, memory of a standard personal computer becomes sufficient to remove the entire data set from entire area of substrate. Also, measurement time is significantly reduced.

In this method, neighboring regions are not superimposed on each other (there is no overlap in 30%). On the contrary, there are significant "voids" in the data (Figure 4). In this regard, it is necessary to carry out the data recovery procedure in the intermediate areas. An example of implementation using the Veeco Wyko NT9300 optical profilometer software is shown in Figures 5 and 6. After processing the data, a complete map of the surface can be observed, in view of which it is possible to estimate the curvature of the substrate and observe local surface irregularities.

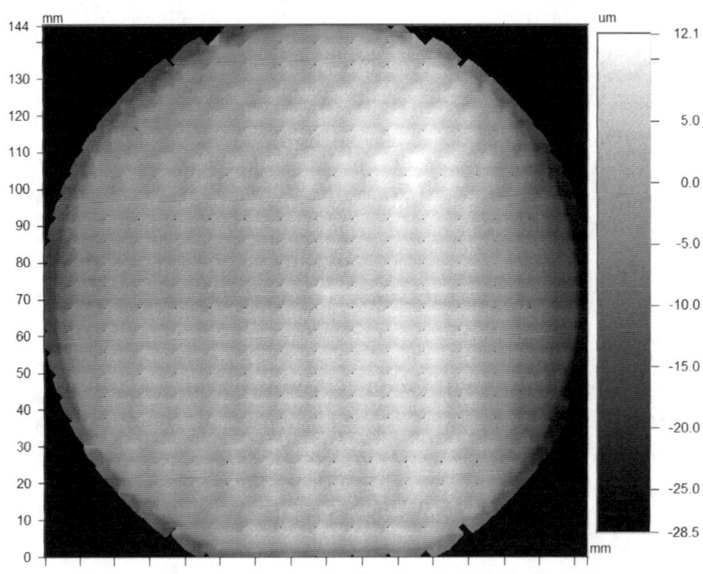

Figure 5. Restored surface map-2D.

IOP Conf. Series: Materials Science and Engineering **289** (2017) 012007 doi:10.1088/1757-899X/289/1/012007

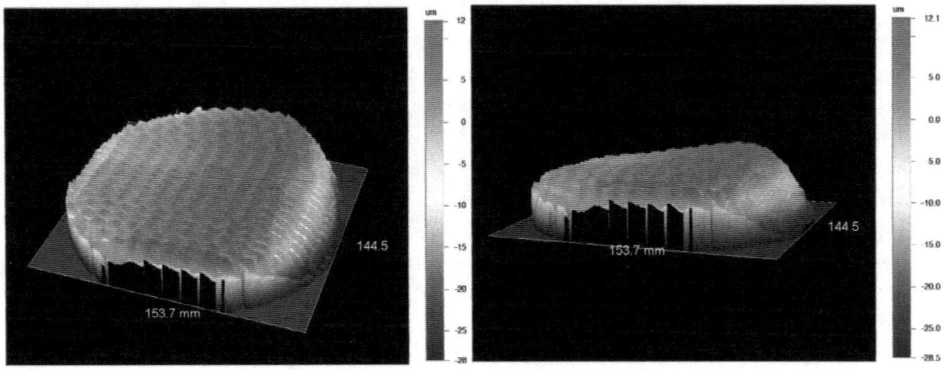

Figure 6. Restored surface map-3D.

A feature of this technique is the unevenness of the resulting image of the surface. Characteristic "saws" are present on 2D and 3D maps (Figures 7, 8). This artifact is related to the data extraction feature and cannot be removed using the Veeco Wyko NT9300 optical profilometer software. However, after recovering the data and saving the profile (Figure 7) in .txt format, this sawtooth distortion can be easily removed using software, for example, Matlab.

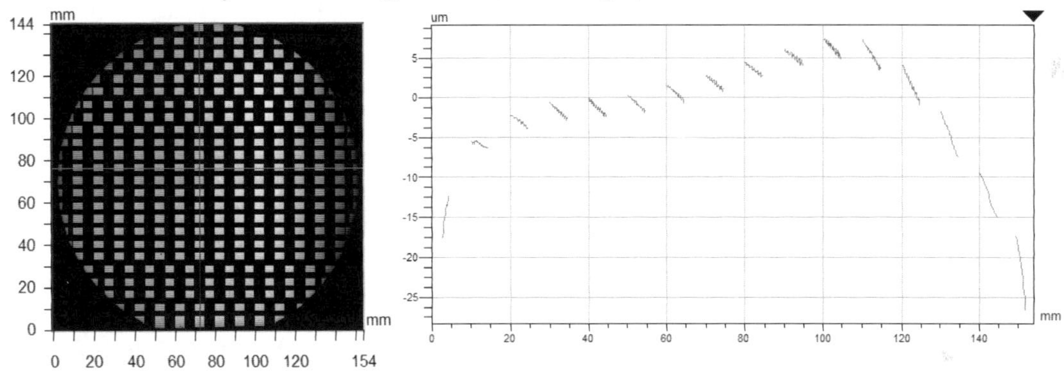

Figure 7. Surface profile of surface along the red line.

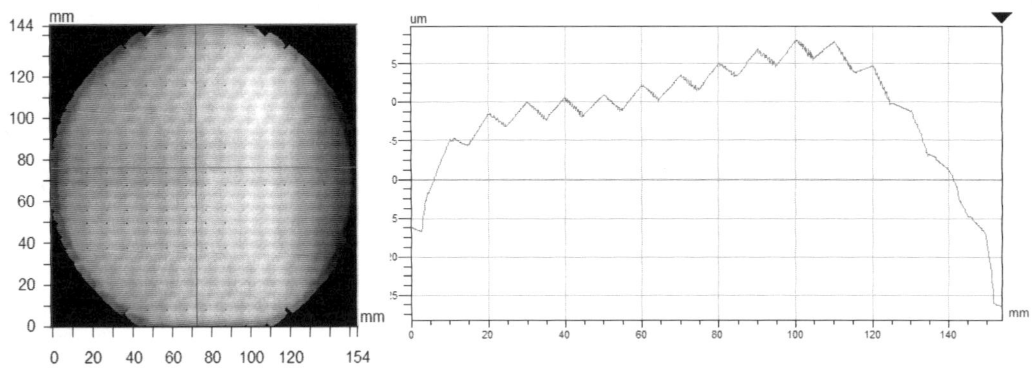

Figure 8. Reduced profile of surface along the red line.

4. Estimation of the accuracy of measurements of mechanical stresses

We are estimated technique on the basis of experimental data obtained by the authors in the previous paper [5]. The thickness of the materials is determined by the contactless method by means of an ellipsometer. The thickness of the silicon oxide film is 0.6 ± 0.05 (um), the thickness of the silicon substrate is 670 ± 20 (um). The value of curvature of the surface is calculated from geometrical location of the points determined by means of a contactless profilometer and is 0.0015 ± 0.0010 (1/m). Let the biaxial modulus of elasticity of the silicon substrate is 1.8×10^{11}(Pa) [6]. Previously, the average value of the mechanical stress in SiO_2–210 (MPa) is determined. Taking into account the variation of the measured values across the plate, the measurement error is -16.2 (MPa), i.e. 7.7%.

5. Conclusion

The developed technique for mapping the surface of structures makes it possible to obtain information about the entire surface of the plate and to localize the irregularities. In the process of applying this technique it is possible to obtain accurate and visual information about the surface of samples with a diameter of about 150 mm or more, which is inaccessible by most other methods. The accuracy of determining the value of mechanical stresses according to the Stony formula increases due to non-destructive measurements of the set of parameters (curvature of the surface, film thickness and substrate thickness) in the local areas of the investigated structure increased by 7.7%.

Acknowledgments

The work was performed on the equipment of MIET R&D center "Microsystem technique and the bases of electronic components", supported by the Ministry of Education and Science of the Russian Federation (state contract No. 14.594.21.0012, unique identifier of the project RFMEFI59417X0012).

References

[1] Astashenkova O N Physico-technological basis of management mechanical stresses in thin film compositions of micromechanics, Thesis, Ph. D., 2015.

[2] Gusev E E, Gryazneva T A, Negencev A V, Makhiboroda M A 2017 *Proceedings of the 2017 IEEE Russia Section Young Researchers in Electrical and Electronic Engineering Conference, ElConRus 2017* 1397–1400 doi: 10.1109/EIConRus.2017.7910831

[3] Wu Q,Wang J, Hou Y; Zhu F 2016 *Method for changing curvature of silicon wafer by stress of silicon dioxide* Patent CN №105448666

[4] Johnson T, Darwin M 2013 *Local Stress Measurement* Patent USA №8534135

[5] Djuzhev N A, Dedkova A A, Gusev E E, Makhiboroda M A, Glagolev P Y 2017 *IOP Conference Series: Materials Science and Engineering* 189 012019 doi: 10.1088/1757-899X/189/1/012019

[6] Chen K S 2006 *Handbook: Techniques and Applications* (Springer Science & Business Media), pp. 263 – 340

IOP Conf. Series: Materials Science and Engineering **289** (2017) 012008 doi:10.1088/1757-899X/289/1/012008

Feasibility of using T-shaped feedback in teraohmmeters

N I Yermoshin[1] and EV Yakimov[2]
[1]Postgraduate, National Research Tomsk Polytechnic University, Tomsk, Russia
[2]Associate Professor, National Research Tomsk Polytechnic University, Tomsk, Russia
E-mail: ermoschin.nik@yandex.ru

Abstract. The paper investigates the feasibility of using T-shaped feedback in teraohmmeters. Theoretical and experimental dependences of the output voltage of the T-shaped feedback converter on the measured resistance and circuit parameters are obtained. The use of T-shaped feedback is found to decrease the reference resistance rating from 10 GOhm to 100 Ohm that indicates 100-fold reduction (with an error of less than 1%).

1. Introduction

The methods of testing, control and measurement employed to determine the state of insulation for current-carrying parts of electrical equipment depend on physical nature of isolation. Any insulation (dielectric) used in electrical machines and apparatus is a capacitor with a complex medium. The plates are external components of the apparatus (housing and core) and current-carrying components (cables, wires and buses), and the medium is an insulating material with the structure that depends not only on the material used, but also on its state – the presence of defects [1].

Electrical resistance is the main indicator of the insulation material state, and its measurement is obligatory for testing of all types of electrical equipment and electrical circuits. Teraohmmeters are used to measure the resistance of insulating materials [2].

The main problems to challenge during teraohmmeter development are the stability of the device to industrial frequency noise, error, and the settling time for indicating the insulating material resistance if the object possesses electrical capacity, or the measuring device is supplied with filtering circuits [3–5].

Teraohmmeters based on an operational amplifier (op amp) and supplied with a circuit of a resistance-to-voltage converter with inverse scale have been widely used (Figure 1). The teraohmmeters operating on this principle are advantageous since they operate at a zero common-mode signal and have good noise immunity [6].

The main requirement to such teraohmmeters is low input current of the op amp to increase the upper measurement limit. The modern operational amplifiers of this type are AD549 and ADA4530-1 (Analog Devices), LTC6268 (Linear Technology), and OPA128 and LMC6041 (Texas Instruments). The input current of all of the above microchips is less than 100 fA that theoretically enables measurement of the resistance up to 1000 TOhm at a reference voltage of 100 V.

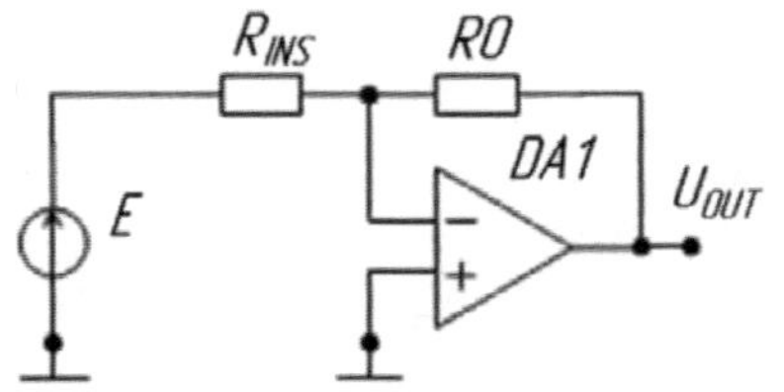

Figure 1. Resistance-to-voltage converter with inverse scale, based on op amp: *DA1* is op amp; *E* is the EMF value of the source; R_{INS} is insulation resistance; $R0$ is reference resistance.

In case of an ideal op amp, the output voltage of the teraohmmeter is [7]

$$U_{OUT(ID)} = -E \cdot \frac{R0}{R_{INS}} \tag{1}$$

2. Problem statement

As seen from equation (1), the larger the resistance $R0$ in the feedback circuit, the greater resistance can be measured. However, typically, the greater the resistance of the resistor, the lower its accuracy. Therefore, to reduce an error in the teraohmmeter, the least possible value for the feedback resistance must be chosen (less than 1 GOhm). For example, high-resistance precision chip resistors Riedon's HVS exhibit an error of 1–2% in the range from 1 GOhm to 100 GOhm. While, in the range from 1 MOhm to 1 GOhm, the error is 0.25–0.5%, which is 4-times less than that in the range 1 GOhm to 100 GOhm. However, the feedback resistance decreases simultaneously with the output voltage amplitude of the resistance-to-voltage converter.

Theoretically, T-shaped feedback can eliminate this contradiction (Figure 2).

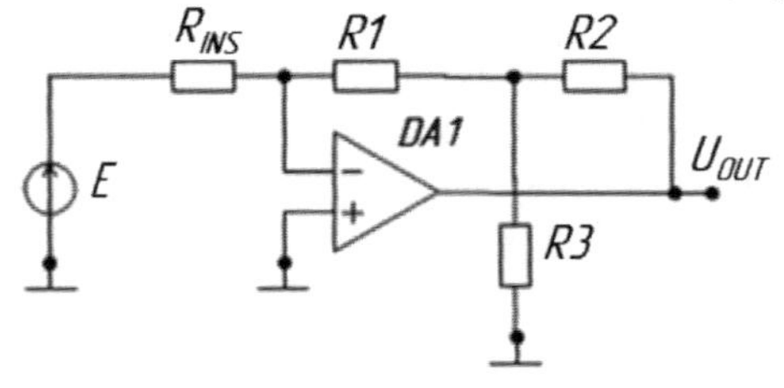

Figure 2. Resistance-to-voltage converter with T-shaped feedback: *R1, R2* and *R3* are resistances of T-shaped feedback.

In this case, we need lower value resistances to obtain a stable useful signal. For an ideal op amp, the conversion equation for this circuit can be written as [8]:

$$U_{OUT(ID)} = -E \cdot \frac{R1+R2+\frac{R1 \cdot R2}{R3}}{R_{INS}} = -E\frac{R_{EQ}}{R_{INS}}, \tag{2}$$

where R_{EQ} is the equivalent resistance of the amplifier feedback and $U_{OUT(ID)}$ is the output voltage of the T-shaped feedback converter for an ideal op amp.

Equation (2) does not take into account many parameters of the op amp (input resistance and current, offset voltage, finite voltage gain, presence of protection from static electricity, etc.), which limit the use of T-shaped feedback to design a teraohmmeter. In order to consider the given parameters of the op-amp, it is advisable to perform the analysis using special programs for electrical circuit simulation.

The study aims to experimentally and theoretically investigate the applicability of T-shaped feedback in teraohmmeters, and to determine the conditions under which the circuit remains operational. All the experiments used a constant voltage source E=100V, and an ADA4530-1 op amp, with less than 20 fA input current at temperatures ranging from –40 to +85° C and differential voltage gain of at least 30 million.

The Altium Designer software was used to determine the theoretical value of the output voltage of the teraohmmeter $U_{OUT(T)}$ to compare it with the experimental results. Equation (2) was used to calculate the theoretical values of the output voltage of the teraohmmeter using an ideal op amp $U_{OUT(ID)}$.

As described in [9], an op amp of the teraohmmeter should have a three-stage structure with an integrator in the first stage to provide protection from noise. Therefore, this circuit diagram was used for the experiments and modeling. Figure 3 illustrates the circuit diagram of the teraohmmeter in the Altium Designer.

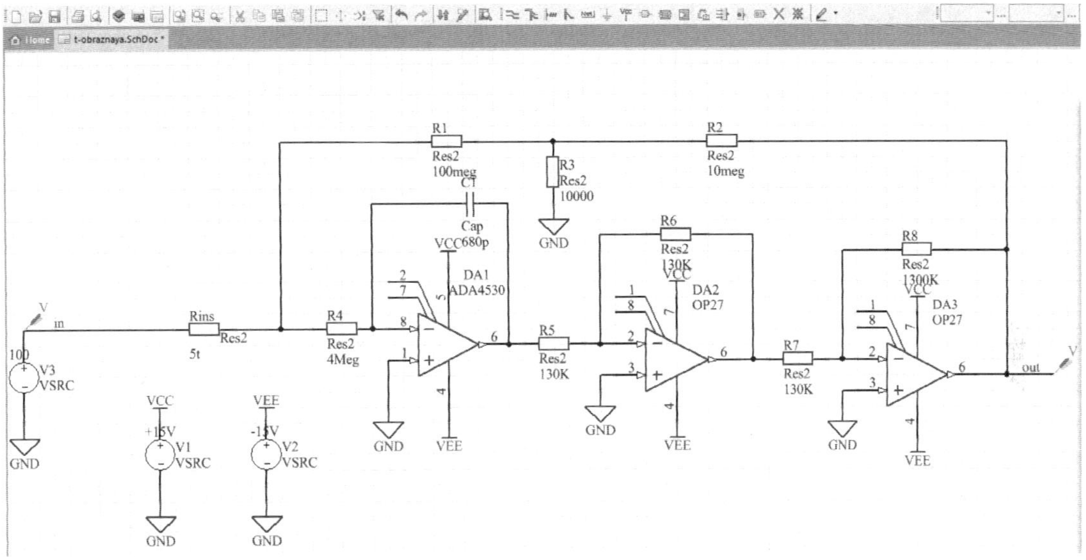

Figure 3. Circuit diagram of the teraohmmeter in the Altium Designer.

3. Investigation of T-shaped feedback

The experimental study revealed that the output voltage of the converter with T-shaped feedback $U_{OUT(EXP)}$ does not depend on the combination of resistances *R2* and *R3*, if they provide the same equivalent resistance R_{EQ}. This can be verified through the analysis of the measurement results at a constant resistance rating *R1* and varying resistances *R2* and *R3* (Table 1).

With this, the changes in resistance rating *R1* affect the output signal. Figures 4 and 5 show the dependences of the output voltage on the insulation resistance at different values of resistance *R1*.

Table 1. Dependences of the output voltage for the T-shaped feedback converter at different combinations of resistances $R2$ and $R3$ ($R_{EQ} = const$).

$R1$, MOhm	$R2$, kOhm	$R3$, Ohm	R_{EQ}, GOhm	R_{INS}, TOhm	$U_{OUT(ID)}$, mV	$U_{OUT(EXP)}$, mV
10	10	101	1	1	100	103
10	100	998	1.012	1	101.2	105
10	1001	10070	1.005	1	100.5	103
100	10	101	10	5	200	205
100	100	998	10.12	5	202	210
100	1001	10070	10.04	5	201	207

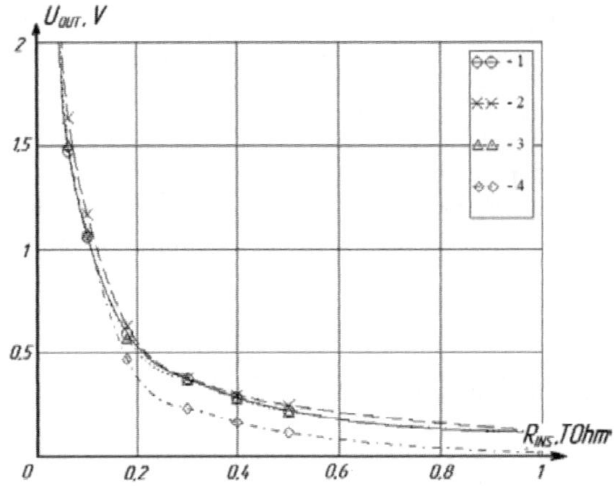

Figure 4. Dependences of the output voltage of the T-shaped feedback converter on the insulation resistance at R_{EQ} = 1 GOhm : 1 is dependence for an ideal op amp $U_{OUT(ID)}$ ($R1$ = $R2$ = 1.001 MOhm, $R3$ = 998 Ohm); 2 is dependence for $U_{OUT(EXP)}$ at $R1$ = 100 MOhm, $R2$ = 100 kOhm, $R3$ = 10.07 kOhm; 3 is dependence for $U_{OUT(EXP)}$ at $R1$ = 10 MOhm, $R2$ = 100 kOhm, $R3$ = 998 Ohm; 4 is dependence for $U_{OUT(EXP)}$ at $R1$ = $R2$ = 1.001 MOhm, $R3$ = 998 Ohm.

The graph in Figure 4 shows that the experimental dependences are equal to the theoretical ones, when reproducing R_{EQ} = 1 GOhm using resistances *R1*=100 MOhm and *R1*=10 MOhm. However, the error is 20% or more at *R1*=1 MOhm and the measured resistances exceeding 200 GOhm. At resistance rating *R1* less than 1 MOhm, the error becomes even greater.

A similar result can be observed in Figure 5, which illustrates the dependences at R_{EQ} = 10 GOhm. As can be seen from the graph, the error is more than 20% at $R1$ = 10 MOhm and the measured resistances exceeding 2.5 TOhm.

The experimental data presented above show that T-shaped feedback enables reduction of the reference resistance rating at least by a factor of 100. With a larger ratio, the measurement error increases, and the output signal of the converter tends to zero at the upper measurement limit when the resistance is reduced by a factor of 1000 or more.

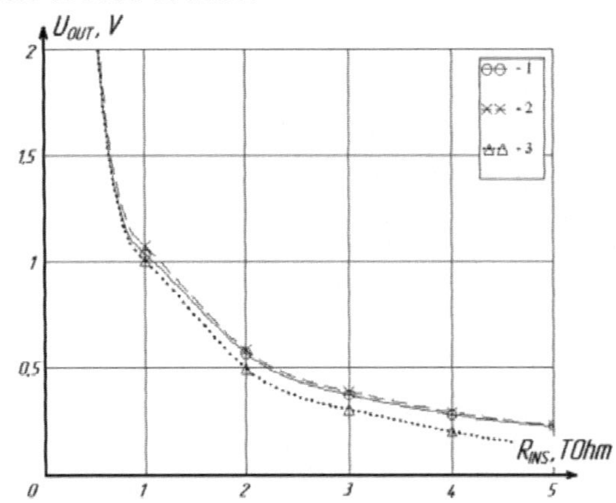

Figure 5. Dependences of the output voltage of the T-shaped feedback converter on the insulation resistance at R_{EQ} = 10 GOhm: 1 is dependence for an ideal op amp at $R1$ = 10 MOhm, $R2$ = 1.001 MOhm, $R3$ = 998 Ohm; 2 is dependence for $U_{OUT(EXP)}$ at $R1$ = 100 MOhm, $R2$ = 1.001 MOhm, $R3$ = 10.07 kOhm; 3 is dependence for $U_{OUT(EXP)}$ at $R1$ = 10 MOhm, $R2$ = 1.001 MOhm, $R3$ = 998 Ohm.

This causes the problem of finding optimal rating for reference resistance *R1* of T-shaped feedback for a certain upper measurement limit.

For this purpose, compare the experimental and theoretical dependences of the output voltage on the changes in reference resistance *R1* of T-shaped feedback at $R_{EQ} = const$ (Figure 6).

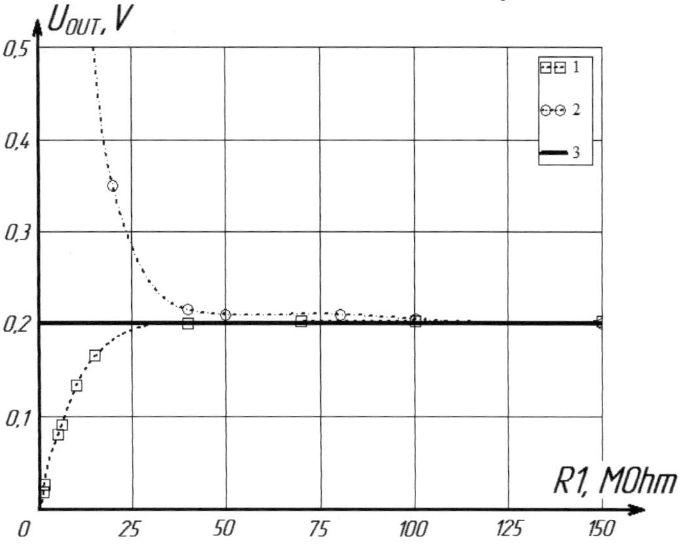

Figure 6. Dependences of the output voltage of the T-shaped feedback converter on reference resistance $R1$ at $R_{EQ} = 10$ GOhm, $R_{INS} = 5$ TOhm: 1 is experimental dependence for $U_{OUT(EXP)}$; 2 is theoretical dependence for $U_{OUT(T)}$; 3 is theoretical dependence for an ideal op amp $U_{OUT(ID)}$.

Based on equation (2), the output voltage of the converter $U_{OUT(ID)}$ should be equal to 0.2 V at $R_{INS} = 5$ TOhm and $R_{EQ} = 10$ GOhm. However, the experimental dependence shows that the use of T-shaped feedback at this measurement limit is possible for *R1* greater than 25 MOhm to make an error less than 10%. Lower values of resistance *R1* cause improper operation of the converter, and make the output voltage tend to zero.

Theoretical dependence is slightly different in nature. At *R1* less than 10 MOhm, the op amp is saturated, and the output voltage of the converter is equal to the supply voltage of the op amp. The converter operates within the range of resistance *R1* from 40 MOhm to 100 MOhm, but there is an additive error of about 10%. In the range from 100 MOhm, the converter output voltage corresponds to the calculated value with an error of less than 1%.

The results of both experimental and theoretical studies show that at resistance *R1* in the range from 100 kOhm to 40 MOhm the op amp of the teraohmmeter cannot provide proper operation of the teraohmmeter based on an ideal op amp.

4. Conclusions

The feasibility of using T-shaped feedback has been studied for teraohmmeters. The ADA4530-1 op amp with the input current of 20 fA was used, which theoretically enabled measurements of the resistance up to 5000 TOhm at a reference voltage of 100 V.

It was experimentally and theoretically proved that the dependence of the output voltage of the T-shaped feedback converter does not depend on the combination of *R2* and *R3* resistances, if they have the same equivalent resistance $R_{eq} = const$, and resistance *R3* ranges from 100 Ohm to 10 kOhm. At the same time, the changes in resistance *R1* significantly affect the output signal. According to the experimental data, the use of T-shaped feedback enables 100-fold reduction of the reference resistance

IOP Conf. Series: Materials Science and Engineering **289** (2017) 012008 doi:10.1088/1757-899X/289/1/012008

rating. For example, resistor *R1* with a resistance not less than 100 MOhm is required to reproduce equivalent resistance $R_{EQ} = 10$ GOhm with an error of less than 1%.

References

[1] Leonov A P *et al* 2015 *IOP Conf. Ser.: Mater. Sci. Eng.* **81** 012094 doi: 10.1088/1757-899X/81/1/012094

[2] Leonov A *et al* 2014 *IOP Conf. Ser.: Mater. Sci. Eng.* **66** 012004 doi: 10.1088/1757-899X/66/1/012004

[3] Yakimov E *et al* 2016 *MATEC Web of Conferences* **79** 01069 doi: 10.1051/matecconf/20167901069

[4] Goldshtein A E, Fedorov E M 2010 *Russian Journal of Nondestructive Testing* **46** 424–430 doi: 10.1134/S1061830910060069

[5] Yakimov E *et al* 2017 *J. Phys.: Conf. Ser.* **783** 012061 doi: 10.1088/1742-6596/783/1/012061

[6] Fedorov E M, Bortnikov I D 2015 *Technical Physics* **60** 1689–1692 doi: 10.1134/S1063784215110110

[7] Popov V P 2007 *Fundamentals of circuit theory* (Moscow: High School)

[8] Bessonov L A 2014 *Theoretical Foundations of Electrical Engineering. Electrical circuits* (Moscow: Jurajt)

[9] Yakimov E V, Zhukov V K 2003 *Measurement Techniques* 35–39

Photovoltaic properties of DSSC with composite counter electrodes based on Pt and SLGO

N Ibrayev[1], E Seliverstova[2,4], R Dzhanabekova[3] and T Serikov[2]
[1]Professor, Institute of Molecular Nanophotonics, Buketov Karaganda State University, Karaganda, Kazakhstan
[2]Senior lecturer, Institute of Molecular Nanophotonics, Buketov Karaganda State University, Karaganda, Kazakhstan
[3]Master student, Institute of Molecular Nanophotonics, Buketov Karaganda State University, Karaganda, Kazakhstan
[4]Senior research fellow, Institute of Applied Mathematics, Karaganda, Kazakhstan
E-mail: niazibrayev@mail.ru

Abstract. Langmuir-Blodgett (LB) films based on graphene oxide were synthesized and studied. Structure and optical properties of the films are studied. At using of the films as counter electrodes in DSSC cells, it is shown that pure LB graphene oxide films, despite the advantages in electrotransport properties, have a rather high resistance. This is lead to the decreasing in the number of charge carriers injected into the semiconductor layer from the dye molecules, and in a complex decrease in the efficiency of solar cells. However, when LB films are deposited over the Pt layer, the electron transport characteristics of the counter electrodes are improved, which is manifested in an increase in the efficiency of solar cells compared to cells with a conventional Pt electrode.

1. Introduction

Graphene and its modified derivatives, like other nanostructures, show the dependence of properties on size and morphology. 2D-graphene shows different properties in comparison with its 1D form, i.e. carbon nanotubes. The 2D structure of graphene makes it more compatible with planar substrates, which can be a viable alternative for silicon substitution. In addition, the 2D structure of graphene layers and their atomic dimensions give them unusual properties [1]. High conductivity and a high ratio of surface area to volume are the most outstanding characteristics of layered graphene nanostructures [2]. In addition, it was shown in [1, 3] that graphene possesses a high specific surface area, high electron mobility and a higher oxidation potential than Pt, which is the traditional material used in dye-sensitized solar cells (DSSC). Due to the variety of existing forms of graphene, it is possible to improve the necessary parameters of the counter electrodes in solar cells. For example, reduced graphene oxide has surface lattice defects, which implies a higher catalytic activity than a completely reduced defect-free graphene [4, 5].

Given the transparency of the single- or multilayer graphene structure and its conductivity, and the fact that the cost of producing of graphene can be cheaper than for other carbon nanomaterials, makes it a very promising material in the production of photovoltaic devices.

There are several methods for obtaining graphene nanostructures. Typically, CVD technology, aerosol spraying and polyione assembling are used for deposition of graphene layers [6, 7]. Polyione assembling and CVD are complicated procedure, that requires additional preparation of substrates and

IOP Conf. Series: Materials Science and Engineering **289** (2017) 012009 doi:10.1088/1757-899X/289/1/012009

template for growing of graphene films. Furthermore, there is the problem of controlling of thickness of synthesized coatings, which is a negative factor.

Langmuir-Blodgett (LB) technique allows to control the thickness of films as well to predict the orientation of the particles within it. It means that the combination of possible variations in number of layers and the distance between the individual graphene sheets enable control not only the conductivity and resistance of the films, but also to adjust the surface area of the film. This opens up the possibility of creating a highly efficient catalyst graphene layer with high conductivity for using in photovoltaic cells.

In this paper, the results of an investigation of the photovoltaic characteristics of DSSC with counter electrodes based on graphene oxide LB films and Pt are presented.

2. Experiment

For the preparation of LB films of graphene oxide (SLGO, Cheaptubes), dispersions in dimethylformamide (DMF) were used. The solvent (analytical grade) were purchased from Sigma-Aldrich and used without further purification. The concentration of SLGO in the solution is 0.03% by weight. Graphene solutions were processed for 60 minutes in an ultrasonic bath.

The investigated monolayers were formed by spreading from a solution in a Langmuir bath (KSV Nima). The isotherms of compression of graphene oxide monolayers at the water-air interface were measured and investigated. To stabilize the monolayer, it was held for 40 minutes. Deionized water purified using the AquaMax water purification system was used as a subphase. The specific resistance of water was equal to 18.2 MΩ/cm. The surface pressure was 72.8 mN/m at pH = 5.6 and temperature of 22 °C.

Measurements of the π-A isotherms showed that, the SLGO monolayer under compression passes from the gaseous to the liquid-stretched [8, 9] state at small volumes of the sprayed solution on the surface of the subphase. In this case, the surface pressure increases smoothly without the collapse of the monolayer. An increase in the number of particles in the monolayer leads to a shift of the π-A isotherm curve toward smaller areas, and the length of the gaseous and liquid phases in this case decreases. The value of the surface pressure was chosen for the transfer of films, when the monolayer is in a more condensed liquid state – at 8 mN/m. Solid films were deposited by the LB method according to the X-type transfer, since it was shown in [8, 10] that such films are more homogeneous in structure than films obtained according to the Z-type transfer. The rate of monolayer transfer was equal to 8 and 19 mm/min, and the compression rate of the monolayer was equal to 5 mm/min. For the deposition of SLGO films, glass plates coated with a FTO layer were used. 1 monolayer of graphene oxide was deposited on the substrates. For the layered films SLGO was deposited over the layer of Pt catalysts. The platinum catalyst was electrochemically applied from an ethanol solution of H_2PtCl_6 (Sigma Aldrcih).

The preparation and assembling of solar cells was carried out according to the procedure detailed in [11, 12]. To prepare of solar cells, glass substrates coated with a FTO layer (Sigma-Aldrich, surface resistance of ~ 10 Ohm/sq) were used. Blocking layer of TiO_2 was deposited on the surface of the FTO. Semiconductor films were prepared by the doctor-blading method. To prepare the films, TiO_2 with a nanoparticle size of 25 nm was used (Sigma-Aldrich). The thickness of the resulting films was measured by the SEM (Tescan Mira-3) and was equal to 10 – 12 microns. Sensitization of semiconductor electrodes with ruthenium dye (N719, Sigma Aldrich) was carried out by immersion of TiO_2 films in an ethanol dye solution with concentration equal to 4×10^{-4} mol/l for 20 hours. Before measurement, the films were kept in an oven at a temperature of 80 °C for at least of 1 hour. The number of adsorbed dye molecules in all samples was controlled by the change in the optical density of the solutions before and after sorption.

The current-voltage characteristics of DSSC were measured with Solar Cell Tester CT50AAA (Photo Emission Tech. Inc., USA) under illumination of the xenon lamp with light intensity of 100 mW/cm^2. The measurements were performed at room temperature in photovoltaic mode. From the obtained values of the open circuit voltage, short-circuit current and fill factor, the efficiency of the cells was determined. Electric transport properties of solar cells based on porous titania films with various opposite electrodes were studied by measuring the impedance of an electrochemical cell. The measurements were carried out on an Z-500PRO impedance meter (Elins, Russia) under irradiation of cells with standard simulated solar radiation (Air Mass (AM) 1.5). The amplitude of the applied sinusoidal signal was 20 mV, and the frequency varied from 1 MHz to 100 MHz.

3. Results

SEM images of the obtained LB films are shown in Figure 1. It can be seen from the Figure 1 that the number of particles with a large number of layers, as well as areas with wrinkles and folds, is small in SLGO LB films. When analyzing the area of the of the LB film on the surface of the substrate, it was found that the monolayer occupies about 80 % of the total area.

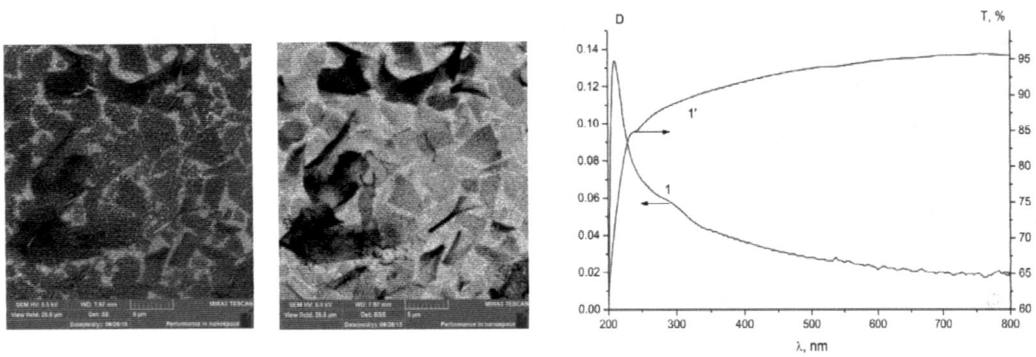

a) b) c)

Figure 1. SEM images of the graphene oxide LB film (*a, b*). Image (*b*) was obtained with the reflected-electron sensor. (*c*) Absorption (1) and transmission (1') spectra of graphene oxide LB films.

The maximum of the absorption spectrum of the LB film exhibits at 230 nm with an weak shoulder about 300 nm (Figure 1*c*). The transparency of the films in the visible spectral range varies from 90 to 95 %.

The measured I-V characteristics are shown in Figure 2 and in Table 1. As it was shown, the efficiency of the cells decreases when the graphene oxide LB films is used instead the platinum catalysts. Such a drop, first of all, is caused by a decrease in the value of the cell fill factor by 4 times. Reducing the efficiency of solar cells using graphene is a well-known fact [13, 14].

In addition, a large resistance was recorded for the cell with the SLGO electrode, compared to the value typical for the traditional Pt electrode (Table 2). These parameters indicate the low electron transport properties of graphene oxide, despite its high photocatalytic activity. However, the use of composite layered electrodes Pt+SLGO allows to solve this problem. As can be seen from the Table 1, for this type of electrode, the efficiency values is higher than for a cell with a platinum catalyzing layer.

Thus, as shown by the results of measurements, the use of graphene oxide films in combination with a platinum catalyst is more promising for photovoltaic cells.

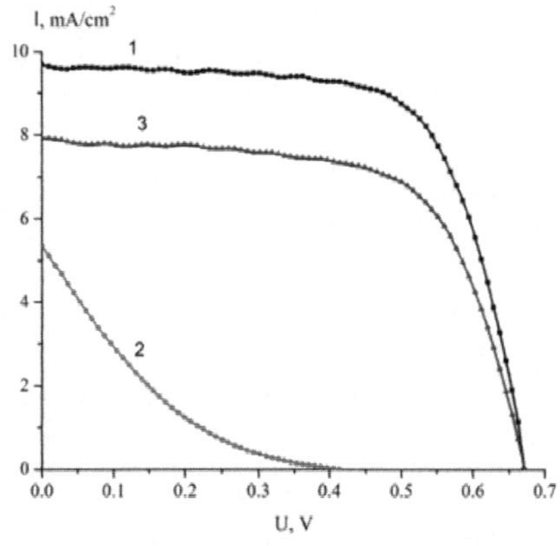

Figure 2. I-V characteristics of solar cells with different counter electrodes: 1 – Pt; 2 – SLGO; 3 – Pt+SLGO.

Table 1. Photovoltaic parameters of solar cells.

Counter electrode	U_{oc}, V	J_{sc}, mA/cm^2	FF	Efficiency, %
Pt	0.67	9.7	0.67	4.4
SLGO	0.41	5.3	0.13	0.3
Pt+SLGO	0.67	7.9	0.65	3.4

Figure 3 shows the hodographs for solar cells with different electrodes. The measurements showed that the magnitude of both the real and imaginary part of the resistance of the electrochemical cell increased significantly when using graphene oxide. The cell resistance is reduced to 40 Ohms and 45 Ohms for platinum electrode and a Pt + SLGO composite electrode respectively.

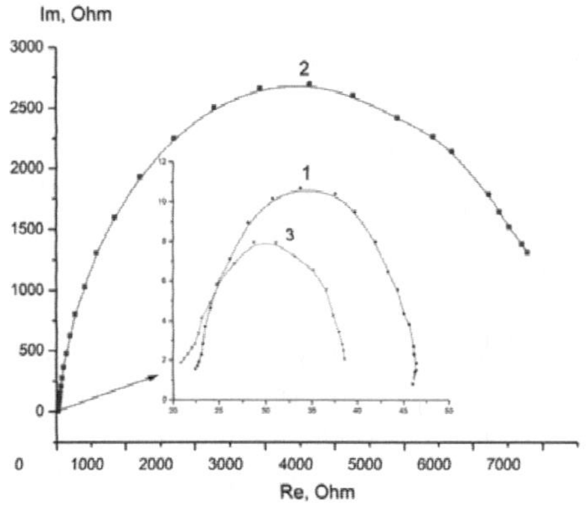

Figure 3. The impedance hodograph curves of DSSC with different electrodes: 1 – Pt; 2 – SLGO; 3 – Pt+SLGO.

Based on the hodographs obtained, the effective diffusion coefficient of electrons D_{eff}, the effective recombination rate k_{eff}, the effective electron lifetime τ_{eff}, the resistance of electron transport in a titanium dioxide film R_w, the charge transfer resistance R_k, associated with electron recombination, were calculated from the central curve of the impedance spectra [15, 16] (Table 2).

It was shown, that in graphene oxide films, the diffusion length of electrons is 2 orders of magnitude higher than in platinum electrodes.

Table 2. Electrotransport parameters of solar cells.

Counter electrode	D_{eff}	k_{eff}, s^{-1}	τ_{eff}, s	R_k, Ohm	R_w, Ohm
Pt	3.3×10^{-5}	28.7	0.03	17.5	20.5
SLGO	1.4×10^{-3}	3.3	0.3	6980.0	22.5
Pt+SLGO	3.8×10^{-5}	22.4	0.04	24.5	20.0

However, R_k of these films is more than three orders of magnitude larger. This means, as indicated above, despite the advantage of graphene oxide as a catalyst in comparison with Pt, SLGO film resistance is greater with comparison to the traditional platinum electrodes which is expressed in high values of D_{eff} and τ_{eff}. This, in turn, is reflected in a decrease in the number of electrons injected into the semiconductor layer from the dye molecules, which do not have a time to reduce. As a result, a very small number of electrons will diffuse into the semiconductor, which is expressed in a decrease in the current values generated by the solar cells.

4. Summary

Thus, the conducted studies have shown that for photovoltaic cells the use of graphene oxide films in combination with a platinum catalyst is more promising. Graphene oxide LB films, despite the advantages in electrotransport properties, have a high resistance, which is a result in the decreasing of the number of charge carriers injected into the semiconductor layer from the dye molecules and in a complex decrease in the efficiency of solar cells.

Acknowledgments

This work was performed at financial support of Ministry of Education and Science of the Republic of Kazakhstan.

References

[1] Guo C X, Guai G H , Li C M 2011 *Adv. Energy Mater.* **1** 448–452 doi: 0.1002/aenm.201100119
[2] He M, Jung J, Qiu F et al. 2012 *J. Materials Chem.* **22** 24254–24264 doi: 10.1039/C2JM33784C
[3] Roy-Mayhew J D, Bozym D J, Punckt C et al 2010 *ACS Nano* **4** 6203–6211 doi: 10.1021/nn1016428
[4] Whitby R, Gun'ko V M, Korobeinyk A et al. 2012 *ACS Nano* **6** 3967–3973 doi: 10.1021/nn3002278
[5] Whitby R, Korobeinyk A, Gun'ko V M et al. 2013 *J. Phys. Chem. C* **117** 11829–11836 doi: 10.1021/jp4022213
[6] Cruz R, Pacheco Tanaka D A, Mendes A 2012 *Sol. Energy* **86** 716–724 doi: 10.1016/j.solener.2011.11.010
[7] Meng T-T, Zheng Z-B, Wang K-Z 2013 *Langmuir* **29** 14314–14320 doi: 10.1021/la403428q
[8] Seliverstova E V, Ibrayev N Kh, Dzhanabekova R Kh 2016 *Nanosystems: phys., chem., math.* **7** 65–70 doi: 10.17586/2220-8054-2016-7-1-65-70
[9] Dzhanabekova R Kh, Seliverstova E V, Ibrayev N Kh 2017 *Bulletin of the University. Physic series* 1(85) 8–14
[10] Seliverstova E, Ibrayev N Kh, Dzhanabekova R Kh 2017 *Rus. J. Phys. Chem. A* **91** 1761–1765 doi: 10.1134/S003602441709028X
[11] Ibrayev N Kh, Seliverstova E, Nuraje N et al. 2015 *Mat. Sci. Semicon. Proc.* **31** 358–362 doi: 10.1016/j.mssp.2014.12.006
[12] Sedghi A, Nourmohammadi M H 2015 *Int. J. Electrochem. Sci.* **10** 3354–3362
[13] Li Zh Y, Akhtar M Sh, Kuk J H et al. 2012 *Mater. Lett.* **86** 96–99 doi: 10.1016/j.matlet.2012.07.006
[14] Wan L, Wang Sh, Wang X et al. 2011 *Solid State Sci.* **13** 468–475 doi: 10.1016/j.solidstatesciences.2010.12.014
[15] Adachi M, Sakamoto M, Jiu J et al. 2006 *J. Phys. Chem.* **110** 13872–13880

doi: 10.1021/jp061693u
[16] Thomas S, Deepak TG, Anjusree GS et al. 2014 *J. Mat. Chem.* **2** 4474–4490 doi:10.1039/C3TA13374E

IOP Conf. Series: Materials Science and Engineering **289** (2017) 012010 doi:10.1088/1757-899X/289/1/012010

Steel surface TiCrN, TiMoNcoatings structural phase state change features after low-energy alpha particles irradiation

S B Kislitsin[1], A I Potekaev[2], V VUglov[3], A AKlopotov[4], V D Klopotov[5], Yu F Ivanov[6] and A T Parpiev[7]

[1]Head of the Laboratory of Applied and Theoretical Materials Science, NuclearPhysics Institute, Almaty, Kazakhstan,

[2]Director of the SFTI, National Research Tomsk State University, Tomsk, Russia,

[3]Professor of BSU, Byelorussia State University, Minsk, Byelorussia,

[4]Professor of TSABU, Tomsk State Architecture and Building University, Tomsk, Russia,

[5]Assistant professor, National Research Tomsk Polytechnic University, Tomsk, Russia,

[6]Leading Researcher of the Institute of High Current Electronics SB RAS, Tomsk, Russia

[7] Postgraduate, National Research TomskPolytechnicUniversity, Tomsk, Russia

E-mail: vdklopotov@mail.ru

Abstract.The low-energy alpha particles irradiation impact on structural phase composition and 321S31 steel surface TiCrN and TiMoN coatings morphology have been summarized. The samples irradiation has been carried out in the DTs-60 (the Dubna cyclotron) heavy-ion accelerator by the $^4He^{+2}$ low-energy ions (40 keV) into a $1.0 \cdot 10^{17}$ ion/cm^2 fluence. It has been established thatthe 321S31 steel alpha particles implantation having energy of 40 keV results in the austenite initial γ-structurepartial transfer, stimulated by the irradiation, to the martensiteα-structure. The low-energy alpha particles irradiation within the fluencies range from 10^{16} to 10^{17} ion/cm^2 for the TiCrN and TiMoN coatings formed by the ionic bombardment condensation method doesn't result in the phase transfers with a coating crystal lattice type change.

1. Introduction

Now days the great efforts are directed at obtaining the more powerful reactor facilities [1]. These reactor facilities product life must be equal to 60 years. This product life increasing is accompanied by a thermal and irradiation embrittlement damaging effect increasing. There are various methods to compensate this effect, such as the steels using with the higherthermal and irradiation embrittlement resistance; the neutrons flow decrease during operation; the products metal shortness initial critical temperature reduction; the using of coatings on a reactor body sideon the basis of ceramics [2, 3]. For the final method using, it's necessary to produce the special coverings, had high physic-mechanical properties and used for the main material protect from the irradiation, thermal and other influences. Often the transition metals nitrides are used as coatings [4, 5]. These compounds combine ceramics and metal properties and have increased hardness, wear resistant and corrosion stability, thermal stability [6–8]. The transition metals TiCrN, TiZrN, TiMoN and other nitrides coatings are considered to be perspective as protective coatings [9, 10]. However these covering properties stability problem under the irradiation influence is not sufficiently explored [9–12].

The investigations results of the TiCrN and TiMoN coatings surface structural-phase composition and morphology changing caused by the alpha particles irradiation modeled the transmutative helium accumulation in the nuclear reactors constructional materialsare presented in this paper.

2. Samples and investigation methods

2.1. The covering production method

The coverings based on Ti-Cr-Nhavebeen applied on a carbon steel substrate surface (< 0.2 %) by the vacuum arc condensation method with the ionic bombardment (CIB). The substrates are capacitance discs 15 mm in diameter and 1 mm wide. The vacuum deposition modes have been the following: Ti and Cr cathodes, nitrogen residual pressure ($P_N = 10^{-1}$ Pa), stress shear 60 V, cathode current 100 A, deposition time 60–120 s. As result the samples series 50÷150 mm wide with the TiCrN coatings with the approximately equal titanium and chromium content have been obtained. The coatings deposition procedure is set out in great detail in [6, 8] papers.

The Ti-Mo-N coatings have been also formed by the CIB method at the titanium and molybdenum flows combination under a nitrogen residual atmosphere. The 321S31 steel plates with dimensions of 15×15×1 mm have been used as a substrate.Before the depositionthe substrate surface ionic bombardment have been studied by titanium ions for 60 s, at potential on the substrate 1 kV, at titanium cathode arc current 100 A and at pressure within the chamber 10^{-3} Pa, that gave the surface cleaning and its heating to a temperature of 450÷500 °C.The coatings havebeen deposited in N_2 gas ambient at pressure 10^{-1} Pa within the chamber, at potential on the substrate is equal to 120 V and at the titanium and molybdenum cathodes arcing currents 100 A and 180 A, respectively. The titanium and molybdenum plasma flows combination have been made by two cathodes allocation at an angle of 60° to each other at a distance equal to 120 mm from the substrate. As a result the samples series with the TiMoN coatings 2 µm wide have been formed.

The X-ray diffraction analysis has been made on the basis of a BRUKER company D8 ADVANCE diffractometer with using the copper radiation with the diffracted beam monochromatorin Brega-Brentano survey geometry. In some cases the survey in X-ray beam small-angle drop geometry (at an angle of 2°) within the angles range $2\theta^0 \in 30^0 \div 46^0$ has been used.

2.2 Alpha particles low-energy irradiation

The low-energy alpha particles irradiation has been carried out on the basis of a DTs-60 (the Dubna cyclotron) heavy-ion accelerator low-energy channelingECR (electron cyclotron resonance) at the Nuclear Physics Institute branch in Astana. The $_4He^{+2}$ ions energy was 40 keV (20 keV/charge). The selected irradiation fluencieswerebetween $1 \cdot 10^{16}$ and $5 \cdot 10^{17}$ ion/cm^2.The irradiated surface square was 10 mm^2. The irradiation temperature didn't exceed 150 °C.

According to the calculations (under the SRIM program) the 321S31 steel alpha particles projective range having energy of 40 keV is approximately 150 nm, the 321S31 steel straggling is approximately 75 nm, the TiCrN projective range is approximately 190 nm, the TiCrN straggling is approximately 83 nm, the TiMoN projective range is approximately 150 nm, the TiMoN straggling is approximately 80 nm. Therefore the implanted helium accumulates at the covering-substrate boundary in the carbon steel TiCrN coating samples. The implanted helium accumulates in the coating in the TiMoN samples.

3. Ti-Cr-N system structural phase states

In the Ti-Cr-N system the chromium and the titanium are related to transition metals [13, 14]. The titanium with the chromium in the liquid state are indefinitely soluble in each other.At the temperature decrease the alloys crystallize with forming a β(Ti,Cr) solid solutions continuous series (Figure 1). The main transformations occur in the solid state. The intermediate stages with the Laves phase structure form near the $TiCr_2$ composition.

IOP Conf. Series: Materials Science and Engineering **289** (2017) 012010 doi:10.1088/1757-899X/289/1/012010

The Cr-Ti-N system third element (nitrogen) significantly differs by chemical properties from the metals, forming this compound. The nitrogen with transition metals can form compounds with an ordered structure. These compounds form from the solid solutions and can produce the variable composition phases. The phases and different polymorphic transformations wide variety also exists in the nitrogen compound with the titanium and the chromium [13]. This is caused by these compounds chemical bond various character. The existing typically metallic bond in the metals solid solutions with the nitrogen in some phases and compounds begins to move in a covalent and an ionic bond. Also the bonds intermediate types manifestation is discovered (metallic and covalent combination, covalent and ionic combination and other).

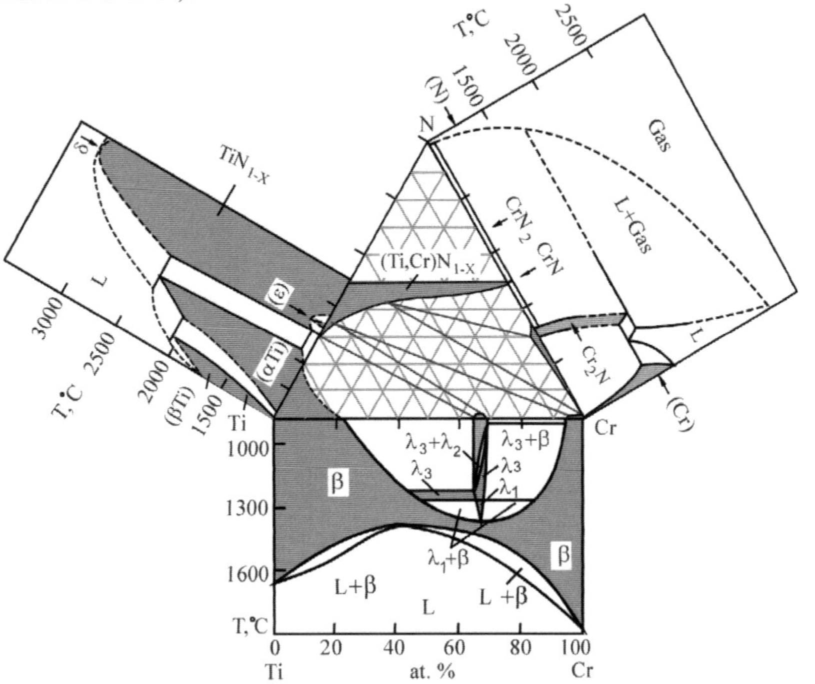

Figure 1. Cr-Ti, Cr-N, Ti-N systems binary diagrams [13] and Cr-Ti-N ternary system isothermal cross section at 1000 °C [14]. The single-phase areas are filled with grey color.

The TiN and CrN nitrides are fully mutually soluble and form solid solutions continuous series (Figure 1). The $(Ti,Cr)_XN_{1-X}$ ternary compound homogeneity region is wide. The TiN, CrN and $(Ti,Cr)N_{1-X}$ nitrides have the B1 cubic structure (NaCl prototype). The TiN_X compounds are related to the wide homogeneity region unordered nonstoichiometric nitrides group [15]. InTiN_X compounds the Ti atoms forms metallic face-centered cubic sublattice may containing N interstitial atoms or vacancies in its octahedral interstices. The nonstoichiometric disordered compound crystal structure type saving at the structural vacancies concentration change causes the homogeneity region presence. In our case the CrN nitride narrow homogeneity region means that the structural vacancies appearance leads to the crystal structural type change.

4. Ti-Mo-N system structural-phase state

There are the solid solutions continuous series in the Ti-Mo system (β-Ti,Mo) (Figure 2*a*). The titanium alloying by the molybdenum leads to the α⇔β polymorphic transformation temperature decrease. You can see at the presented diagram (Figure 2*a*) that the solid solution decomposition occurs to β_1 and β_2 solid solutions and the phase separation area are asymmetrical and shifted to the Ti side. The Mo solubility in the (α-Ti) is weak and isn't above 14 % (at.) at the monotectoide transformation temperature.

A significant negative departure from the Vegard law is found in a (β-Ti, Mo) solid solution *a* lattice parameter concentration dependence (Figure 2*b*), reflecting the strong lattice compression and as a consequences should results in interstitial solution forming problem. It really happens (Figure 3).

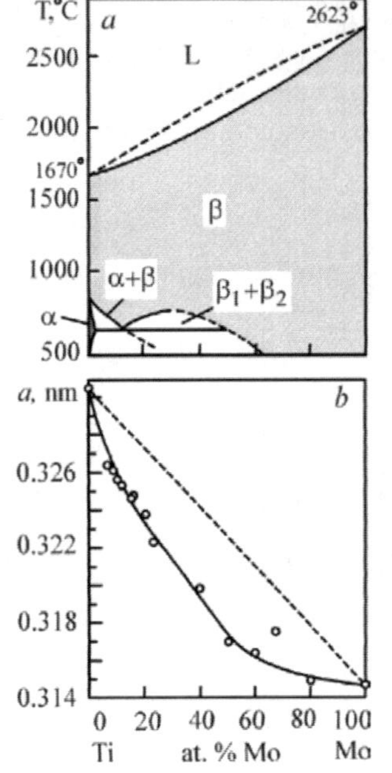

Figure 2. The Mo–Ti system state diagram (*a*) and the solid solution lattice parameter concentration dependence (*b*) [13].

You can see in the ternary constitutional diagram that the nitrogen atoms solubility in the solid solution (β-Ti, Mo) is very limited and it happens in a very ternary compound narrow homogeneity region (β-Ti, Mo, N) in the ternary diagram (Figure 3).

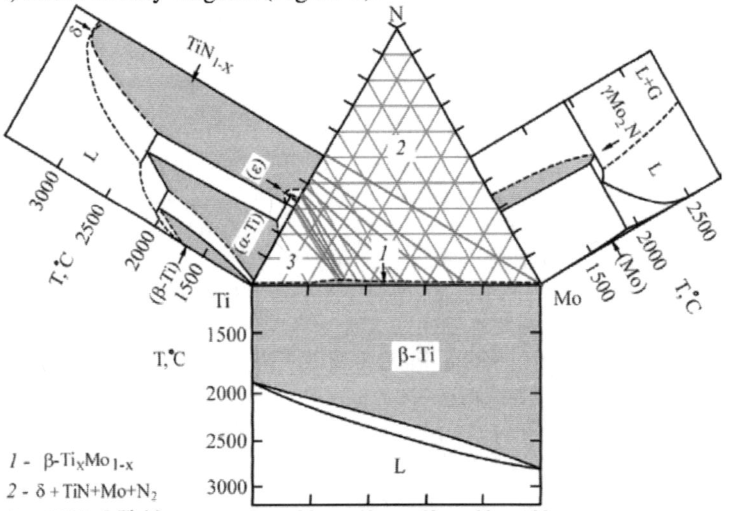

Figure 3. The Mo-Ti, Mo-N, Ti-N systems binary diagrams [13] and the Mo-Ti-N ternary system isothermal cross-section at the 1000 ºC [14]. The single-phase regions are indicated by grey color.

There are a few molybdenum nitrides in the Mo-N system [13]. There are a high temperature γ-Mo_2N (L′1 structure type, Fe_4N prototype) and a low temperature β-MoN_2 (L′6 structure type) in the Mo_2N composition region. The nitride in the δ-MoN forms in the composition area at 50 % (at.) N with the hexagonal syngony (*P6₃/mmc, B_h* structure type) and is unstable at the high temperatures.

The Mo-Ti, Mo-N, Ti-N systems binary diagrams [13] and the Mo-Ti-N ternary system isothermal cross-section at the 1000 °C are presented in Figure 3 [14]. The presented Mo-Ti-N ternary system isothermal cross-section at 1000 °C shows that Mo-Ti-N system ternary compounds practically don't form, except for β-Ti_XMo_{1-X}solid solution narrow homogeneity region with the dissolved nitrogen little amount (region 1, Figure 3).

5. Results and considerations

5.1 The 321S31 steel

It has been established by the X-ray structure analysis method, that the steel in the initial state is two-phase with the γ-Fe content to 85 % on the base of the fcc lattice and α-Fe content about 15 % on the basis of the bcc lattice. The irradiation by using of the low energy alpha-particles to the fluence of $1\cdot10^{17}$ has been led to the α-phase content increasing in the steel to approximately 25 %. Consequently the γ-Fe content has been increased to 75%. Therefore, the irradiation by using the low energy alpha-particles leads to the significant 321S31 austenitic steel structural-phase state change caused by the irradiation through an austenite-martensite phase transition [11].

5.2 The TiCrN coating on the carbon steel

According to [6], Ti-Cr-N coatings, deposited on the metal substrate by using the CIB method at temperatures 450 °C – 500 °C, are the solid solution (Ti, Cr)N with the fcc lattice. The TiCrN coatings formed by using the plasma flows condensation, having titanium and chromium atoms, in nitrogen atmosphere, are the phases mix having the content range from (Ti, Cr)N + Cr to TiN + (Ti,Cr)N with the relative concentrations increasing of the titanium to the chromium. This means that the chromium content increasing leads to the phases mix formation in the covering, such as (Ti, Cr)N solid solution and Cr, and the titanium content increasing leads to the other phases mix formation, such as (Ti,Cr)N solid solution and TiN. The elements content analysis made by means of the energy dispersed X-ray microanalysis (EDXM) has been shown that the coating is the titanium, chromium and nitrogen composite with the titanium and chromium content of 52.51 ± 0.07 at % and 47.49 ± 0.07 at % respectively. The substrate consists of 99 at. % Fe, 0.6 at.% Cr and 0.4 at.% Si. Complementary Bragg reflexes weren't indicated by the steel samples X-ray investigations with the coating on the base of TiCrN. May be this is caused by the small coating thickness (50÷150 nm). The alpha-particles ions irradiation having energy of 40 keV to the fluence of $1\cdot10^{17}$ ion/cm^2 has been led to some diffraction pattern change. The background has been significantly increased. The reflexes from the bcc lattice plates have became wide and the weak reflexes occur on the irradiated samples diffraction pattern at the angles are equal to 37° and 62°. The analysis showed that complementary reflexes relate to the (Ti,Cr)N_{1-X} compound with the B1 cubic structure.

Thereby the carbon steel surface TiCrN coating irradiation by using of the helium ions leads to the structure defects formation. This causes the disordered nonstoichiometric (Ti,Cr)N_{1-X} nitrides formation with the B1 structure [15].

5.3 The 321S31 steel TiMoN coating

It's established by using the X-ray analysis method that the near-surface layer generally consist of the δTiN titanium nitride (cubic syngony with the Fm-3m space group, NaCl prototype) and a (β-Ti,Mo) solid solution little amount. It's established that the δTiN titanium nitride lattice parameter is equal to 0.4298±0.0006 nm. The (β-Ti,Mo) solid solution lattice parameteris equal to 0.3149 nm. As it's seen onthe presented (β-Ti,Mo) solid solution lattice parameter concentration dependence, the Mo alloying by using the titaniumto 20 at.% only slightly change the lattice parameter (Figure 2*b*).

The X-ray photography usingin the grazing beam geometry has allowed indicating the (β-Ti, Mo) solid solution weak reflexes. Besides the (β-Ti, Mo) solid solution content isn't uniform along the coating thickness and increases with the distance increasing from the surface.

The low energy alpha-particles irradiation to the fluence of $1 \cdot 10^{17}$ hasn't been led to the significant diffraction pattern change. As well as to the irradiation, the (β-Ti, Mo) solid solution, δTiN titanium nitride and alpha-phase compound reflexes have been indicated from the steel substrate.

6. Summary

The low energy alpha-particles irradiation influence on the 321S31steel and TiCrN and TiMoN coatings surface crystal structure and morphology by using the 10^{16} – 10^{17} ion/cm^2 fluencesis investigated by using the X-ray analysis method, scanning electron and atom-power microscopy.

It's presented that in the 321S31steel the alpha-particles implantation having energy of 40 keV leads to the austenitic initial γ-structure partial conversation into martensite α-cstructure caused by the irradiation. The $\gamma \rightarrow \alpha$ transformation grate depends on the irradiation fluence, i.e. the helium supersaturation grate in the straggling region, reflecting ions ranges dispersionin the ion stopping region, i.e. in the projective range region. For the TiCrN and TiMoN coatings formed by using the ionic bombardment condensation method the low energy alpha-particles irradiation within the fluence range from 10^{16} to 10^{17} ion/cm^2 doesn't lead to the phase conversations with the crystal lattice type change.

Acknowledgements

This work was supported by Russian Science Foundation (project number 14-29-00091).

References

[1] Zotova A O et al 2014 *Materials technology problems* **79** 22–33
[2] Murty K L and Charit I 2008 *J. Nucl. Mater* **383** 189–195 doi:10.1016/j.jnucmat.2008.08.044
[3] Kurata Y et al 2010 – 2012 *Structural Materials for Innovative Nuclear Systems. Workshop Proceeding* 177–188
[4] Ciriello A et al 2007 *J. Nucl. Mater* **371** 129–133 doi:10.1016/j.jnucmat. 2007.05.024
[5] Belous V A et al 2012 *Proceedings of XX International conference on physics of radiation phenomena and radiation material science* 333–334
[6] Uglov V V et al 2005 *Surface & Coatings Technologies* **200** 178–181 doi: 10.1016/ j.surfcoat. 2005.02.136
[7] Ivanovskii A L 1996 *Chemistry Progress* **65** 499–518
[8] Uglov V V 2007 *Vacuum* **81** 1345–1347 doi:10.1016/ j.vacuum.2007.01.042
[9] Gavarini S 2007 *J. Nucl. Mater* **362** 364–373 doi:10.1016/j.jnucmat.2007.01.239
[10] Uglov V V 2010 *Surface & Coatings Technologies* **204** 2095–2098 doi: 10.1016/ j.surfcoat. 2009.11.023
[11] Kislitsin S 2010 *Structural Materials for Innovative Nuclear Systems. Workshop Proceedings* 87
[12] Kislitsin S B 2013 *Nuclear Science and Technology Protblems. Radiation Damages and Radiation Material Technology Physics Series* **84** 17–22
[13] Lyakishev N P 1996–2000 *Double Metallic Systems Phase Diagrams* (Moscow: Publishing house «Mashinostroenie») **1–3**
[14] Holleck H 1984 *Transition Metalls Binary and ternary Carbid- and Nitrid system* (Berlin Stuttgard: Gebrüder bornteager)
[15] Gusev A I and Rempel A A 2001 *Non-Stoichometry, Disorder and Order in Solids.* (Ekaterinburg: UB RAS)

IOP Conf. Series: Materials Science and Engineering **289** (2017) 012011 doi:10.1088/1757-899X/289/1/012011

The influence of the distance between the donor-acceptor groups of polymethine dyes on their photovoltaic properties

E Seliverstova[1,2] and N Ibrayev[3]
[1]Senior lecturer, Institute of Molecular Nanophotonics, Buketov Karaganda State University, Karaganda, Kazakhstan
[2]Senior research fellow, Institute of Applied Mathematics, Karaganda, Kazakhstan
[3]Professor, Institute of Molecular Nanophotonics, Buketov Karaganda State University, Karaganda, Kazakhstan
E-mail: genia_sv@mail.ru

Abstract. Spectral-luminescent and photovoltaic properties of polymethine dyes of various structures are studied. It is shown that an increase in the length of the methylene chain between the active chromophores leads to a red-wave shift of the absorption and fluorescence spectra. Significant changes in the absorptivity and lifetime of fluorescence do not occur in this case. The best photovoltaic parameters have cells sensitized with shorter dye molecules. It is shown, that for a longer dye the resistance associated with electron recombination on the TiO_2/electrolyte surface is much higher than the electron transfer resistance in the semiconductor, which reduces the efficiency of electron transfer in the solar cell, sensitized with longer dye molecules.

1. Introduction

About solar cells sensitized with organic dyes (DSSC) usually mentioned as an alternative to solar batteries based on inorganic semiconductors. DSSCs are less expensive and harmless than inorganic semiconductors. Despite the potentially high quantum efficiency, solar cells based on organic dyes have a number of drawbacks that prevent the commencement of large-scale industrial production of such solar cells. Their main drawback is the low efficiency of conversion of solar energy, which, usually, does not exceed of 7 – 13% [1, 2]. For effective operation of an organic solar cell, a strong absorption of light by the working layer, an optimal difference in the energy levels of the donor and acceptor of the charge carriers, sufficient mobility of electrons (holes) in the acceptor (donor) phase and ohmic contacts of the acceptor (donor) with the cathode (anode) are necessary. As was shown in [3, 4], polymethine dyes have extremely high extinction coefficients. The possibility of reconstructing the absorption spectra in the entire visible and near infrared region makes polymethine dyes a perspective object for use in solar cells.

Because of their unique optical properties, polymethines are used in nonlinear optics, in recording and storage devices, like fluorescent probes in biomolecules or as contrast agents in optical images of tissues. Oxidation-reduction properties of polymethine dyes have also been studied from the point of view of their role as a sensitizer or desensitizer for silver halides. A wide spectrum of oxidation-reduction potentials allows polymethines to act both as electron donors and as electron acceptors in photoinduced electron transfer processes [4]. A literary review in the field of cells based on polymethine dyes shows that the efficiency of dyes of this class varies in a wide range from 2 to 7% with a current density more than 21 mA/cm^2 [5]. The dyes with functional groups (SO_3, -COOH, -

OH), which promote more effective adsorption of the dye on the semiconductor surface, have the highest photovoltaic parameters [6].

In this paper, the results of a study on the influence of the structure of polymethine dyes on their spectral-luminescent and photovoltaic properties are presented.

2. Experiment

The dyes with different lengths of polymethine chain between the terminal donor-acceptor groups were chosen as the object of the study and shown in Figure 1. Dye absorption and fluorescence spectra were measured on a Cary-300 (Agilent) spectrophotometer. The lifetimes of the excited state of the dye molecules were measured using a pulsed spectrofluorimeter with a picosecond resolution and registration in a time-correlated photon count mode (Becker&Hikl, Germany). The fluorescence quantum yields (φ_{fl}) of the dyes were measured relative to the standards – ethanol solutions of methylene blue ($\varphi_{fl} = 0.01$) and oxazine 17 ($\varphi_{fl} = 0.70$).

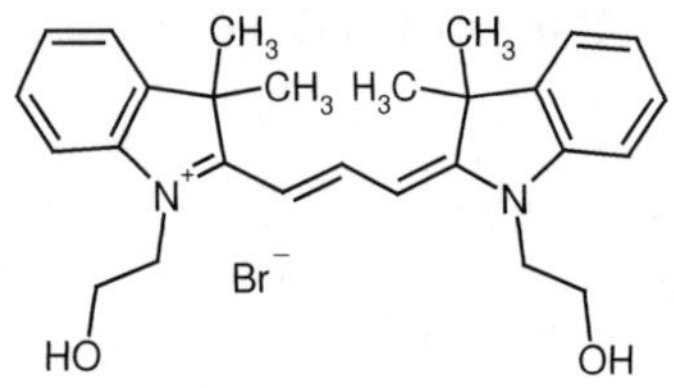

K1

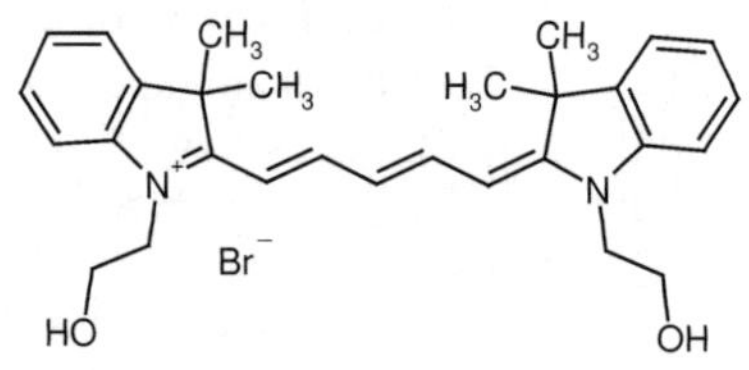

K2

Figure 1. Structure of dyes.

The preparation and assemblying of solar cells was carried out according to the procedure detailed in [7, 8]. To prepare of solar cells, glass substrates coated with a FTO layer (Sigma-Aldrich, surface resistance of ~10 Ohm/sq) were used. Blocking layer of TiO_2 was deposited on the surface of the FTO. Semiconductor films were prepared by the doctor-blading method. To prepare the films, TiO_2 with a nanoparticle size of 25 nm was used (Sigma-Aldrich). The thickness of the resulting films was measured by the SEM (Tescan Mira-3) and was equal to 10 – 12 microns. Sensitization of semiconductor electrodes with polymethine dye K1 or K2 was carried out by immersion of TiO_2 films in an ethanol dye solution with concentration equal to 4×10^{-4} mol/l for 20 hours. Before measurement, the films were kept in an oven at a temperature of 80 °C for at least of 1 hour. The number of adsorbed dye molecules in all samples was controlled by the change in the optical density of the solutions before and after sorption.

The current-voltage characteristics of DSSC were measured with Solar Cell Tester CT50AAA (Photo Emission Tech. Inc., USA) under illumination of the Xe lamp with light intensity of 100 mW/cm^2. The measurements were performed at room temperature in photovoltaic mode. From the obtained values of the open curcuit voltage, short-circuit current, fill factor, efficiency of the cells were determined. Electric transport properties of solar cells were studied by measuring the impedance of an electrochemical cell. The measurements were carried out on an Z-500PRO impedance meter (Elins, Russia) under irradiation of cells with standard simulated solar radiation (Air Mass (AM) 1.5). For the analysis of the data obtained, the ZView 3.2b program and the equivalent solar cell circuitry (Figure 2), presented below, were used.

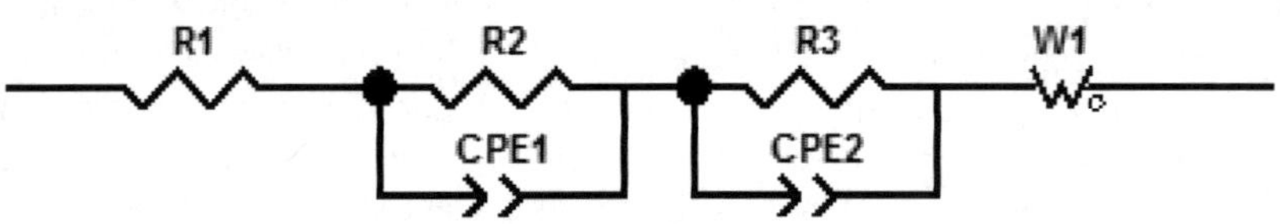

Figure 2. Equivalent solar cell circuitry.

3. Results

Measured absorption and fluorescence spectra and spectral-luminescent parameters of dyes in ethanol solutions are presented in Figure 3 and Table 1.

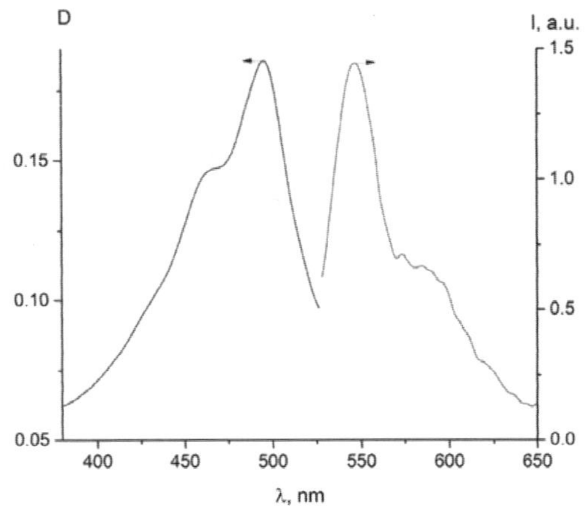

Figure 3. Absorption (a left) and florescence (at right) spectra of K1 dye in ethanol solution, $C_{dye}=10^{-5}$ mol/l

Table 1. Spectral parameters of K1 and K2 dyes.

Dye	K1	K2
${}^{a}\lambda_{1\ max}$, nm	550	646
${}^{a}\lambda_{2\ max}$, nm	520	598
$\Delta\lambda^{a}_{1/2}$, nm	54	42
$\varepsilon*10^4$, l/mol cm	0,8	1,4
${}^{f}\lambda_{max}$, nm	568	670
${}^{f}\lambda_{max}$, nm	–	724
$\Delta\lambda^{f}_{1/2}$, nm	26	40
τ_f, ns	2.6	2.45
φ_f	0.05	0.25

It can be seen from the Figure and the data that as the length of the polymethine chain increases, the absorption spectrum shifts to the red region of the spectrum. The extinction coefficient (ε) of the K2 dye is higher than that of K1, as the quantum yield of fluorescence.

The results of measurements of the I-V characteristics of the dyes are presented in Table 2.

Table 2. Photovoltaic parameters of solar cells sensitized with K1 and K2

Dye	J_{sc}, mA/cm^2	U_{oc}, mV	*FF*	η, %
K1	0.4575	378.5	0.264	0.11
K2	0.36	272.0	0.300	0.02

It can be seen that the dye K1 has better photovoltaic parameters, in comparison with K2, in spite of that the fill factor of the K2 dye is higher. This can be explained by the fact that the absorption spectrum of K2 is shifted to the region of red wavelengths. As is known, in this area the intensity of sunlight is lower than in the shorter-wave region (at green region of spectrum).

The electrotransport and amplitude-frequency characteristics of the prepared solar cells were investigated (Figure 4).

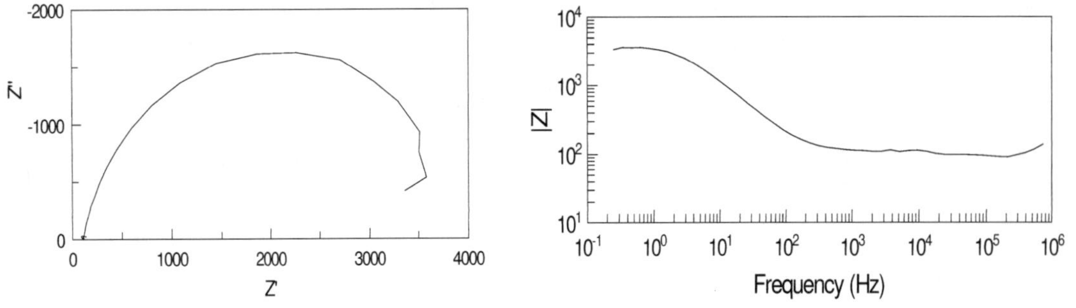

Figure 4. Impedance (a) and amplitude-frequency characteristics (b) of a solar cell sensitized by K1.

Table 3. The value of the elements of the DSSC circuit.

Dye	R_1, mOhm	R_2, Ohm	C_1, μF	$\tau = R_2 * C_1$, s
K1	106	3955	22.4	0.093
K2	392	4740	13.7	0.072

From Table 3 it is seen that the diffusion time of the charge carrier for the dye K2 is less than for K1. This, in turn, reduces the probability of electron transfer to a semiconductor film, which affects the magnitude of the current generated by the cell [9]. In addition, this parameter is affected by the value of the active resistance R2 (in Figure 3 – parameter Z') and it is have the value of the order of several thousand Ohms. Hence, the resistance associated with the recombination of electrons on the TiO_2/electrolyte surface is much higher than the resistance of electron transfer in a semiconductor [10]. This is the result of the fact that the lifetime of an electron is much higher than the time of electron transfer, and the diffusion length is greater than the thickness of the cell [11].

4. Summary

Thus, when studying the influence of the length of a polymethine chain between chromophores in a dye molecule, it was shown that the cells sensitized with dye molecules K1 possess the best photovoltaic parameters. This can be explained by the fact that the efficiency of absorption of sunlight in a dye with a long chain decreases, due to a shift in the absorption spectrum of the dye to the region of red wavelengths. In studying the electrophysical parameters of solar cells, it was shown that for K2 dye the resistance associated with electron recombination on the TiO_2/electrolyte surface is much higher than the electron transfer resistance in the semiconductor, which reduces the electron transfer efficiency within the solar cell.

Acknowledgments

This work was performed at financial support of Ministry of Education and Science of the Republic of Kazakhstan.

References

[1] Gratzel M 2003 *J. Photochem. Photobiol. C.* **4** 145–149 doi: 10.1016/S1389-5567(03)00026-1
[2] O'Regan B, Gratzel M 1991 *Nature* **353** 737–741 doi: 10.1038/353737a0
[3] Ishchenko A A 1998 *Theoret. Experiment. Chem.* **34** 191–210 doi: 10.1007/BF02523249
[4] Ishchenko A A 2002 *Polym. Adv. Technol.* 13 744–752 doi: 10.1002/pat.269
[5] Mishra A, Fischer M K R, Bäuerle P 2009 *Angew. Chem. Int. Ed.* **48** 2474–2499 doi: 10.1002/anie.200804709

[6] Ibrayev N Kh, Seliverstova E V, Ishchenko A A et al. 2017 *J. Photochem. Photobiol. A: Chem.* **346** 570–575 doi: 10.1016/j.jphotochem.2017.06.029
[7] Ibrayev N Kh, Seliverstova E, Nuraje N et al. 2015 *Mat. Sci. Semicon. Proc.* **31** 358–362 doi: 10.1016/j.mssp.2014.12.006
[8] Sedghi A, Nourmohammadi M H 2015 *J. Electrochem. Sci.* **10** 3354–3362
[9] Adachi M, Sakamoto M, Jiu J et al. 2006 *J. Phys. Chem.* **110** 13872–13880 doi: 10.1021/jp061693u
[10] Andrade L, Cruz R, Aguilar Ribeiro H et al. 2010 *Int. J. Hydrogen En.* **35**. 8876–8883 doi: 10.1016/j.ijhydene.2010.06.020
[11] Bisquert J 2002 *J. Phys. Chem.* B. **106** 325–333 doi: 10.1021/jp011941g

IOP Conf. Series: Materials Science and Engineering **289** (2017) 012012 doi:10.1088/1757-899X/289/1/012012

Application of CAE-modeling for the study of the influence of the sensor location on the flow-through water heater operation

A G Yakunin
Head of chair, Altai State Technical University, Barnaul, Russia
E-mail: yakunin@agtu.secna.ru

Abstract. The article deals with issues related to increasing the efficiency of the system of automatic maintenance of the temperature of liquid media entering the pipes to the place of consumption. For this purpose, a flowing water heater model is proposed, made in the SolidWorks environment, the construction parameters of which can be changed using the appropriate macro and screen form. It is shown that the choice of the location of the temperature sensor has a significant effect on such parameters of the device as the accuracy of maintaining a given temperature regime and the duration of the transient process caused by a change in the temperature of the liquid entering the heater. On a concrete example, it is shown that by changing the distance between the sensor and the heating module, it is possible to achieve minimum temperature fluctuations of the heat-transfer-agent at the heater outlet.

1. Introduction

Automatic control and regulation systems have recently become increasingly important, since the quality of the operation of such systems largely influence on the quality of products, the amount of energy consumption, the ease of operation of equipment and a number of its other technical, consumer and economic properties [1, 2]. The issues of developing such systems and increasing the efficiency of their work in a variety of applications have been the subject of a large number of works since the middle of the last century, for example [3–8]. Among the whole variety of such systems and the devoted to them publications, it is especially worth mentioning the works devoted to the control and regulation of temperature processes, because the temperature is one of the main factors, the sustentation of which at a proper level is important both in production and in an everyday life. And among them, in turn, can be distinguished regulators of the temperature of liquid media, and water first of all, since a water is used both in everyday life and at enterprises as a heat-transfer-agent or like a component of the technological process.

Typically, temperature control systems are implemented using microcontrollers and PID algorithms to ensure the rapid completion of transient processes and to minimize temperature fluctuations during regulation [9], but in the simplest systems, threshold regulators are also used. At present, industry produces a wide range of such regulators for liquid media [10], including regulators controlling the operation of flow heaters [11–14]. Their feature is both the instability of the flow of the liquid passing through the heater due to a change in the degree of water consumption, and the changeability of the water temperature entering the heater. In addition, the spreading in the space the heat source, the temperature sensor and the place of water consumption leads to the appearance of a so-called transport delay, as a result of which the use of pure PID algorithms for maintaining the temperature at a constant level becomes ineffective [15]. For this reason, the calculation of such regulators becomes extremely complex and approximate, because it is practically impossible to take into account all features of the structural design of the heating module analytically, as proposed in [16, 17].

In this regard, it becomes urgent to develop a methodology that allows to ensure high quality of temperature control for flow heaters of various designs by choosing optimal parameters that define the geometry and the relative position of components of the heater. In particular, the aim of this work was the development of a technique based on the application of a computational experiment on the CAE model of flow heaters constructed in the environment of the corresponding simulators of dynamic temperature processes occurring in liquid and solid media.

2. Environment and Stages of Modeling

Among the many software products that allow solving the tasks of fluid dynamics and heat transfer on personal computers, the CFD (Computational Fluid Dynamics) packages from AutoDesk [18] and ANSYS [19], as well as the Flow Simulation module [20] from Dassault Systèmes SolidWorks [21] are widely distributed. Although SolidWorks is a CAD system, the use of the included CFD module for the given task solved was recognized as the most suitable, because it provides the optimal combination of performance, interface simplicity and sufficient functionality on limited computing resources of a personal computer. This conclusion is confirmed, in particular, by the results presented in [22–24].

The entire process of conducting research in the Flow Simulation environment is divided into four stages. The first one creates the actual model of the object under study, that is the assembly consisting of individual components that are included in it. This stage can be performed completely manually, using the appropriate interface means. In this case all or part of the components of the assembly can be imported from other development environments that support formats for the exchange of CAD data. An alternative solution is to use your own macro that automates the process of creating a model in accordance with specified requirements. At the second stage, the mesh is superimposed on the model and the shape of the cells is chosen, since the solution of the problem is reduced to the solution of non-linear inhomogeneous partial differential equations by the finite element method. At the next stage, the environmental conditions and parameters of various external factors that are stationary or dynamically affecting the simulation object are set. The process of simulation and obtaining of temperature distributions and distributions in space and time of other parameters of interest are performed at the last, final stage.

Since the aim of the work was to develop a technology for designing and researching water heaters, at the first stage, to automate the process of creating a model, an appropriate macro was developed, launched from the user's dialog box, allowing you to vary the geometric dimensions of the structural elements of the heating block, for example, to choose there optimal relation for the given criterion of quality. As a basis for the design of this unit, the variant shown in Figure 1*a* was chosen, which is most often found in models of different manufacturers. Appearance of the form for specifying the geometric dimensions of the structure is shown in Figure 1*b*, and in Figure 1*c* the variants of assemblies obtained using the developed macro are presented. The algorithm of its operation is shown in Figure 2. From it, as well as from Figure 1*b*, it can be seen that with the help of this macro, the basic details of the heating module are separately constructed for such parameters as the diameter of the inlet and outlet nozzles, the diameter and the length of the tank: the tank body (cylinder), its covers, heating element and its base. In this case, the dimensions of the details that are not specified in the dialog box (except for the base of the heating element having a fixed value) were set proportionally to the dimensions set by the interface. The final construction of the assembly of the heating module is performed by pressing the button "Construct the tank". For convenience, a separate additional button was added to the SolidWorks toolbar for calling the macro.

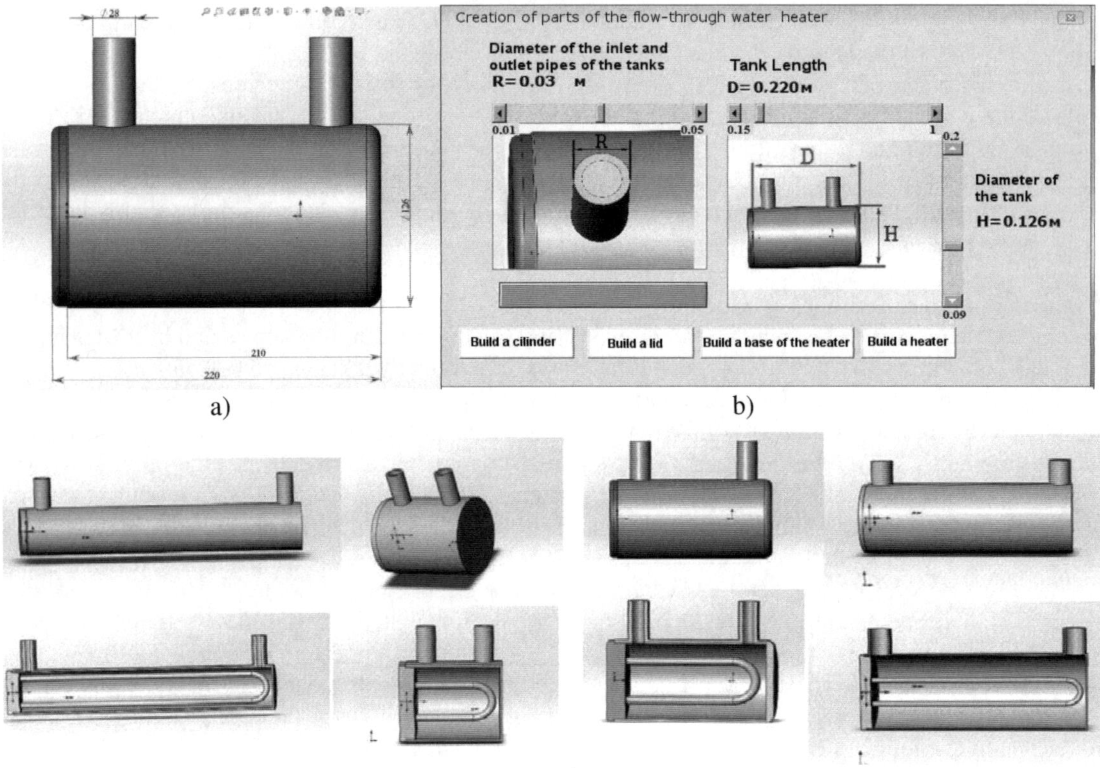

Figure 1. The design of the heating module (*a*), the dialog form of the macro for entering the parameters of its assembly (*b*), and the various variants of such assemblies, constructed with this macro (*c*).

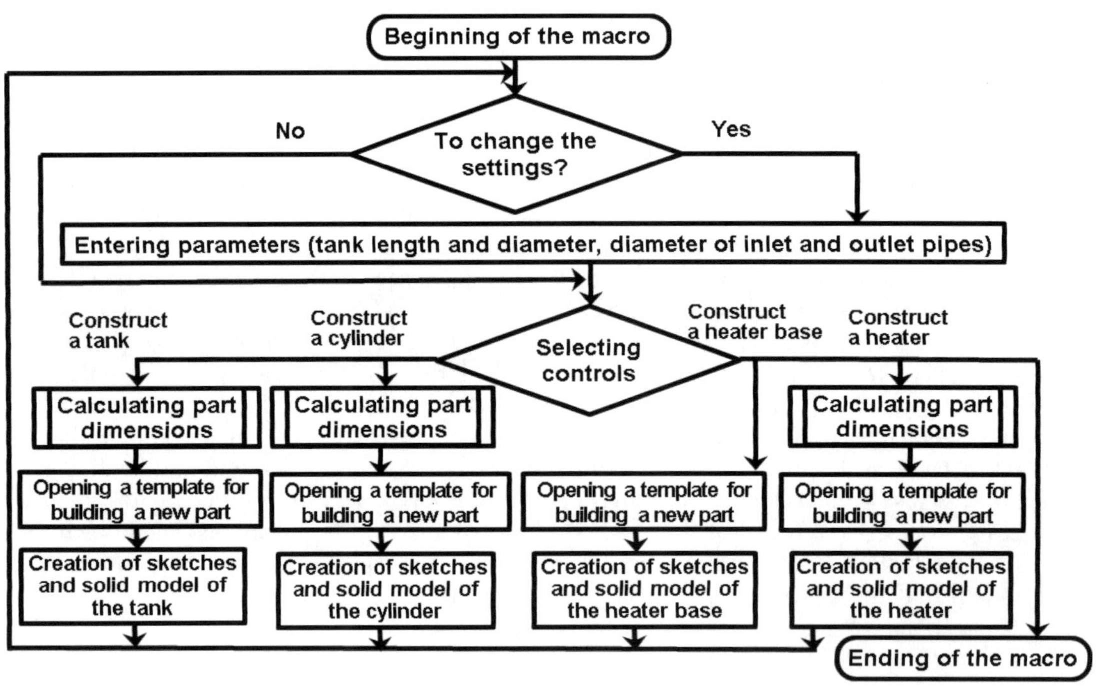

Figure 2. Simplified algorithm for the operation of the macro for forming the assemblies of the heating module.

IOP Conf. Series: Materials Science and Engineering **289** (2017) 012012 doi:10.1088/1757-899X/289/1/012012

The main material of the assembly was stainless steel. For the base of the heating element an insulating material with low thermal conductivity "insulator" was used, and for the case of the heating elements themselves, either copper or also stainless steel. The maximum dimensions of the parts shown in Figure 1*b* were chosen based on the maximum allowable power of the 15 kW heater. The construction of a larger tank would require a change in its design in connection with the need to install several heating elements. In the future it is planned to realize this possibility, as well as to expand the number of different configurations of flow water heater assemblies for other models of flow heaters.

At the next stage the model was placed in the design area, the size of which was set approximately 5–10 times larger than the size of the heating module. The shape of the 3D cells was set rectangular everywhere with using the adaptive property. Their size for solid elements was set 5–10 times smaller than the cells in the calculation area (Figure 3).

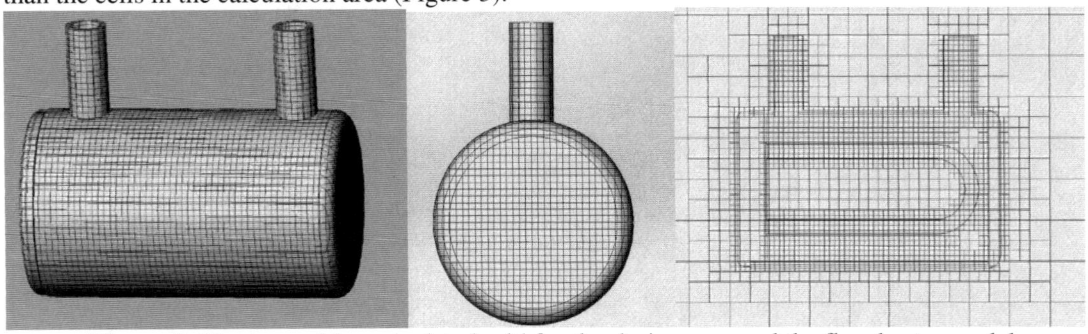

Figure 3. The type of the calculated grid for the design area and the flow heater model.

After the superposition of the grid, the subregions of the flow and the initial and boundary conditions were specified for the model. For the surrounding water heater, air was chosen as the fluid, and water was selected for the internal cavity of the flow water heater model. The temperature of the air and the walls of the design area was set at 22 ° C, and the water temperature was set at 15 ° C. The pressure in all cases was 101325 Pa. The mass flow of water at the inlet branch was set at 0.15 kg / s, and the heating element power was 15kW. The Appearance of the model after performing all the listed procedures is shown in Figure 4. To study the change in the temperature of the heated water, the length of the outlet branch pipe was increased, the design area was stretched in a vertical direction. On this area are marked points for which temperature plots were calculated. On this step the preparation for study a flow heater can be considered complete.

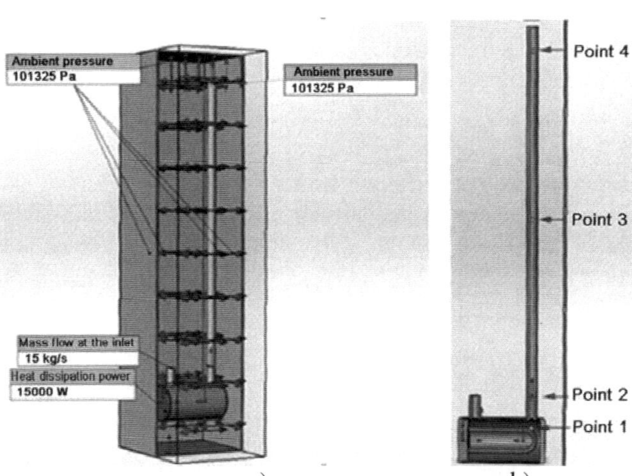

Figure 4. Type CAE model for the study of heat distribution in a flow-through water heater with specified boundary and initial conditions (*a*) and placement of points on it for monitoring the temperature of heated water (*b*).

3. Investigation of the operation of the flow-through heater in the static mode

Within the framework of this work two kind of computational experiment were carried out. The first one was to assess the inertial properties of the heating module and to analyze the heat distribution along the volume of the heating module and in the outlet branch pipe. The second was the research of operation of the heater using various methods of temperature control.

The result of the computational experiment on the study of the dynamics of the temperature field changes during static heating is shown in Figure 5. It can be seen from this that the temperature inside the volume of the heater is highly inhomogeneous. This heterogeneity remains in the outlet branch pipe and the temperature in it is equalized at a distance greater than the linear dimensions of the heater itself. In this case, the termination of the transient and the establishment of the static regime occurs in the interval from 50 to 90 s after the heater is turned on. °

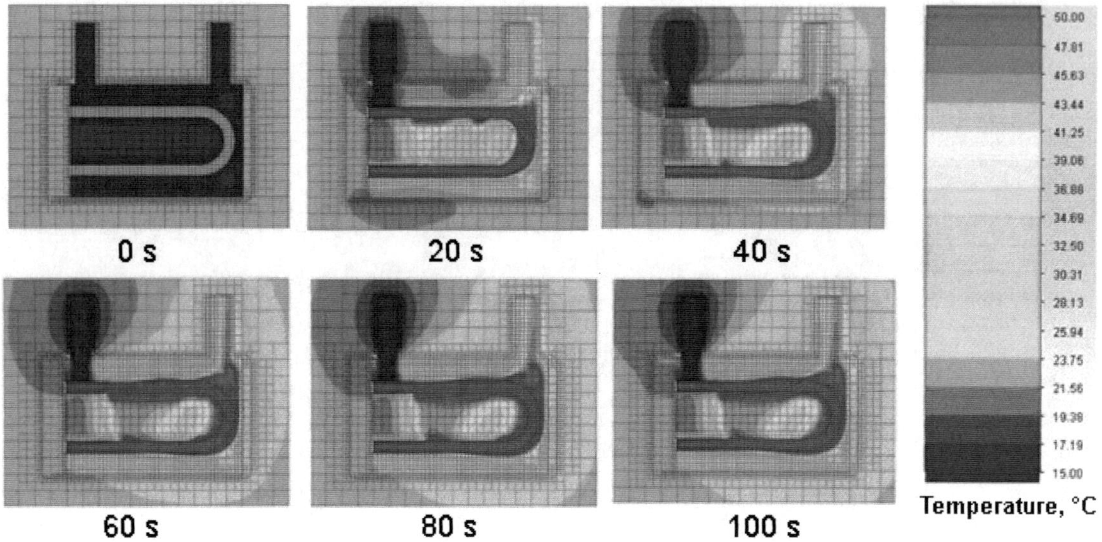

Figure 5. Dynamics of temperature fields change in the flow heater after the heating element is switched on.

To more accurately establish the time constant and determine the distance after which the temperature of the liquid in the outlet tube could be considered homogeneous, temperature measurements were made at four points on the outlet pipe, as shown in Figure 4. The plots of dependence of the temperature vs. time at these points are shown in Figure 6.

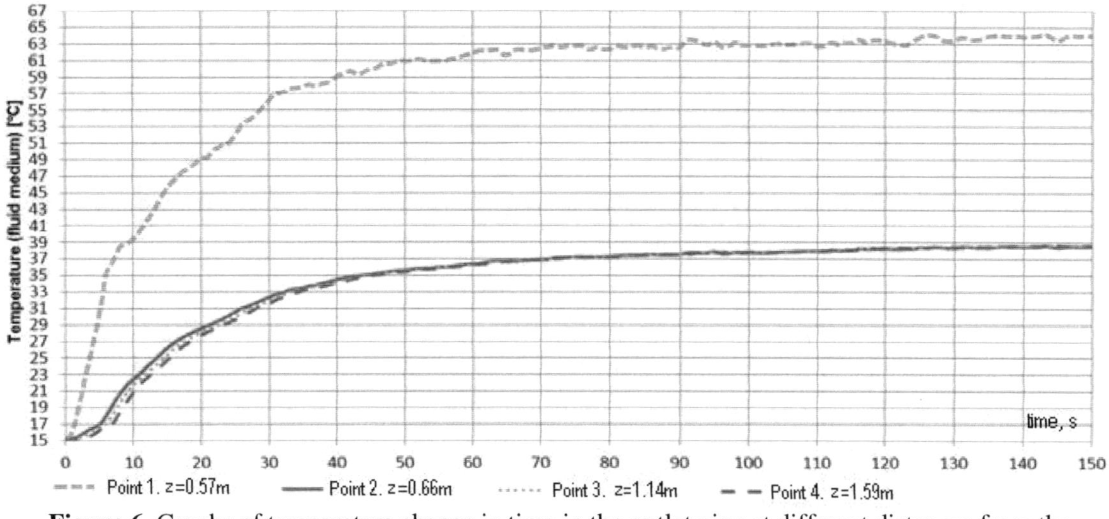

Figure 6. Graphs of temperature change in time in the outlet pipe at different distances from the heater with static heating. In all cases, $x = 0.969$m, $y = 0.988$m.

It can be seen from this that the installation of a temperature sensor in the immediate vicinity of the heater outlet does not at all reflect the temperature of the water entering to consumer. And besides, because of convection processes, this temperature is subject to significant fluctuations, which makes it difficult to use point 1 to place the temperature sensor of the automatic control system. For these purposes, other points more remote from the heater for a distance of 0.5 m or more are more suitable.

4. Investigation of the operation of the flow-through heater in the thermal control mode

At this stage of computational experiments, the operation of a flow-through heater in the circuit of the automatic temperature control system was investigated. Two control modes were considered: threshold and proportional. In the threshold mode, the heating element was switched on when the temperature of the water exceeded the set threshold value. In linear mode with proportional control, the heating element was switched off when the set temperature was reached the set threshold value, and if it did not exceed this value, the power supplied to it linearly depended on the difference between the preset and the actual temperature.

Figure 7*a* shows the temperature of the heated water as a function of time for different locations of the temperature sensor. The cutoff threshold of the heating element in this experiment was set equal to 30 ° C, and the inclusions to 29 ° C. From these graphs it is seen that with a smaller distance from the sensor to the heater, the amplitude of the oscillations decreases, which is explained by the smaller value of the transport delay. To study the transient processes due to the change in the flow velocity, the flow rate was linearly reduced from 0.15 kg / s to 0.06 kg / s in the boundary condition "mass flow rate at the inlet" from 80 to 81 seconds, and from 90 to 91 s is linearly returned to the original value. The graph of the corresponding transient process is shown in Figure 7*b*.

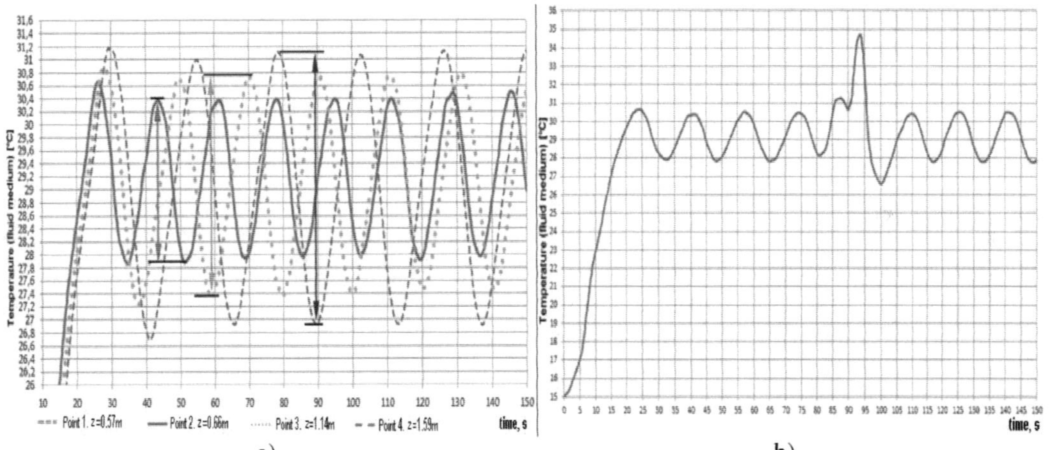

a) b)

Figure 7. The time dependences of the temperature changes during the operation of the thermostat system in the threshold mode for different locations of the temperature sensor (*a*) and the transient diagram when the flow rate is changed (*b*). The sensor is placed in the position $z = 0.66$m.

Similar studies were carried out with the operation of a temperature control system in proportional control mode. In this mode, the power input to the heating element varied from 0 to a maximum value of 16 kW, with a temperature drop of 30 to 28 ° C. The results of such studies are shown in Figure 8. Comparing the results shown in Figures 7 and 8, it can be seen that the use of linear control has made it possible to reduce the amplitude of the oscillations approximately 4-fold. At the same time, the duration and amplitude of the oscillations of the transient process with a change in the water supply rate in both cases turned out to be approximately equal, which, apparently, is due to the fact that the transport delay is more affected on them.

5. Conclusions

The use of the developed methodology based on carrying out computational experiments using CAE models allows to significantly reduce the time and volume of financial expenses for carrying out studies on optimization of algorithmic and constructive solutions in the development of flow heaters.

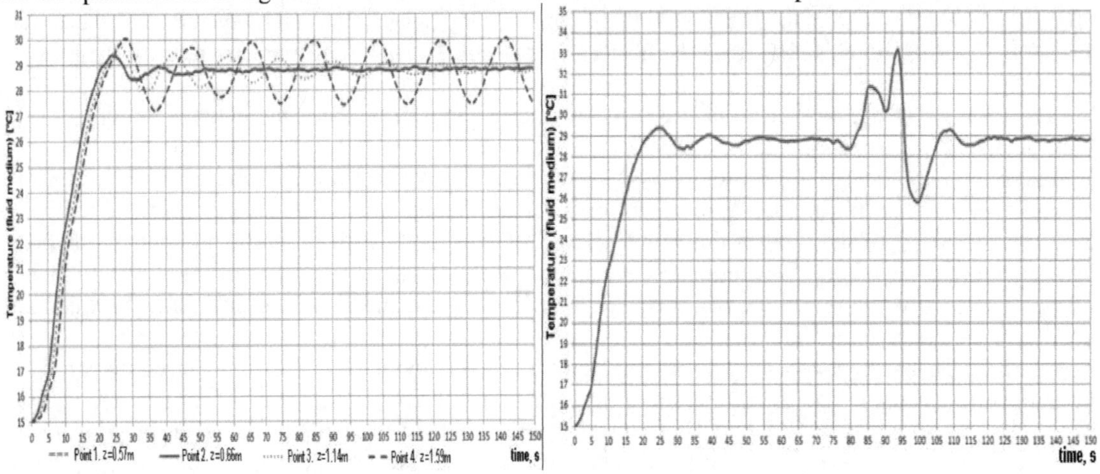

a) b)

Figure 8. The graph of the temperature changes during the operation of the thermostat system in the proportional control mode for various locations of the temperature sensor (*a*) and the transient schedule when the flow rate is changed (*b*). The sensor is placed in position z = 0.66m.

Its use in solving a particular development task for the heater design shown in Figure 1*a* has made it possible to optimize the location of the temperature sensor and estimate the value of the pulsations of the heated water for different control regimes of the power supplied to the heater [24] and in case of changing in the water flow rate. Further development of the work is supposed to be done in the direction of expanding the set of structural designs of heating modules and in studying the operation of heaters in other modes. Thus, it can be assumed that the introduction of an additional sensor directly on the housing of the flow heater in combination with the implementation of the algorithms described in [15–17] will significantly reduce the amplitude and duration of the transient processes caused both by the change in the water flow rate and its temperature fed to the input connection of flow heater. A definite effect can also be provided by the transition to the use of algorithms with proportional-differential-integral control.

Acknowledgement
The author expresses his deep gratitude to the leadership of the Altai State Technical University of I.I. Polzunov (AltSTU) for financial support of the publication of the article, OOO Promavtomatika for technical assistance in the preparation of the article and the alumnus AltSTU, the master Vorobyov D.S. for the develop of the model of the water heater and performing calculations in the SolidWorks.

References

[1] Zhelyazova B 2005 *Elsevier's Dictionary of Automation Technics 1st Edition* (Sofia: Elsevier Science)
[2] Paraskevopoulos P N 2001 *Modern Control Engineering (Automation and Control Engineering)* (New York: Marcel Dekker, Inc)
[3] Elliott Brothers (London) Ltd. 1946 *J. Sci. Instrum.* **23** 409 doi: 10.1088/0950-7671/23/3/409
[4] Suchkova L I at all 2013 *Kontrol'. Diagnostika* **13** 95–101 (in Russian)
[5] Suchkova L I, Hussain H M, Kunz RV, Yakunin A G 2013 *Izvestiya of the ASU* **1/1** 210-214 (in

Russian)
[6] Suchkova L I, Hussain H M, Yakunin 2013 A G *Polzunovsky almanakh* **1** 75–82 (in Russian)
[7] Yakunin A G, Hussein H M 2017 *Journal of Physics: Conf. Series* **881** 012028 doi:10.1088/1742-6596/881/1/012028
[8] Automatic Control in Aerospace 1989: Selected Papers from the IFAC Symposium (*Tsukuba, Japan*)
[9] Grubic M and Würz U 1978 *J. Phys. E: Sci. Instrum.* **11** 692–696 doi: 10.1088/0022-3735/11/7/025
[10] Liquid heaters [Electronic resource] URL: http://www.directindustry.com/industrial-manufacturer/liquid-heater-215840.html
[11] Tankless water heating [Electronic resource] URL: https://en.wikipedia.org/wiki/Tankless_water_heating
[12] Technical characteristics of flowing water heaters [Electronic resource] URL: http: //www.stelmarket.ru/teh_inf/39.htm
[13] Tempra® Whole House Tankless Electric Water Heaters [Electronic resource] URL: http://www.stiebel-eltron-usa.com/products/tempra-series-whole-house-tankless-electric-water-heaters
[14] DHC Single Sink Point-of-Use Tankless Electric Water Heaters [Electronic resource] URL: http://www.stiebel-eltron-usa.com/products/dhc-point-of-use-tankless-electric-water-heaters
[15] Shilin A A, Bukreev V G 2014 *Thermal Engineering (English translation of Teploenergetika)* **61** 741–746 doi: 10.1134/S0040601514100115
[16] Shilin A A , Bukreev V G 2014 *Thermal Engineering (English translation of Teploenergetika)* **61**671–678 doi: 10.1134/S0040601514080126
[17] Shilin A A, Bukreev V G 2014 *Communications in Computer and Information Science* **487** 387–396
[18] Autodesk knowledge network: AutoDesk CFD, Knowledge Overview [Electronic resource] URL: https://knowledge.autodesk.com/support/cfd
[19] ANSYS: Fluid [Electronic resource] URL: http://www.ansys.com/products/fluids
[20] SOLIDWORKS Flow Simulation [Electronic resource] URL: http://www.solidworks.com/sw/products/simulation/ flow-simulation.htm
[21] Oficial site of Dassault Systèmes SolidWorks Corporation [Electronic resource] URL: http://www.solidworks.com/
[22] Kurowski P.M. 2017 Thermal Analysis with SOLIDWORKS Simulation 2017 and Flow Simulation 2017 (*SDC publications)*
[23] Hong-yang Zhang,Tong-chun Li, Zong-kun Li 2009 *Water Science and Engineering* **2** 95–102 doi: 10.3882/j.issn.1674-2370.2009.01.009
[24] Vorob'ev D S, Yakynin A G 2017 *The 21st Century: Fundamental Science and Technology: Proceedings of the Conference* 75–77

IOP Conf. Series: Materials Science and Engineering **289** (2017) 012013 doi:10.1088/1757-899X/289/1/012013

A technique for predicting steel corrosion resistance

V F Novikov[1], R A Sokolov [2], D F Neradovskiy [3] and K R Muratov [4]
[1]Head of the department, Tyumen Industrial University, Tyumen, Russian Federation
[2]Specialist, Tyumen Industrial University, Tyumen, Russian Federation
[3]Senior lecturer, Tyumen Industrial University, Tyumen, Russian Federation
[4]Assistant professor, Tyumen Industrial University, Tyumen, Russian Federation
E-mail: falcon.rs@mail.ru

Abstract. Research works were carried out to develop a technique with the aim to increase the lifetime of steel items used in corrosive media. The possibility to monitor corrosion parameters of steel samples is analyzed on the basis of magnetic properties obtained by means of a magnetic structuroscope DIUS-1.15M designed by the Institute of Metal Physics of the Ural Branch of the Russian Academy of Sciences (IMP UB RAS).

1. Introduction

There are many modern metal corrosion inspection and monitoring techniques, such as weight loss analysis, metallographic evaluation, ultrasonic testing and radiography (metal thickness measurements); magnetic sensing (magnetic field measurements); optical sensing (light intensity measurements) and others. Corrosion prediction is normally based on determining the corrosion rate of coupons by means of direct or indirect methods [1, 2]. This approach is destructive, time-consuming and does not provide 100 per cent efficiency of monitoring. New monitoring parameters should be stated to assure high reliability of corrosion rate prediction by means of nondestructive accelerated techniques [3].

2. Experimental procedures

In magnetic structuroscopy, magnetic hysteresis loops are generated to monitor strength properties of steels. The loops involve many experimental points, but only some of these points are used for nondestructive monitoring. Among them we can name coercive force, magnetic permeability, relaxation coercive force, differential permeability, etc. An objective of this research was representing a quasi-static magnetic hysteresis loop as a periodic signal and calculating harmonics of this signal.

Quasi-static hysteresis loops for ferromagnetics were generated by means of a modern structuroscope (and a hardware/software module DIUS-1.15M designed and produced at the Institute of Metal Physics, Ekaterinburg).

The material under investigation was 45KH (45X) steel which was temperature (950ºC) quenched with the sequence of tempering procedures that followed.

Magnetic hysteresis loops for 45KH (45X) steel generated by means of the hardware/software module DIUS-1.15M are shown in Figure 1.

To enhance possibilities of the entire hysteresis loop we designed and applied the harmonic analysis for a quasi-static hysteresis loop [4].

For this, time parameter in the harmonic analysis was replaced with the magnetic field strength

$$t = H/k, \tag{1}$$

since magnetizing and demagnetizing current in the device used and, therefore, magnetic field strength, are described by a ramp function. To form a periodic function, we imaged the descending (upper) branch of the loop symmetrically with respect to the vertical line going through the H_{max} point (Figure 2).

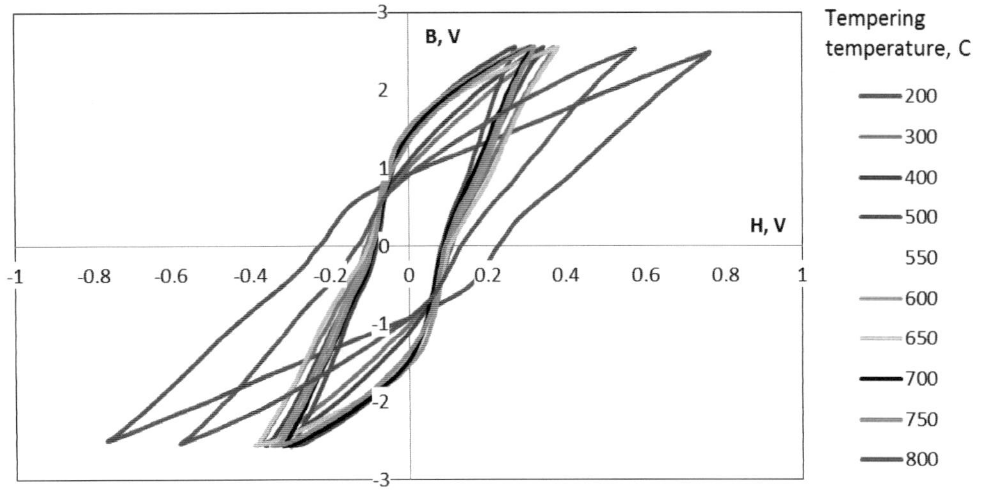

Figure 1. Hysteresis loops of 45KH (45X) steel tempered at various temperatures.

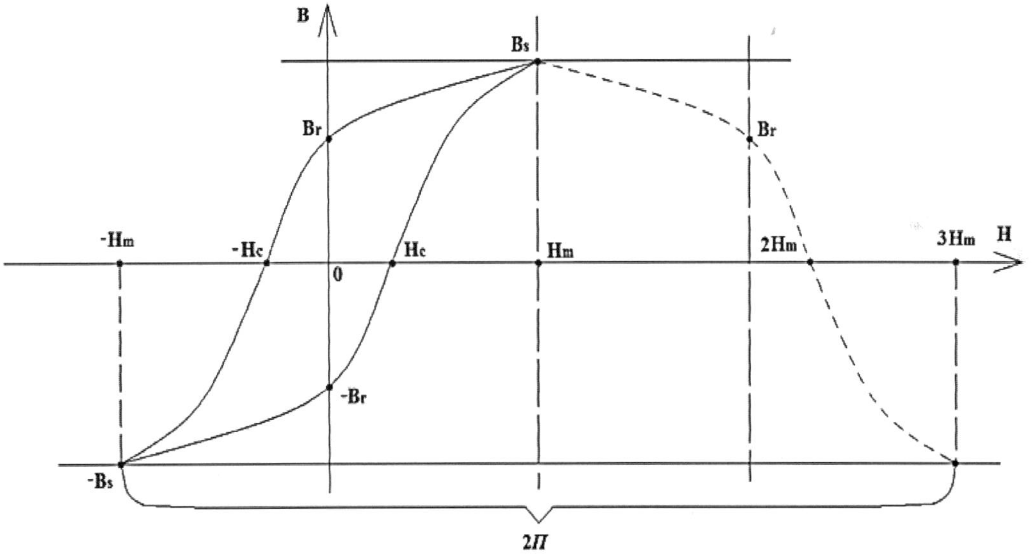

Figure 2. Imaging an involute of the magnetic hysteresis loop.

As a result, we obtained the representation of the hysteresis loop as a periodic function

$$f(H) = f(kt) \tag{2}$$

3. Results and discussion

For this function, the discrete Fourier transform can be applied [5, 6]. With the discrete Fourier transform, the amplitudes of the 1st, the 3rd, the 5th and the 7th harmonics of the hysteresis loops were calculated for 45KH (45X) steel, depending on tempering temperatures given in Figure 3. In this chart, A constitutes dimensionless amplitudes of Fourier spectrum harmonics. As one can note, the peak values of harmonic components endure a non-monotonic change all through the temperature range, what is connected to the structural changes taking place in the test steel.

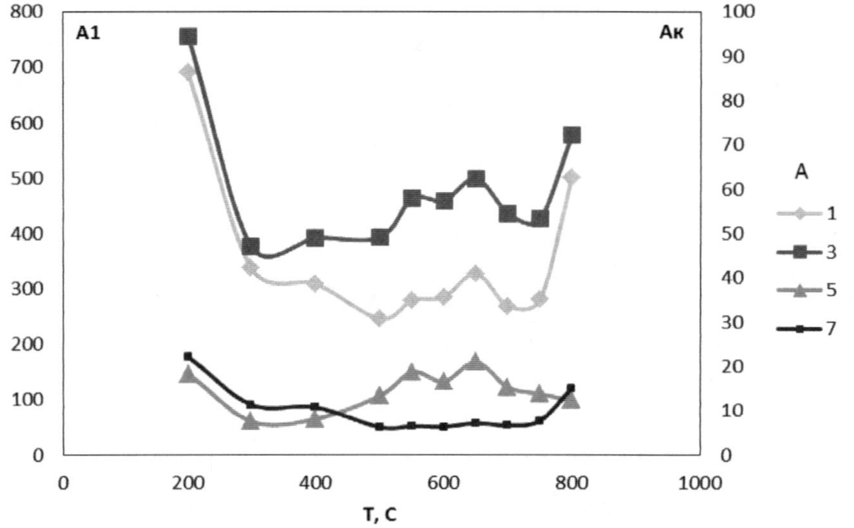

Figure 3. Tempering temperature dependence of A_1, A_k amplitudes (k=3, 5, 7) of the odd Fourier spectrum harmonics in the hysteresis loops for 45KH (45X) steel.

To define the reliability of the readings taken by means of the hardware/software module DIUS-1.15M, the comparison of the readings coercive force was carried out by means of the coercitimeter KRM-Ts-K2M (Figure 4).

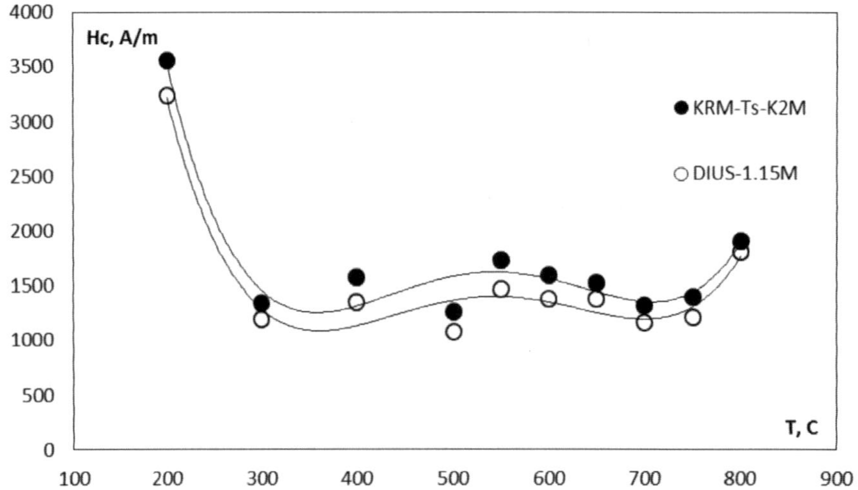

Figure 4. Tempering temperature dependences of the coercive force Hc for 45KH (45X) steel, which were generated by means of the coercitimeter KRM-Ts-K2M and the hardware/software module DIUS-1.15M.

The test steel samples tempered at various temperatures were subjected to corrosion testing in the 5% solutions of hydrochloric and sulfuric acids at room temperature (Figure 5).

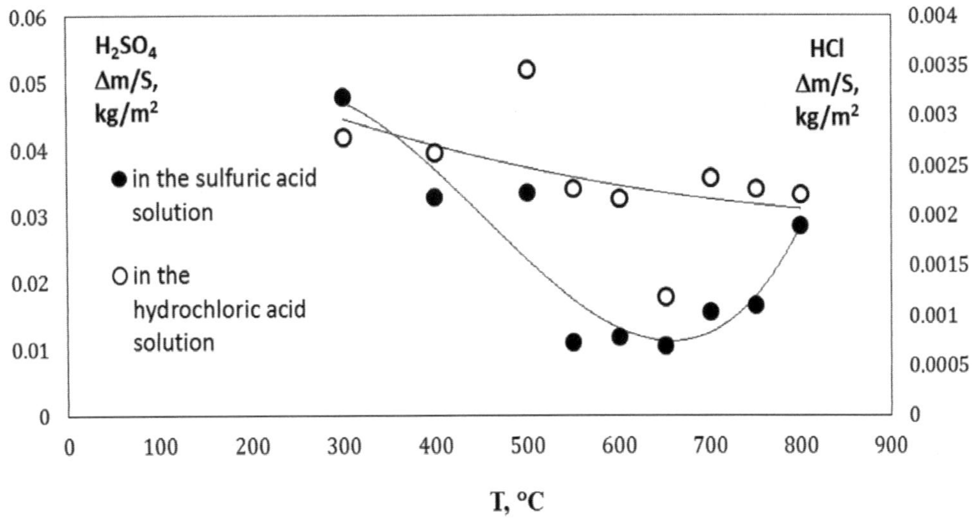

Figure 5. Tempering temperature dependence of the samples weight loss for 45KH (45X) steel in the 5% solutions of hydrochloric and sulfuric acids.

According to the inclusion theory of coercive force, the increased contents of carbide components of carbon lead to the increased coercive force. The structure dispersion results in the increased number of micro-galvano pairs and, therefore, in the increased rate of corrosion processes.

The work [7] shows that at the temperatures over 200°C the retained austenite transforms to the tempered martensite, and internal strains decrease, leading to the decreased coercive force.

At tempering within the temperature range of 300…450°C, carbon is completely released from the supersaturated solid solution, and the martensite transforms to the ferrite. The carbide Fe_2C transforms to the cementite Fe_3C [8]. A structure is formed, which is called the tempered troostite and consists of the ferrite with small cementite particles regularly distributed. All these processes reduce the corrosion susceptibility of steel (Figure 4). As the magnetic hysteresis loops correlate directly with the magnetic parameters of steel, the changes of magnetization harmonic spectra on their basis will also be susceptible to the changes of the steel structure and phase composition. In Figure 3 the amplitudes of the 3rd, the 5th and the 9th magnetization harmonics increase, while those of the 1st and the 7th decrease.

At tempering within a higher temperature range (500…650°C), the cementite particles coagulate – small particles merge into bigger ones and, as a result, their number decreases [9]. After such tempering the steel consists of a coarse mixture of the ferrite and cementite, called the granular, or tempered, pearlite. With the tempering temperature increasing, the structure approaches equilibrium to an increasing extent [10, 11]. The properties of steel also change considerably, including corrosion properties (the increased corrosiveness in the sulfuric acid solution). The coercive force has a low peak value within the temperature range of 500…600°C, which is mostly related to the presence of the alloyed cementite or special carbides with smaller inclusions. The amplitudes of magnetization harmonics increase and the peak values for the amplitudes of the 1st, the 3rd and the 5th harmonics can also be observed due to some concentration of carbide particles and their reaching a "critical size" through the coagulation [8, 10].

At tempering within a yet higher temperature range (600…800°C), carbides are released. The amplitudes of the harmonics stay monotonic up to 700°C, but with a higher temperature the amplitudes of the 1st and the 7th harmonics increase, while the amplitude of the 5th harmonic decreases [12, 13]. The carbide release results in the formation of micro-galvano pairs which lead to the increased corrosiveness.

As it is shown in Figure 4, the minimal corrosiveness of 45KH (45X) steel can be observed within the tempering temperature range of 450–600°C. This arises from the decreased structure heterogeneity, the material reaching equilibrium.

The comparison of the harmonic spectrum components and the weight loss in the abovementioned corrosive media allowed to obtain the dependences given in Figure 6.

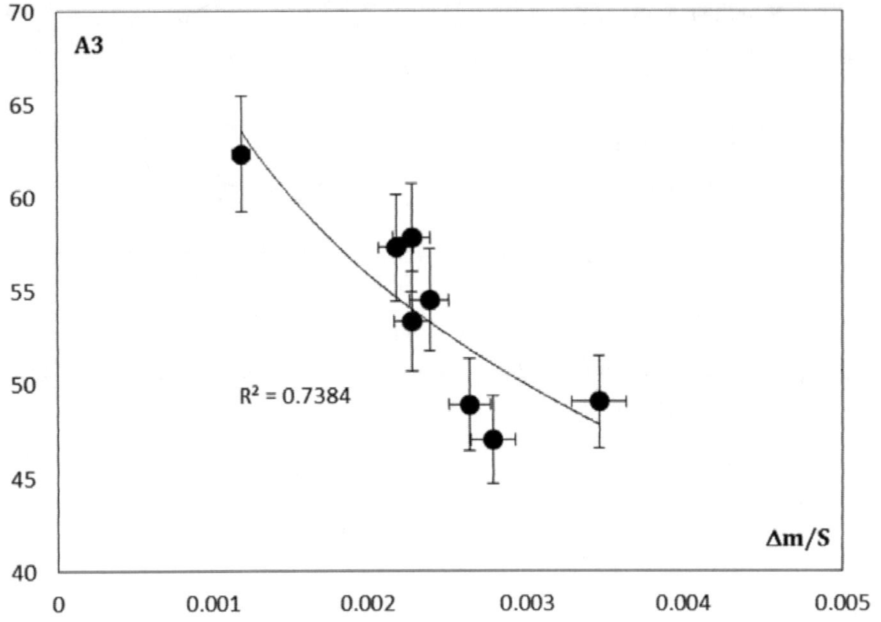

a)

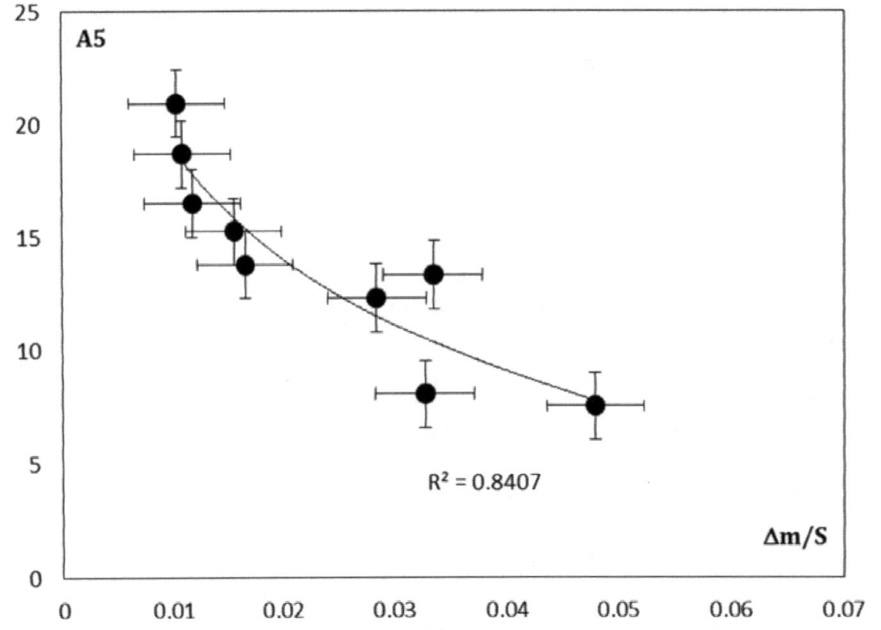

b)

Figure 6. Dependences of the harmonic amplitudes A_3 and A_5 on the relative corrosion weight loss m/S of the heat-treated 45KH (45X) steel samples: (*a*) dependence of the 3rd harmonic amplitude on the weight loss in the hydrochloric acid solution; (*b*) dependence of the 5th harmonic amplitude on the weight loss in the sulfuric acid solution.

4. Conclusions

Thus, we obtained the dependences for the harmonic spectrum and the relative weight loss, described by the logarithmic function with 74% and 84% reliability.

To sum up, the research work resulted in the following conclusions:

1. The correlation between the corrosion weight loss and the changes of the harmonic spectrum amplitudes for quenched and tempered 45KH (45X) steel samples is determined with 75% and 80% rate.

2. The coercive force for 45KH (45X) steel is proved not to correlate with the corrosion in the solutions of hydrochloric and sulfuric acids.

3. The dependence obtained with the use of the harmonic analysis constitutes a promising research direction to develop a technique for predicting 45KH (45X) steel corrosion resistance.

References

[1] Klyuev V V 2003 *Nondestructive Monitoring and Diagnostics* (Moscow: Mashinostroyeniye)
[2] 2004 *GOST 9.008-85.YeSZKS. Metals and Alloys. Techniques to Define Corrosion and Corrosion Resistance Indicators* (Moscow:Izdatelstvo Standartov)
[3] Novikov V F, Ustinov V P, Radchenko A V, Yevko V P 2015 Method and device to monitor excessive corrosion of steel Patent RU 2570704
[4] Novikov V F, Neradovskiy D F, Sokolov R A 2016 *Herald of Perm National Research Polytechnic University. Machine-Building, Material Science* **18** 38–50 (in Russion).
[5] Belashov V Y, Chernova N M 1997 *Effective Algorithms and Programs of Computational Mathematics* (Magadan: NEISRI)
[6] Ponomaryov Y F 1986 *Russian Journal of Nondestructive Testing* **7** 67–78
[7] Malakhov A I and Zhuko A P 1989 *Bases of Material Science and Theory of Corrosion* (Moscow: Vysshaya Shkola).
[8] Mikheyev M N and Gorkunov E S 1993 *Magnetic Methods of Structural Analysis and Nondestructive Monitoring* (Moscow: Nauka)
[9] Novikov I I 1974 *Theory of Metals Heat Treatment* (Moscow: Metallurgiya)
[10] Mogutnov B M, Tomilin I A, Schwartzman L A 1984 *Thermodynamics of Iron Alloys* (Moscow: Metallurgiya)
[11] Mikheyev M N, Morozova V M, Noskova N I 1981 *Structure and Physical and Mechanical Properties of Steels* (Sverdlovsk: Institute of Metal Physics of Ural Scientific Center of Academy of Sciences Russia)
[12] Bida G V, Nichipuruk A P 2005 *Magnetic Properties of Heat-Treated Steels* (Ekaterinburg: the Ural Branch of the Russian Academy of Sciences)
[13] Kurdyumov V G, Utevskiy L M, Entin R I 1977 *Transformations in Iron and Steel* (Moscow: Nauka)

IOP Conf. Series: Materials Science and Engineering **289** (2017) 012014 doi:10.1088/1757-899X/289/1/012014

Radial-piston pump for drive of test machines

A I Nizhegorodov[1], A N Gavrilin[2], B B Moyzes[2], A I Cherkasov[3], O M Zharkevich[4], G S Zhetessova[5], N A Savelyeva[6]
[1]Professor, Irkutsk National Research Technical University, Irkutsk, Russia
[2]Associate professor, Tomsk Polytechnic University, Tomsk, Russia
[3]Senior lecturer, Tomsk Polytechnic University, Tomsk, Russia
[4]Associate professor, Karaganda State Technical University, Karaganda, Kazakhstan
[5]Professor, Karaganda State Technical University, Karaganda, Kazakhstan
[6]Doctoral student, Karaganda State Technical University, Karaganda, Kazakhstan
E-mail: nastromo_irkutsk@mail.ru

Abstract. The article reviews the development of radial-piston pump with phase control and alternating-flow mode for seismic-testing platforms and other test machines. The prospects for use of the developed device are proved. It is noted that the method of frequency modulation with the detection of the natural frequencies is easily realized by using the radial-piston pump. The prospects of further research are given proof.

1. Introduction

The destructive effect of vibration on various technical systems is due to the occurrence of resonance phenomena in the devices elements as well as dynamic loads which lead to mechanical failure and faults of the objects [1–4] during a standard operation. As a rule, several elements of the system simultaneously resonate resulting a synergetic effect. There are many methods and means of testing for vibration and vibration resistance, for example [5, 6]. Hydrostatic vibro-exciters are referred as a special type of test facilities which are capable to excite spectrally saturated vibrations at very high carrying capacity. Also they are distinguished by the ability to control the vibration mode in a wide range [7–10]. During the seismic resistance tests the large-scale mock-ups were performed by loading the fragments of buildings and structures which give the adequate to real seismic effects [3, 4]. In the result there were worked out several approaches to seismic resistance. This requires the installation of test platforms of much higher carrying capacity. These tests platforms must be capable to simulate the true conditions for the test of dynamic load occurring during earthquakes. Test platforms equipped with a hydrostatic vibratory drive are the most suitable for such purposes as they feature a high specific power and control capability inherent in volumetric hydraulic drives [1, 2].

Due to the impact of seismic vibrations on construction objects the irreversible deformations of elements and components of their structures occur in the result of resonance phenomena. It is experimentally proven that any structure is characterized by a spectrum of natural frequencies, which, as a rule, is in the range of 0.5...15 Hz [3]. As seismic oscillations have a continuous spectrum with the largest harmonic levels in the interval of 0.1...10 Hz, there is always a correspondence between the excitation frequencies and the natural frequencies, which leads to the appearance of resonances.

Another factor of seismic resistance is the duration of such an impact. Short-term load of 4...5 even at a 10-point of earth shock is not dangerous, as the resonances do not have time to develop, and a small-amplitude earthquake which is lasting for several tens of seconds leads to serious destruction [3, 11–16].

The method of a time sweep frequencies is based on those factors which include the preliminary scanning and follow-up tests at resonance frequencies. As the oscillation spectrum is simulated sequentially, the test conditions are inadequate to real seismic impacts when the resonances of the elements and their construction parts do not occur simultaneously.

If we select the amplitude levels from the spectrum of random ground vibrations and average them to the third-octave ranges, then the spectrum can be represented as an imitating line spectrum which ratio of the harmonic levels is close to the spectrum of amplitude-modulated (AM) oscillations. This result is not accidental, as the actual ground vibrations during an earthquake are the natural modulation process [3] (Figure 1), although it is not described by mathematic harmony. Thus, the process of amplitude or amplitude-frequency modulation is the most adequate oscillation process simulating seismic oscillations during test. The envelopes of AM oscillations can be defined by different functions, that allows to simulate the oscillation processes which are different in harmonic composition.

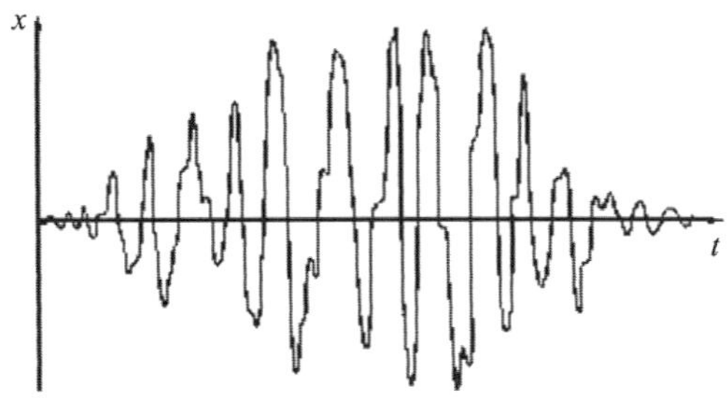

Figure 1. The ground vibrations during an earthquake.

2. Design and operation of a radial piston pump

Figures 2 and 3 show a radial piston pump comprising a stator *1*, a rotor *2* and a phase regulator *3* with an eccentric *4* and an eccentric ring *5* mounted coaxially in the body. The shafts of the phase regulator and the rotor are located in the bearings *6*, *8* and *9*. The pistons *11* are located in the rotor cylinders *10*, which are linked by the ends with the eccentric *4*, and the transverse slits a with the eccentric ring *5*. The channels *12* and *13* (Figure 3) with a width smaller than the width of the rotor are built in the pump body. There are pressure clamps *15* with springs *16* located in the ducts *14*. The cavities above the clamps are communicated through valves *23* and *24* with hydraulic lines *A* and *B* (Figure 3), which make tight contact of the jumpers with the rotor and provide high tightness of the channels *12* and *13*.

There are bypass valves *17* with springs 18 (Figure 4) in the pressure clamps. Holes *19* link them with the cylinders *10*, when the latter are under the jumpers. The above-valves spaces are connected to the channels *13* and *12* through the holes *20* and *21*. The fixed sealing discs *25* and *26* with seals *27* are installed on both sides of the rotor (Figure 2). Due to annular ducts *28* which are in the body and connected to the hydraulic lines *A* and *B* through the return valves *29* and *30*, the discs *25* and *26* are pressed to the ends of the rotor providing high hermeticity of the channels *12* and *13* (Figure 3).

One of the drawbacks of radial piston pumps and rotary hydro-pulsators [1, 5] is the precision of a complex sequence of couplings that closes on the spool valve. The sequence includes the rotor bore, its mounting surfaces for bearings, their base mounting bushings and the collector The gaps in the hydro-mechanical couplings sequence such as the «spool – rotor» and «spool-collector» are designed within 0.02...0.03 mm in order to ensure the sufficient tightness. Therefore the other couplings must be within the tolerances of 0.005...0.05 mm [1].

The flow distribution system in the pump eliminates gaps and provides a higher hermeticity due to the float position of the jumpers and end disks. When the fixed phase regulator (phase angle φ=0) is at the position of the eccentric 4 (Figure 3), then the hydraulic unit operates as a radial-piston pump with a constant and maximum flow (m^3/s):

$$Q = 2ezs_{\mathrm{p}}f_{\mathrm{r}} = \frac{ezs_{\mathrm{p}}\omega_{\mathrm{r}}}{\pi},$$

where e – the eccentricity of the phase regulator, m, z – the number of pistons, s_p – the area of the piston, m^2, f_r – the rotor speed, Rev/s, and ω_r – the angular velocity of rotor rotation, rad/s.

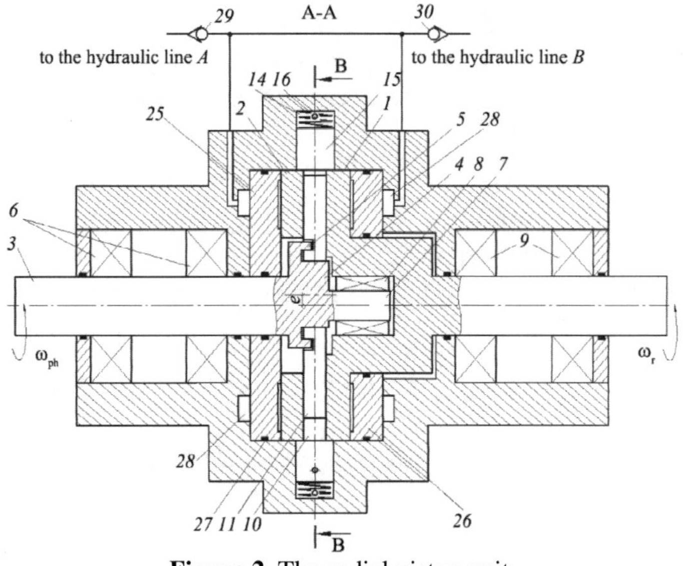

Figure 2. The radial piston unit.

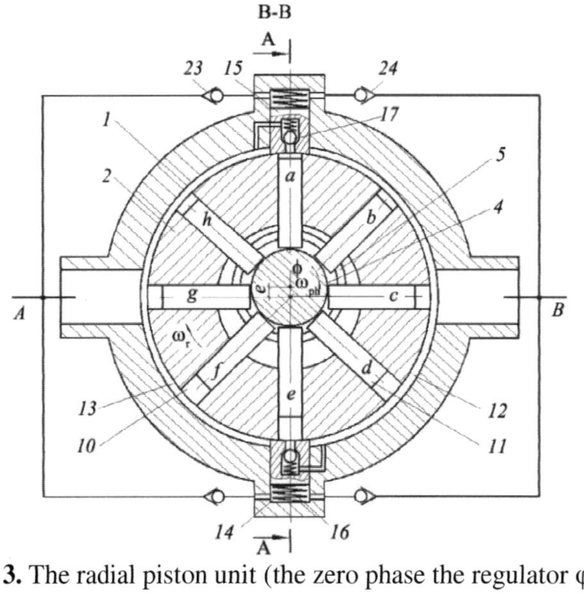

Figure 3. The radial piston unit (the zero phase the regulator φ=0).

When the rotor rotates clockwise, the pistons are retracted by the eccentric ring *5* (Figure 2). Then pistons pass from position *a* through stages *b, c* and *d* to position *e*, sucking liquid through channel *12* from hydroline *B*. The remaining pistons contact with channel 13, go over their positions *e*, through stages *f, g* and *h,* to position *a* and push the liquid into hydroline *A*. When the cylinder passes under the jumper *15* (Figure 4), the piston moves from position I (indicated by dash) into the position II of the «dead point» and completes the displacement of the liquid into channel *13* until the point *i* on the edge of the cylinder is aligned with point *j* on the edge of the jumper. When the cylinder is cut off from the channel *13*, a liquid compression arises in the over-piston cavity and if it exceeds the working pressure of the pump, the valve *17* will transfer the trapped liquid into the channel *13* through the apertures *20* and *21*. In the first phase the regulator ($\varphi=\pi/2$) is positioned anti-clockwise (Figure 5), the pistons return from position *c* through the stage *d* to position *e*, and suck liquid from channel *12*. The other pistons pass from position *e* through stage *f* to position *g* and force liquid into channel *12*. The volumes of the sucked and displaced liquid are compensated, there is no supply. The mutual compensation of the volume occurs the same way: when the pistons move from position *a*, through stage *b* to position *c* and then move from the position *g* through stage *h* to position *a*. Thus, the first phase position corresponds to the time of reversed flow in the hydrolines *A* and *B*.

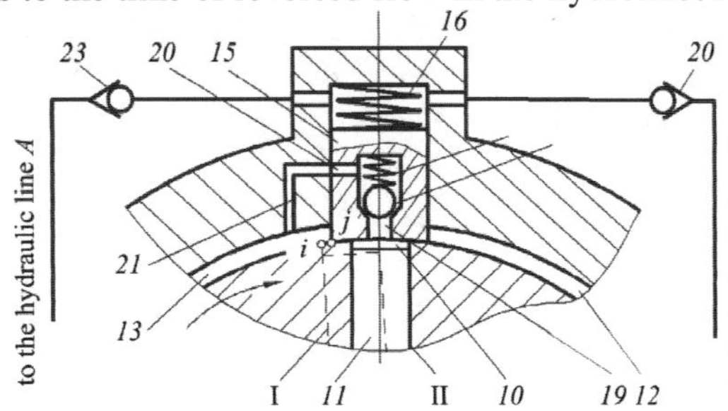

Figure 4. The design of jumper and bypass valve.

When the cylinders pass under the lower jumper *15* (Figure 4) the piston (Figure 5) moves from position I to position II at maximum speed. And when point *i* is aligned with point *j*, the piston will displace the larger volume of the trapped liquid through valve *17* and its compression will also be eliminated.

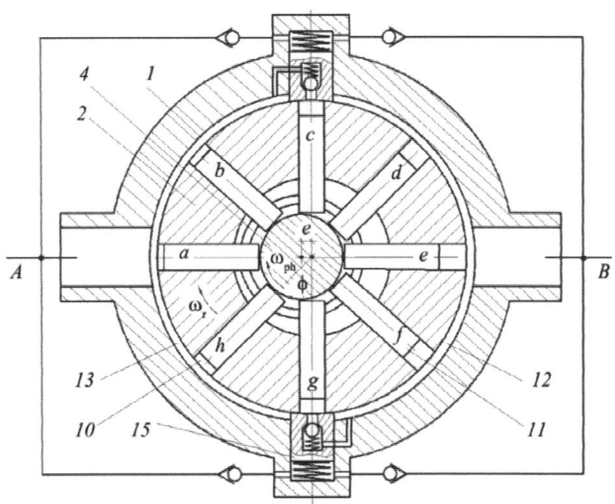

Figure 5. The radial piston unit (the first phase the regulator φ=π/2).

When the regulator is transferred to the second phase position (φ=π), the feed reaches maximum again, but the liquid is displaced into channel *12* and the hydroline B, and is sucked from channel *13* and hydroline A. The third phase position (φ=3π/2) fully corresponds to the first – there is no feed.

Oscillation mode of the phase regulator can be used for hydraulic drives of the machines which are used to test structures for alternating loads. Oscillation mode is to be set between zero and the second phase positions in the range of φ=0...180°. The angular oscillations frequency determines the frequency of load of the test structure. The level of structure deformation determines the holding time *t* of the regulator in the extreme phase positions and, accordingly, the stroke of the hydraulic cylinder, (*m*):

$$h = \frac{ezs_{\mathrm{p}}\omega_{\mathrm{r}}}{\pi}\frac{t}{F},$$

where *F* – the effective area of the piston of the two-rod hydraulic cylinder, m^2, which is connected to the pump.

When the phase regulator rotates with a constant angular velocity ω_{ph}, for example, towards the rotor rotation (Figure 3) its phase position changes continuously:

$$\varphi=\omega_{ph}\cdot t,$$

and reversed feed occurs in one revolution. The pump goes into rotary pulsator mode, when a cyclic, sign-variable fluid motion is created in the hydrolines *A* and *B*.

3. Test platform operation

Figure 6 shows the schematic diagram of the platform hydraulic drive for seismic testing of the objects based on the radial piston hydraulic unit *1*. The unit creates oscillatory fluid flows in pipelines *2* and *3* with an angular frequency ω_{ph} given by the phase regulator. The hydraulic cylinder *4* excites the oscillations *x* of the platform *10* at a frequency of Hz:

$$f_{ph}=\omega_{ph}/2\pi.$$

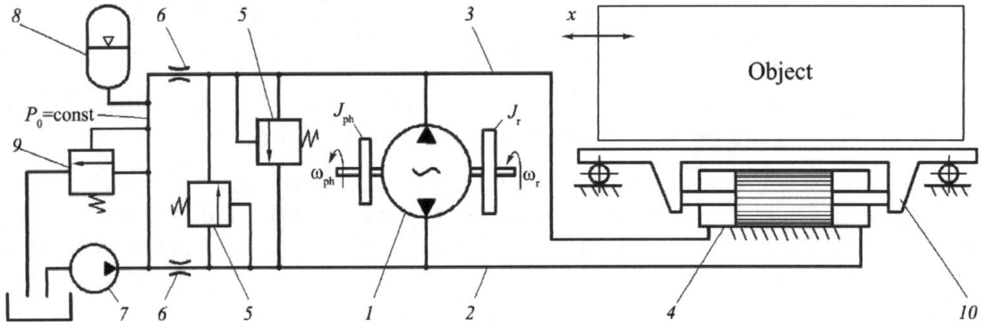

Figure 6. The platform hydraulic drive based on the radial piston hydraulic unit.

The drive operates at the initial pressure P_0 in the system supported by make-up pump *7*, the accumulator 8 together with the throttle *6*. Valve *9* provides adjustment P_0, and valves *5* eliminate overload of the hydraulic system. In this case, the rotor shafts and the phase regulator are equipped with flywheels with inertia moments J_r and J_{ph}, which smooth out the pulsation of their torque moments.

One of the seismic test methods is a smooth sweep frequency of the excitation and the preliminary scanning of the test object in the 0.5 to 15 Hz range (sometimes more). The method also allows to detect natural frequencies and subsequent tests in the local resonance mode. When scanning the test object, the rotor preset the constant angular speed ω_r, and the phase regulator drive – variable $\omega_{ph}(t)$, with constant acceleration $d\omega_{ph}/dt$=const, and provides a slow sweep frequency of the excitation f_{ph} in the interval of $\omega_{ph(min)}$–$\omega_{ph(max)}$ (Figure 7).

In contrast to rotary hydro-pulsators with a controlled valve [1], here the liquid supply to pipelines *2* and *3* (Figure 6) is determined taking into account the angular velocity of phase regulator rotation:

$$Q = Q_{amp}\cdot\cos\omega_{ph}\cdot t,$$

where Q_{amp} – feed amplitude, m^3/s:

$$Q_{amp}=\frac{ezs_p}{\pi}\left(\omega_r-\omega_{ph}\right), \quad (1)$$

the vibration law and vibration velocity for the test platform are described by the follow equations:

$$x=\frac{ezs_p}{\pi F}\left(\frac{\omega_r}{\omega_{ph}}-1\right)\cdot\sin\omega_{ph}t, \quad (2)$$

$$v=\frac{ezs_p}{\pi F}\left(\omega_r-\omega_{ph}\right)\cdot\cos\omega_{ph}\cdot t. \quad (3)$$

The spectral characteristic of the test stand, deployed in a two-decadal range $\omega_{ph(min)}\ldots\omega_{ph(max)}$ (Figure 7). The lower graph corresponds to the unidirectional rotation of the regulator and the rotor ω_{ph}–ω_r, the upper one to multy-directed rotation ω_{ph}+ω_r. Frequency-modulated oscillations form a continuous spectrum. The «lifetime» of the continuous spectrum harmonics depends on the change rate of the modulating function $\omega_{ph}(t)$. The oscillation spectrum is sequential and the mobility of the process does not create the right conditions for the object excitation, as the corresponding reactions cannot be established in it. Therefore, the frequency sweep speed should be low. However, the intensity of the related levels, for example, of the vibration velocities taken in the third-octave ranges of the frequency axis in the interval $\Delta\omega=\omega_2-\omega_1$, will have a final value and a real physical result will be sufficient to excite the resonances of the structural elements of the object. The spectral characteristic (Figure 7) is applied within working range which corresponds to local resonances (0.8…16 Hz) of the most building structures. We aim to exclude the influence of the resonance of the platform-object system. Therefore the stiffness c of the oscillatory system is determined mainly by the liquid volumes enclosed in the hydraulic cylinder chambers and determined by the expression:

$$c = 2F^2 \frac{E_0}{W_0},$$

where E_0 – the modulus of fluid elasticity, n/m^2, W_0 – the fluid volume in the hydraulic cylinder chambers, m^3.

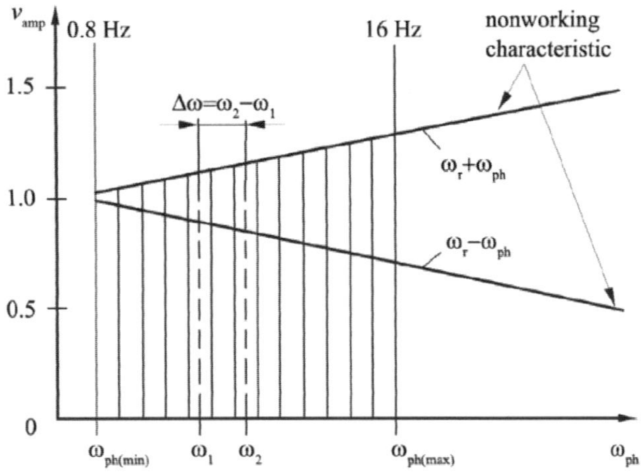

Figure 7. The spectral characteristic of the seismic-testing platforms.

The stiffness should be maximum. The operating volume of the hydraulic unit 1 and other elements of the hydraulic drive circuit in Figure 6 are small, compared to W_0, however, the length of the pipelines should be minimized, and all hydraulic equipment is assembled into a single hydroblock. Taking into account attached mass of test object the natural oscillation frequency of the platform f_0, must not be less than an order of magnitude higher than the upper limit of the frequency range $f_{ph(max)}$ (16 Hz):

$$f_0 = \frac{1}{2\pi}\sqrt{\left(\frac{1}{(m_{pl}+\kappa m_{to})}\right)\cdot\left(c_{eq}+\frac{2F^2E_0}{W_0}\right)} > 10 f_{ph(max)},$$

where m_{pl} – the platform mass, kg; m_{to} – the mass of the test object, kg; k – the coefficient that takes into account the fraction of the attached object mass; c_{eq} – the equivalent stiffness of the «hydraulic springs» of the hydrolysines and hydraulic devices, reduced to the piston of the loading hydraulic cylinder.

4. Results

The angular velocity of the phase regulator ω_{ph} is successively set to the corresponding resonances and maintained for a regulated time in the seismic test mode at the detected natural frequencies of the object. The angular velocity of the rotor ω_r is changed by means of an adjustable drive if it is necessary to adjust the platform-object system, for example, to the required vibration amplitudenor vibration speed (2, 3).

5. Summary

The method of frequency modulation with the detection of the natural frequencies of the test object is the simplest and easily realized by means of the considered pump. But the potential of the method is limited, as there is no simultaneous occurrence of resonances and the synergetic effect mentioned above is absent. To expand the functionality of the radial piston hydraulic machine and implement amplitude-frequency modulation during testing, in addition to the phase adjustment, it is necessary to provide volumetric regulation, which requires further improvement of the hydrounit.

The development of hydrostatic random oscillations generators which are capable of exciting narrow-band vibration with a high level of spectral density of vibration acceleration [1, 5] is the further development of the concept of volumetric hydraulic machines for the initiation of test dynamic systems. Such hydraulic units have already been created, they can be used not only in the seismic testing technique, but also in the testing of various instruments and equipment for vibration and vibration resistance and will be considered in future publications of the authors.

References

[1] Kljuev V 1982 *Test engineering: reference book 1,2* (Moscow: Mashinostroeniye)
[2] Plotnikova I V, Efremova O N, Tchaikovskaya O N 2017 J. Ponte **73** 158–162 doi: 10.21506/j.ponte.2017.4.45
[3] Zhunusov T 1972 *Vibrations of buildings and ground during the explosions and earthquakes* (Alma-Ata: Nauka)
[4] Leonov A P, Red'ko V V, Red'ko L A et al. 2015 IOP Conference Series: Materials Science and Engineering **81** doi:10.1088/1757-899X/81/1/012094
[5] Kljuev V 2010 *Devices and systems for measuring vibration, noise and shock. Reference book. Book 2* (Moscow: Mashinostroeniye)
[6] Gavrilin A, Moyzes B, Cherkasov A *et al* 2016 *MATEC Web of Conferences* **79** 01078 doi: 10.1051/matecconf/20167901078
[7] Gavrilin A, Krauin'sh P, Perestoronin P 1990 *Vibration diagnostics of machines and mechanisms* 64–65 (in Russian)
[8] Nizhegorodov A, Gavrilin A, Moyzes B 2015 *Key Engineering Materials* **685** 320–325 doi: 10.4028/www.scientific.net/KEM.685.320
[9] Nizhegorodov A, Gavrilin A, Moyzes B et al 2016 *Journal of Vibroengineering* **18** 3734–3742 doi: 10.21595/jve.2016.16994
[10] Nizhegorodov A I, Gavrilin A N, Moyzes B B *Journal of Physics: Conf. Ser.* **671** 012037 doi: 10.1088/1742-6596/671/1/012037
[11] Redko V V, Starikova N S, Redko L A, Vavilova G W 2015 *IOP Conference Series: Materials Science and Engineering* **81** 012083 doi: 10.1088/1757-899X/81/1/012083
[12] Goldshtein A E, Vavilova G V, Mazikov S V 2016 *MATEC Web of Conferences* **79** 01009 doi: 10.1051/matecconf/20167901009
[13] Gomboeva S.G, Shishkin I F, Hamhanova D N, Vasendina E A, Plotnikova I V and Efremova O N 2016 *Key Engineering Materials* **685** 463–466 doi: 10.4028/www.scientific.net/KEM.685.463
[14] Kalinichenko N P, Kalinichenko A N, Lobanova I S, Borisov S S 2013 *Russian Journal of Nondestructive Testing* **49** 668–672 doi: 10.1134/S1061830913110041
[15] Fedorov E M, Bortnikov I D 2015 *Technical Physics* **60** 1689–1692 doi: 10.1134/S1063784215110110
[16] Fedorov E, Koba A 2016 *MATEC Web of Conferences* **79** 01082 doi: 10.1051/matecconf/20167901082

Risk assessment for stonecutting enterprises Accidental risks in the course of petroleum production and stone extraction

A J Aleksandrova [1], S S Timofeeva[2]
[1] Graduate student, Irkutsk National Research Technical University, Irkutsk, Russia
[2] D. Sc., Professor, Irkutsk National Research Technical University , Irkutsk, Russia
E-Mail: aleksandrova.angelina1993@yandex.ru

Abstract. The paper is devoted to the assessment of accidental risks occurring at the works engaged in stone extracting and petroleum production. Two basic kinds of accidents common for stone extracting and petroleum production have been chosen to be discussed in the part under consideration. The most dangerous accidental situation characteristic for a stone milling line is an unsanctioned explosion, UE, of blasting agents used for the development of stone deposits. The analysis of a risk occurrence in certain accidental situations is to be carried out. With reference to petroleum extraction, a combustibles and lubricants (C & L) explosion is the most dangerous of characteristic accidental situations. To reveal the most probable causes of accidental situations to be realized, a graph of cause and effect relations has been constructed for each of the accidental situations most probable causes to real situation of an accident. Disasters of a natural origin are the most probable causes of unsanctioned explosions at the deposits of stone raw materials. Technology related natural disasters are the most probable causes of unsanctioned explosions to be realized at multiple well platforms engaged in petroleum production.

1. Introduction

An accidental risk is a risk of unwanted ecological after-effects technology-related in a production. As opposed to a constant risk, an accidental risk is connected with an uncertainty which is followed by different after-effects. The after-effects depend both on the character of the accident itself (transport accidents, fires, explosions, chemical or biological ones, radioactive emissions, building and construction collapses, hydrodynamics-related accidents, etc.) [1] and on the condition of the environment, climate, season and other factors.

There are plenty of accident and catastrophe causes, but the basic ones are as follows:

– non-observance of a workplace and technological discipline;
– design faults;
– building faults;
– equipment, buildings and constructions deterioration;
– the results of natural disasters.

The lines of petroleum production and stone extraction belong to geological branches, accident incidence rate being in their nature to the utmost due to dangerous natural raw materials (i.e. petroleum and accompanying gas) being extracted and harm full substances and materials (occupational dust, combustibles and lubricants, etc.) being stored [2].

The following kinds of accidents are characteristic for the lines of petroleum production and stone extraction:

– accidents of a natural origin (earthquakes, floods, landslides, mudflows, falls of ground, sloughs, avalanches, slope washes, storms, wilderness fires);

– technology-related accidents (fires, explosions, explosion threats, accidents followed by blowing-out (a threat of blowing-out) chemically hazardous accidental materials; unexpected collapse of buildings, constructions, failure of transport communication line elements).

2. Essence of survey

Two basic kinds of accidents common for stone extraction and petroleum production have been given consideration in the survey performed. In stone extraction, an unsanctioned explosion of the explosives used in the development of stone deposits (EM). In petroleum production, the explosion of combustibles and lubricants is the most common accidental situation. The analysis of these accidental situation risks was made in the course of works performed.

To begin with, stone extracting branch is to be given consideration to. Blasting operations are common in this production and are followed by extracting stone.

The methods of explosive works are as follows: using deep-hole charges (being the most common one for modern open-pits), chambering charges (being used for exploding great solid rock formations with the aim of blowing-out, faulting or fragmenting), using outer and blast hole charges. In the open pits, the last two methods mentioned are not used frequently and, as a rule, they are intended for fragmenting oversized blocks.

The method of breaking with deep-hole charges is the basic one in the open pits. Deep-hole charges are usually located vertically in the rock formation, sometimes they are in an oblique position, whereas their horizontal position can be observed almost never. The main parameters of blast hole drilling, the blast hole being located vertically, are as follows: the diameter of a deep hole, the height of the scarp, the interval between charges, the interval between series, the length of sub drilling, the least resistance line, the tamping height, a charge quantity and its construction [3].

The use of a charge construction characterized by the availability of free-air spaces is an advanced method of an explosion controlling (see Figure 1). It contributes to a higher intensity and evenness of fragmenting (the size of an average piece becomes smaller by a factor of 1.5 – 3, the yield of oversize material becomes several times smaller), a reduced volume of tamping works makes the explosive ratio smaller by 10…30 % simultaneously increasing the yield of broken material, balm stones into the depth of the solid and the seismic effect of the explosion becoming smaller.

A better fragmenting of subsurface rocks as a result of blasting out charges containing air gaps is achieved due to energy rearrangement occurring in an explosion-generated pulse (causing a reduced incidental pressure in the initial period and a longer explosion effect), choking the explosion of the upper charge with light end explosion products.

The main hazards occurring in realization of unsanctioned explosion at the stone deposit are as follows: fragment injuries and injuries caused by spalls of mined rocks and crushable materials along with direct human exposure to shock-induced air waves and to detonating products of explosive charges.

Most accidents and cases resulting in injuries happen due to violation of safety measures. An unsatisfactory level of the personnel's workplace discipline, improper control, irresponsibility, negligence of business managers, inadequate control on the part of the Federal Service for Environmental, Technological and Nuclear Supervision authorities also serve as a background of accident occurring at the productions.

3. Technological aspect

Figure 1 shows a graph of cause and effect relations (a tree of experiences) dealing with accidents having happened in accordance with some probable scripts of event courses. The graph mentioned is related to the deposits given consideration in the given paper as the stone extraction there is performed opencast combined with blast hole drilling.

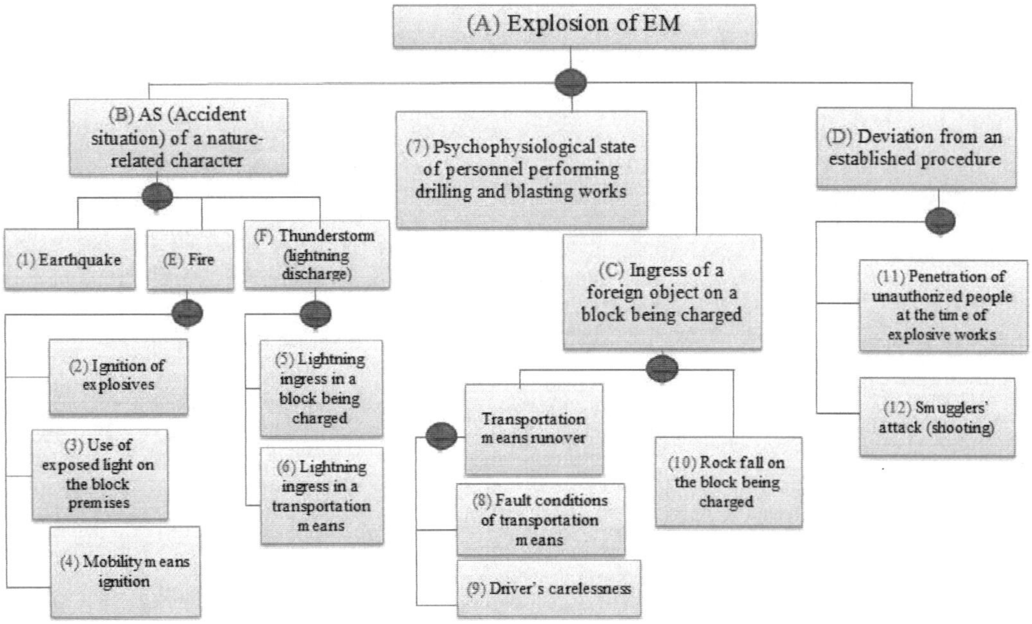

Figure 1. A graph of cause and effect relations (a tree of experiences) dealing with accidents that happen according to probable scripts of event development at stone extraction works.

From the causes of unsanctioned explosions of explosive materials in a block being charged which are represented in the tree, an analysis of static information has been made to reveal the probabilities of their occurrence.

Basing on the information obtained, a probability calculation of the main event, i.e. an unsanctioned explosion [3], was performed in accordance with the following Equation:

$$P(x) = 1 - (1 - P(i_1)) \cdot (1 - P(i_2)) \dots \cdot (1 - P(i_n)) \quad (1)$$

The calculations are indicative of the fact that the probability of an unsanctioned explosion in the site of explosive works is 1.72×10^{-4}. The background (B), namely, "Accident situation of a nature-related character" is the most important factor as concerns the most important event.

Consequently, nature-related disasters are the most probable causes of unsanctioned explosion realization in the deposits of stone raw materials.

Next, an accidental situation characteristic for the line of petroleum production is to be given some consideration, namely, an explosion of combustibles and lubricants. Combustible and lubricant warehouses are located on each multiple-well platform, in the immediate vicinity of other objects. There are such materials in them as petroleum products involving different kinds of combustibles and lubricants, fuels [4] (gasoline, diesel-fuel oil, liquefied-petroleum gas, pressurized natural gas, liquefied natural gas); lubricants (engine, transmission, and special oils, greases), subject-oriented liquids (brake fluids and coolants).

A number of causes, such as a fire, high temperatures, fuel or other combustibles and lubricants leakage, electric equipment failure, faulty transportation facilities, etc. can provoke an explosion in a combustible and lubricant warehouse. Cause and effect relations (a tree of experiences) that are responsible for an accident (explosion of combustible and lubricant materials) are represented in Figure 2 in accordance with a probable script of the event development.

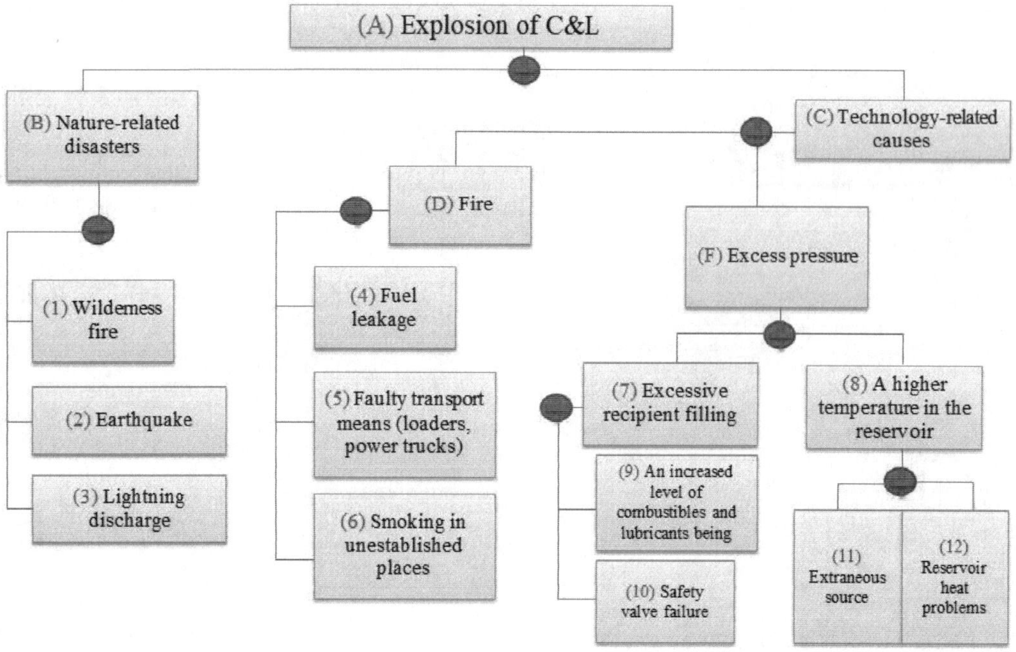

Figure 2. A graph of cause and effect relations (a tree of experiences) inducing accidents in accordance with probable scripts of developing events in petroleum production.

Basing on the causes of an unsanctioned combustible and lubricant explosion that has happened on a multiple-well platform and is represented in the tree of experiences, an analysis of statistical information was made to reveal the probabilities of their appearance.

The probability calculation with regard to the most important event [5], i.e. a combustible and lubricant explosion (Equation 1), results from the information obtained. It follows therefrom that the probability of unsanctioned combustible and lubricant explosion occurrence on a multiple-well platform makes up 6.7×10^{-5}. The background (C) of "Technology-related causes" is most important for the assessment of the main event probability. It follows therefrom that technology-related natural disasters are the most probable cause of realizing unsanctioned explosions on multiple-well platforms intended for petroleum production.

4. Economic disbenefit calculations

If events under consideration get realized at stone extraction and petroleum production works, economic disbenefit follow them resulting from facility accidents. Disbenefit was calculated following realization of accident situations occurring in the both lines and given consideration in the given paper to be compared [6].

An economic disbenefit following an accident at a stone deposit (an unsanctioned blast of explosives in the course of drilling and blasting operations) accounts for 25 mln 959 thou. 311 roubles. The disbenefit following an accident at the multiple-well platform of a petroleum production (combustible and lubricant materials explosion) accounts for 48 mln 613 thou 528 roubles.

A comparative graph of an economic disbenefit following the accident situations realized at petroleum production and stone extraction works is represented in Figure 3.

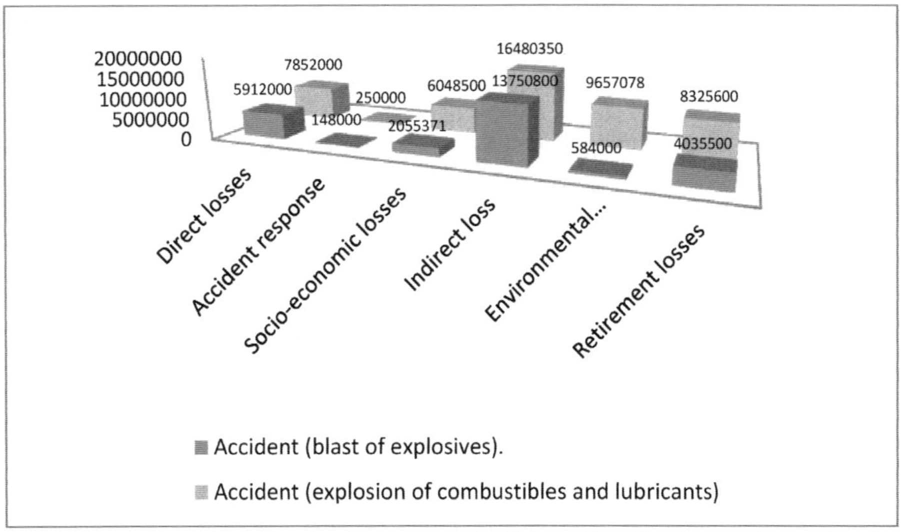

Figure 3. A comparative graph of an economic disbenefit following accident situations realized on production platforms of works under consideration.

5. Practical recommendations concerning making lower accidental risks

To make accidental risks lower in the conditions of stone extraction the following items should be anticipated:

– to arrange the possibility to study regulations and data messaging dealing with blasting work for the personnel performing drilling and blasting operations and employees responsible for a blasting work area;

– to control the knowledge of fire safety regulations;

– to conduct the instructions dealing with explosives transportation from the primary warehouse to work areas for the people performing the transportation;

– to limit the group of people having access to explosives on a need-to-know basis;

– to conduct instructions with drivers, fillers, and securities;

– to control the personnel compliance with fire safety regulations, trucking rules using autotruck transportation, rules of establishment and safe exploitation of production locations in addition to engine-driven primary mining for explosives being used in the services performing explosive works;

– to revise the labor system management in correspondence with the amendments made in the labor legislation.

To make an accidental risk smaller in the conditions of petroleum production the following items should be anticipated:

– to organize regulations and documentary data messaging study dealing with the rules of storing combustibles and lubricants for the personnel responsible for a proper combustible and lubricant storing and for the specialists engaged in work production in the area;

– to control the awareness of fire safety regulations on a regular basis;

– to conduct instructions for people engaged in transportation on the delivery of flammable materials from the primary warehouse to the sites of work performance;

– to limit the group of people having access to combustible and lubricant warehouses on a need-to-have –access basis;

– to conduct instructions with the personnel handling multiple-well platforms for petroleum production;

– to control the personnel observance of fire safety regulations , the rules of freight lift by autotruck transportation, rules of establishing and safe service of production sites in addition to a power-driven preparation for application of flammable materials in the services performing petroleum production works;

– to revise the system of labor safety management and arrange it in correspondence with the amendments made in the labor legislation.

6. Summary

Accidental situations are common for most productions, petroleum production and stone extraction including. The mentioned productions being compared following accidents realized in the production areas, it is evident that the disbenefit resulting from the explosion of combustibles and lubricants is by far greater than the one resulting from explosive materials blast. Most probably this fact is connected with different kinds of the exploding materials and discharge components. Salaries attached to positions and wage pools are also of a considerable significance in this connection. The result of calculations proves that the disbenefit caused by explosion of combustibles and lubricants in a petroleum production area is by the factor of 1.8 greater than the similar index calculated for the blast of explosives.

References

[1] Crisis and Risk Network (CRN): Focal report 2: Risk analysis – Integrated risk management and societal security, Center for Security Studies (CSS), ETH Zürich [Electronic resource] URL:http://www.bevoelkerungsschutz.admin.ch/internet/bs/en/home/dokumente/Unterlagen_Risiken.parsys.0001110.downloadList.77378.DownloadFile.tmp/crnreportfocalreport riskanalysis.pdf

[2] Timofeeva S S 2016 *Methods and technologies for assessing production risks: workshop* (Irkutsk: Publishing House of Irkutsk State Technical University) (in Russian)

[3] Timofeeva S S 2015 *Methods and technologies for assessing production risks: a summary of lectures* (Irkutsk: Publishing House of Irkutsk State Technical University) (in Russian)

[4] Sinkankas J 2009 Guidelines for the Processing of Precious and Semiprecious Stones (Moscow: Mir) (in Russian)

[5] Alexandrova A Yu 2016 *Proc. of Security Problems of the modern World* (in Russian) 7–8

[6] Jerrard R et al 2007 Design management: Exploring fieldwork and applications (London: Routledge) 102–121

IOP Conf. Series: Materials Science and Engineering **289** (2017) 012016 doi:10.1088/1757-899X/289/1/012016

Influence of metallic additives on manganese ferrites sintering

S A Shevelev[1], P A Luchnikov[2], A R Yarullina[3]
[1]Senior Lecturer, National Research Tomsk Polytechnic University, Tomsk, Russia
[2]Leading engineer of the department of metrology and development, Moscow Technological University (MIREA), Moscow, Russia
[3]Researcher, National Research Tomsk Polytechnic University, Tomsk, Russia
E-mail: wevelev@tpu.ru

Abstract. Influence of cuprum nanopowder additive received by electric explosion on the process of manganese ferrites MgFe2O4 consolidating at thermal sintering was researched by dilatometry method. Cuprum nanopowder at a rate of 5 mass % was added into the original commercial-grade powder of manganese ferrite MgFe2O4. Powder mixture was numerously blended with screening for better blending before pressing. Powder compacts were formed by cold one-axle static pressing. It was proved that introduction of cuprum additive caused shrinkage increase at final heating stage. There was abnormal compact enlarging at sintering in the air at isothermal stage; the specified process was not observed in vacuum. This difference can be explained by changes in conditions of gaseous discharge from volume of pores.

1. Introduction

Ceramic materials are widely used for both functional and instrumental purpose. Ceramic materials on the base of zirconium dioxide [1–4] and lithium ferrite [5, 6] stand apart. The main technological stage [7] of ceramic production is powder compact burning, its sintering [8, 9], and also modification [10, 11]. As for ferrites, process of their sintering was mostly determined by interaction of sintered material with surrounding gaseous atmosphere [12, 13]. Moreover, diffusion processes have a great importance and secondary ion mass spectroscopy method is effectively used to study it [14, 15]. Dilatometry method [16–19] and thermal gravimetric analysis [5], grapho-analytical method [20], are widely used to study ceramic materials sintering. Both conventional methods of thermal processing in resistance furnaces and nonconventional methods based on radioactive action of accelerated beam [6, 18], laser and other types of light stimulus can be used for sintering and processing of ceramic materials [21–23]. Method of sintering in resistance furnace is most suitable for manufacturing of ferrites. Furthermore, considerable attention is paid to sintering temperature decrease to reduce final products cost. In this respect blending method proved to be effective [24–27]. Polymers [27], oxides [26] and metals [25] were used as additives. For instance, work [23] presents information about introduction of zinc and niobium into barium hexaferrit formulation which has strong impact on morphology and size of grains.

Providing conditions of uniform diffusion of additive on volume basis of original powder is important at introduction of additives in small amounts in relation to original powder. It means that amount of particles of additive should be compared with amount of particles of original powder ranged from 20–50 micrometers in size. Nanopowders of metals meet the conditions, as they have very high specific surface area and at the same time they have small weight. As a result of it even if they have small weight percent they uniformly diffuse on volume basis of powder.

The present work presents the results of research of influence of additive of nanodisperse powder of cuprum (Cu), made by electric explosion method on the process of consolidating at manganese ferrites sintering $MgFe_2O_4$ made by dilatometry method.

2. Experimental part

Manganese ferrite samples $MgFe_2O_4$, made from original commercial-grade powders according to ceramic processing technology were studied. Nanodisperse powder of cuprum was received by electric explosion method. Average size of particles ranges from 20…60 nm. Specific surface area of nanopowder of cuprum, determined by BET method, was not less than 15 m^2/g.

Influence of cuprum additive on the processes of compacting and sintering of ferrite $MgFe_2O_4$ was carried out. Introduction of additive in number of 5 mass % in original powder was made by multiple blending with screening to provide uniform diffusion of additive on volume basis of original powder.

Samples for sintering were compacted by cold one-axle static pressing in press-form at laboratory hydraulic press PGR-10 (ПГР-10) under a pressure of pressing 155 MPa. Compacts had cylindrical form of 9 mm in diameter and about 3 mm thick.

Research of kinetics of compression of test samples was conducted at highly-sensitive dilatometer DIL402C, company NETZSCH (Germany). Heating and cooling rate was 10 degrees/min.

Research of kinetics of compression by dilatometry method was conducted for compacts, received from original powder without additives at sintering in the air (sample of type A), samples with adding Cu at sintering in the air (sample of type B) and in vacuum (sample of type C).

3. Results

Figure 1 shows kinetic connection of linear shrinkage of all test sample types.

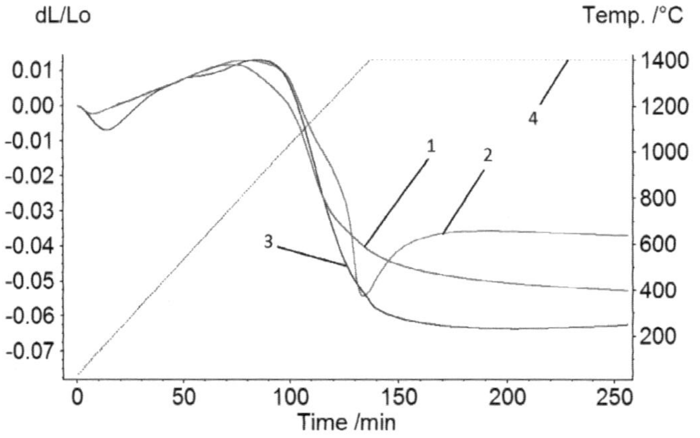

Figure 1. Kinetic connection of linear shrinkage $\Delta L/L_0$: 1, 2, 3 – samples of types A, B and C, correspondingly; 4 – temperature conditions of burning.

According to analysis of connections in Figure 1 it follows that introduction of Cu causes shrinkage increase at heating stage in comparison with samples of type A, as at sintering in the air (curve 2), and in vacuum (curve 3). Determined by first derivative curves (Figure 1), shrinkage of samples of types A, B and C, temperatures corresponding to maximum shrinkage rate are equal to T_m =1126.1; 1330.5; 1145.3^{0}C, correspondingly.

Change of T_m can be explained in the following way. There is oxidizing of Cu with oxide of CuO formation at heating of samples of type B, which causes its enlargement in relation to samples of type A. Liquid phase formation occurs at melting temperature CuO equal to 1242^{0}C, resulting in T_m=1330.5^{0}C. At heating stage of samples of type C there is melting of Cu without oxide formation. As melting temperature of Cu is lower than CuO and is equal to 1085^{0}C, T_m of samples of type C is also lower and is equal to T_m=1145.3^{0}C. Presence of liquid phase in samples of types B and C leads to friction decrease between particles at consolidating.

In accordance with work [17] there are two simultaneous processes at thermal sintering liable for change of linear sizes of a sample. The first process was specified by interparticle interaction, the second process was conditioned by changes of volume of gaseous pores. Presence of liquid phase in samples of types B and C at their sintering prevents from gaseous discharge from volume of pores, which causes their enlargement at the stage of isothermal time. As it can be seen from Figure 1, the process of enlargement mostly occurs at sintering in the air (samples of type B), which causes abrupt enlargement at isothermal time. Gaseous discharge from pores is more effective at sintering in vacuum (samples of type C) because of larger gradient of pressure which causes shrinkage enlargement in comparison with samples of type B.

4. Summary

It was established that consolidating of ferrite compacts $MgFe_2O_4$ with 5 mass% additive of nanopowder Cu, received by electric explosion method, strongly depends on ambient air environment pressure. Gaseous discharge from volume of pore space of a compact is difficult at atmospheric pressure because of blockout of pores of liquid phase of additive. At sintering in vacuum decrease of pores volume and their quantity is more effective, which finally leads to considerable increase of compacts shrinkage. Introduction of low-melting metallic nanopowders should be preferably made at sintering of manganese ferrites in vacuum to obtain nonporous ceramics.

Acknowledgements

This work was financially supported by The Ministry of Education and Science of the Russian Federation in part of the science activity program.

References

[1] Stawarczyk B, Keul C, Eichberger M, Figge D, Edelhoff D, Lümkemann N 2017 *J Quintessence International* **48** 369–380 doi: 10.3290/j.qi.a38057
[2] Taiylor S, Singh M, Doub A V 2016 *Journal of Cluster Science* **27** 1097–1107
[3] Freris I, Riello P, Enrichi F, Cristofori D, Benedetti A 2011 *Optical materials* **33** 1745–1752 doi: 10.1016/j.optmat.2011.06.010
[4] He J, Chen J, Ren L, Wang Y, Teng C, Hong M, Zhao J, Jiang B 2014 *ACS applied materials & interfaces* **6** 2718–2725 doi: 10.1021/am405202d
[5] Surzhikov A P, Frangulyan T S, Ghyngazov S A, Lysenko E N 2012 *Journal of Thermal Analysis and Calorimetry* **108** 1207–1212 doi: 10.1007/s10973-011-1734-z
[6] Surzhikov A P, Lysenko E N, Vlasov V A, Malyshev A V, Nikolaev E V 2013 *Russian Physics Journal* **56** 681–685 doi: 10.1007/s11182-013-0085-6
[7] Vasendina E, Plotnikova I, Levitskaya A, Kvesko S 2016 *IOP Conference Series: Materials Science and Engineering* **110** doi: 10.1088/1757-899X/110/1/012070
[8] Macaigne R, Marinel S, Goeuriot D, Meunier C, Saunier S, Riquet G 2016 *Ceramics International* **42** 16997–17003 doi: 10.1016/j.ceramint.2016.07.206
[9] Marinca T F, Chicinaş H F, Neamţu B V, Popa F, Chicinaş I 2017 *Ceramics International* **43** 14281–14291 doi: 10.1016/j.ceramint.2017.07.180
[10] Ghyngazov S A, Frangulyan T S, Chernyavskii A V, Goreev A K, Naiden E P 2015 *Russian Physics Journal* **58** 188–191 doi: 10.1007/s11182-015-0480-2
[11] Poltavtseva V P, Kislitsin S B. Ghyngazov S A 2016 *Russian Physics Journal* **59** 159–165 doi: 10.1007/s11182-016-0754-3
[12] Surzhikov A P, Peshev V V, Pritulov A M, Gyngazov S A 1999 *Russian Physics Journal* **42** 490–495 doi: 10.1007/BF02508222
[13] Gyngazov S A, Lisenko E N, Petyukevich M S , Frangulyan T S 2007 *Russian Physics Journal* **50** 134–139 doi: 10.1007/s11182-007-0018-3

[14] Surzhikov A P, Chernyavskii A V, Ghyngazov S A, Frangulyan T S 2002 *Russian Physics Journal* **45** 1190–1194

[15] Gyngazov S A, Surzhikov A P, Frangul'yan T S, Chernyavskii A V 2002 *Russian Physics Journal* **45** 753–758

[16] Surzhikov A P, Frangulyan T S, Ghyngazov S A 2012 *Russian Physics Journal* **55** 345–352

[17] Surzhikov A P, Frangulyan T S, Ghyngazov S A 2014 *Journal of Thermal Analysis and Calorimetry* **115** 1439–1445

[18] Surzhikov A P, Frangulyan T S, Ghyngazov S A, Koval N N 2009 *Nuclear Instruments and Methods in Physics Research* **267** 1072–1076 doi: 10.1016/j.nimb.2009.01.144

[19] Melnikov A A, Sigov A S, Vorotilov K A, Davydov A A, Topalova L I, Zhavoronkov N V 1999 *Journal of Crystal Growth* **197**666–669

[20] Gomboeva S G, Shishkin I F, Hamhanova D N, Vasendina E A, Plotnikova I V, Efremova O N 2016 *Key Engineering Materials* **685** 463–466 doi: 10.4028/www.scientific.net/KEM.685.463

[21] Ghyngazov S A, Vasil'ev I P, Surzhikov A P, Frangulyan T S, Chernyavskii A V 2015 Technical Physics 60 128–132 doi: 10.1134/S1063784215010120

[22] Vorotilov K, Sigov A, Seregin D, Podgorny Y, Zhigalina O, Khmelenin D 2013 *Phase Transitions* **86** 1152–1165 doi: 10.1080/01411594.2013.794276

[23] Aleutdinov A D, Ghyngazov S A, Mylnikova T S, Luchnikov PA 2015 *IOP Conference Series: Materials Science and Engineering* **81** 012069 doi: 10.1088/1757-899X/81/1/012069

[24] Kanagesan S, Hashim M, Jesurani S, Kalaivani T, Ismail I 2014 *Journal of Superconductivity and Novel Magnetism* **27** 811–815 doi: 10.1007/s10948-013-2357-3

[25] Flegler A G, Burye T E, Yang Q, Nicholas J D 2014 *Ceramics International* **40** 16323–16335 doi: 10.1016/j.ceramint.2014.07.071

[26] Restivo T A G, Durazzo M, de Mello-Castanho S R H, Moreira A C, Graciano S, Telles V B, Tenorio J A S 2017 *Journal of Thermal Analysis and Calorimetry* doi: 10.1007/s10973-017-6560-5

[27] Li J, Zhang X-H, Cui B-C, Lin Y-H, Deng X-L, Li M, Nan C-W 2017 *Journal of Dentistry* **58** 60–66 doi: 10.1016/j.jdent.2017.01.008

IOP Conf. Series: Materials Science and Engineering **289** (2017) 012017 doi:10.1088/1757-899X/289/1/012017

Detection of insulation defects in the wire through measuring changes in its capacitance

G V Vavilova[1] and A V Ryumkin[2]
[1]Associate professor, National Research Tomsk Polytechnic University, Tomsk, Russia
[2]Student, National Research Tomsk Polytechnic University, Tomsk, Russia
E-mail: wgw@tpu.ru

Abstract. The paper describes the technique to detect local wire insulation defects through measuring the wire capacitance. The operating principle of the CAP-10 device is explained. The principal possibility of this device to detect local defects in wire insulation is shown. The experiments showed that the device can be used to detect defects of different types.

1. Introduction

The cable products are widely used in electrical engineering and electronics. The cable product quality determines the quality of electric power and information signal transmission [1]. The considered single-core electric wire is a part of the cable product; therefore, the wire quality affects the finished product quality [1–3]. One of the main normalized values is the wire capacitance per unit length [4, 5]. This parameter is particularly important for a number of cable products such as communication cables, radio-frequency cables and LAN-cables [4].

To calculate the capacitance, the wire is presented as a cylindrical capacitor (Figure 1). The two electrical components of the capacitor are the wire metal core and the wire polymer insulation surface to which the voltage is applied [6].

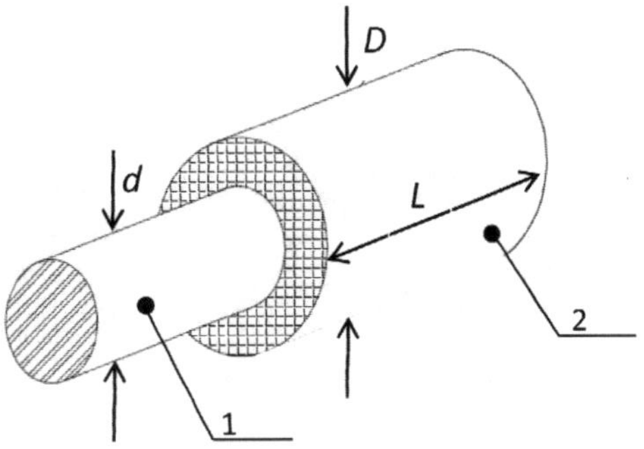

Figure 1. Wire model: core (1); polymer insulation (2).

The capacitance capacitor is calculated using the known equation [7]:

$$C = \frac{2\pi\varepsilon\varepsilon_0 l}{\ln\left(\frac{D}{d}\right)}, \tag{1}$$

where ε is insulating material permittivity; ε_0 is vacuum permittivity; L is wire length; D is insulation diameter; d is core diameter.

The capacitance of the wire depends on its geometric dimensions and insulation electrical properties [5, 8]; therefore, the capacitance change can indicate nominal deviations in geometric or electrical parameters. Consequently, the capacitance change can be the evidence for these deviations [9, 10].

The purpose of the research: to study the capability of the CAP-10 device to detect insulation defects in the wire through measuring the change in its capacitance.

2. Defects of wire insulation

The defects are caused by the breakdown in the technological process of wire manufacturing [11–13]. Therefore, timely detection of wire insulation defects is cost-effective for the cable industry [9, 12, 14]. A defect is product nonconformity to specified requirements [15, 16]. The main types of wire insulation defects are insulation "slicing", cracking, local rupturing, insulation porosity, impurity inclusion, increase or decrease in the outer insulation diameter, etc. [3, 8, 17]. Local defects can cause abrupt changes in geometric dimensions of the wire and electrical properties of the insulation. Consequently, the in-process capacitance changes can be monitored to detect the defects in wire insulation.

3. The operating principle of the CAP-10

CAP-10 can be used to perform in-process testing [18, 19]. It is based on a simple and easy to implement voltmeter-ammeter method. This method implies measurement of the current value to evaluate the wire capacitance value, provided the amplitude and frequency of the applied voltage are known [20, 21].

CAP-10 consists of an electro-capacitive measuring transducer (ECMT), an analogue conversion unit, a digital processing unit and a visualization unit. The physical configuration of CAP-10 is showed in Figure 2 [22].

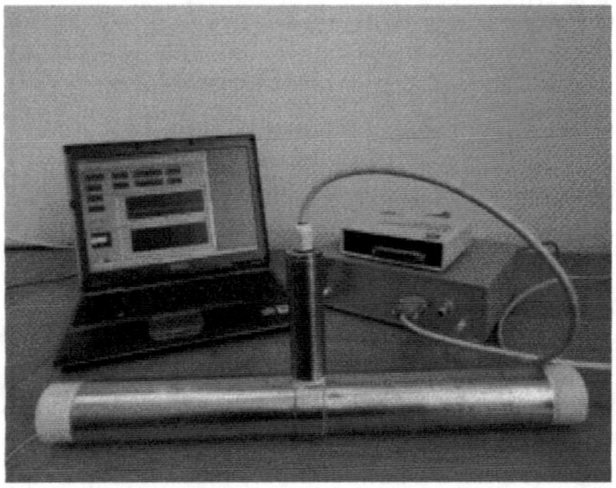

Figure 2. Physical configuration of the CAP-10.

ECMT is a tubular electrode immersed in cooling water of the extrusion line together with the test wire [7, 23]. The ECMT output signal is the current whose amplitude is proportional to the measured value of the test wire capacitance. The analogue conversion unit generates output signals (Re U и Im U) whose values are proportional to the amplitude values of the current in the measuring electrode. The digital processing unit performs filtering, measured data averaging within 1 second and further digital processing. The visualization unit displays the measurement results in the form convenient for the operator [22].

4. Detection of wire local defects using CAP-10

Local defects in wire insulation ("slicing", cracks, rupture, etc) [24] cause abrupt changes in the geometric dimensions and insulation electrical properties [25, 26]. This changes lead to abrupt change in the capacitance value of the test wire [4]. CAP-10 records any slight change in the capacitance (tenths of pF), therefore it can easily detect any abrupt capacitance change.

During manufacturing, the test wire permanently moves inside the ECMT with a known speed [12, 13]. If the wire capacitance is constant, the values recorded by the device remain stable (Figure 3*a*). This indicates that wire insulation is free of defects.

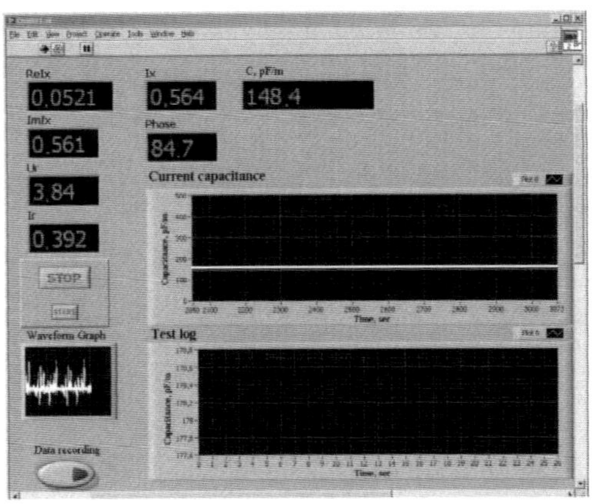

a) Free of defects

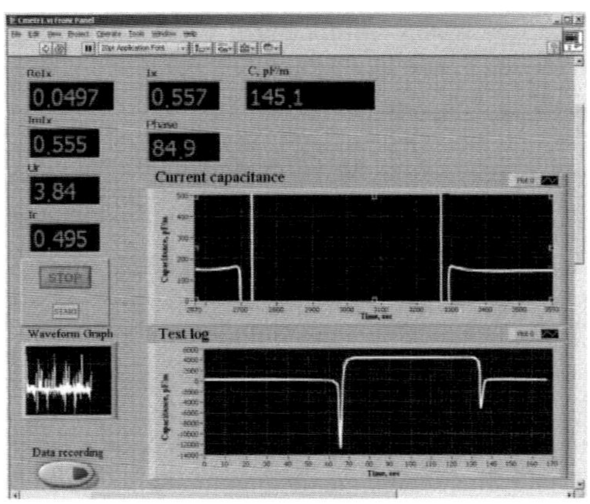

b) Defect detected

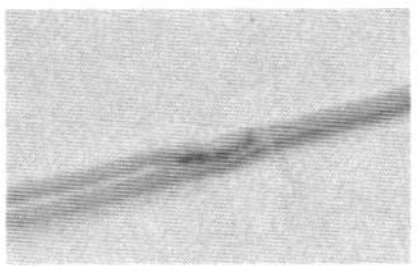

c) Defect

Figure 3. The front panel of the device CAP-10:
a) Test results for the sample free of defects; *b)* Test results for the sample with detected defect, *c)* The sample with defect.

As seen in Figure 3*b*, if the defect (Figure 3*c*) comes into the measuring zone, the wire capacitance abruptly changes. The length of the CAP-10 measuring zone is similar to that of the ECMT measuring electrode (20 cm) [22]. Note that the capacitance change occurs when the defect-containing area is getting in and out of the CAP-10 measuring zone. When the defect-containing area moves through the measuring zone, the capacitance remains constant. When the defect-containing area leaves the measuring zone, the wire capacitance takes its initial value.

Since the test wire permanently moves with a known speed, the defect location can be determined through the time of its detection. The CAP-10 resolution depends on the defect size and the speed of the moving test wire.

The presented capability of CAP-10 extends the scope of its application and provides additional capabilities of spark testing in non-destructive testing in the cable industry [27, 28]. In contrast to spark testing [2, 29], the method employed by the device has an indisputable advantage. Testing can be performed at AC voltage of 3.5 V which is much safer than high voltage used in spark testing [30].

5. Experiment on detection of different types of wire insulation defects

An experiment was performed to test the capability of CAP-10 to detect different types of defects. During the experiment, CAP-10 recorded the capacitance change for different types of wire insulation defects. For this purpose, segments of a single-core electric wire with one brand insulation with a diameter of 1.5 mm were used. The actual capacitance value of the wire segments free of defects was determined in compliance with GOST 27893-88 "Communication Cables. Methods of Tests" [31]. It equals 150 pF/m.

Insulation defects were created artificially on wire segments free of defects. Thus, a set of wire segments with the following types [6, 24, 32] of defects was produced:

- Insulation "slicing" with a length of 1 mm across the total insulation diameter.
- Local insulation rupture with a size of 0.5×0.5 mm.
- A 2.5 mm increased in the outer insulation diameter for the defect with a length of 15 mm.

During the experiment, ECMT with the test wire was immersed in a bath filled with ordinary tap water [33, 34] with a temperature of (22 ± 2)° C. To study the test wire, its movement inside ECMT was simulated at constant speed, and the capacitance value was measured using CAP-10. CAP-10 recorded the capacitance change for different types of defects (Figures 4–6).

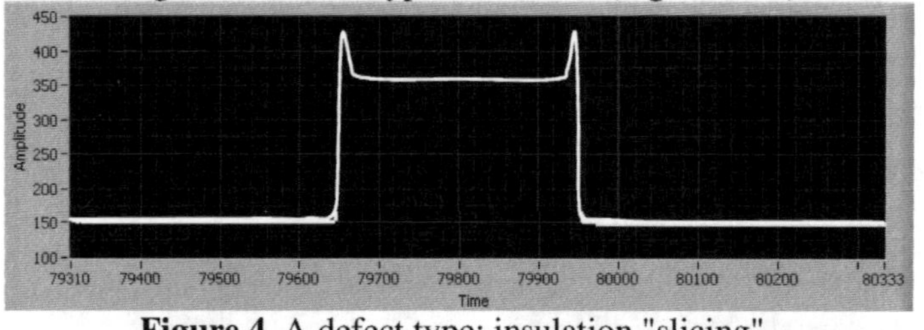

Figure 4. A defect type: insulation "slicing".

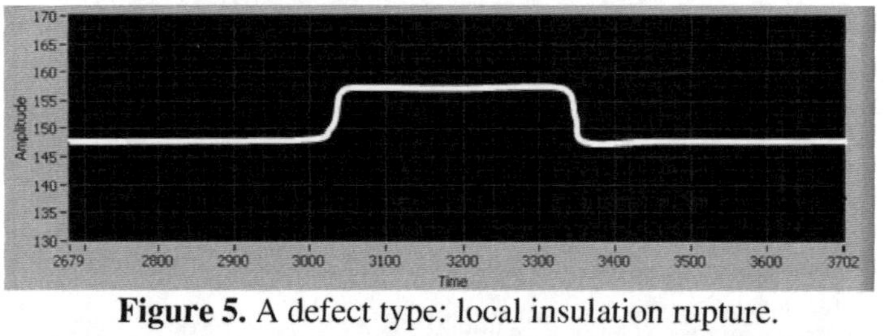

Figure 5. A defect type: local insulation rupture.

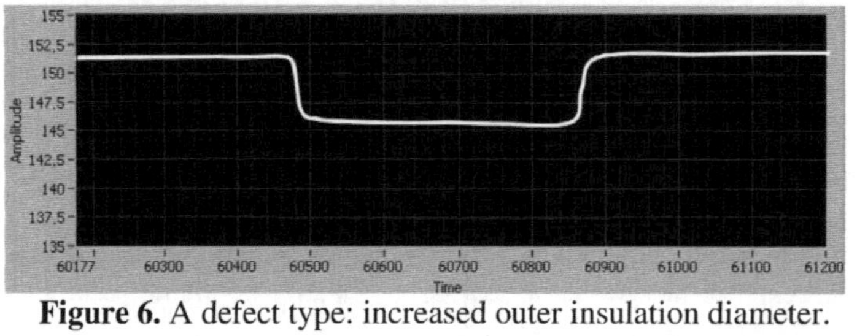

Figure 6. A defect type: increased outer insulation diameter.

The analysis of the results shows that the CAP-10 device identifies different types of wire insulation defects. Further study will aim to determine the CAP-10 sensitivity to geometrical dimensions of the above types of defects.

IOP Conf. Series: Materials Science and Engineering **289** (2017) 012017 doi:10.1088/1757-899X/289/1/012017

6. Conclusion

1. The presence of local defects causes abrupt changes in geometric dimensions of the wire and its insulation electrical properties, and consequently, abrupt changes on its capacitance.
2. CAP-10 records any slight capacitance change; therefore it can be used to detect wire insulation defects through measuring capacitance changes.
3. The capability of CAP-10 to detect different types of defects: insulation "slicing", local rupture of insulation, and increased outer insulation diameter is proved experimentally.
4. More research is required to determine the minimum size of defects causing appreciable wire capacitance change (CAP-10 sensitivity).

References

[1] Horn F W, Bleinberger W E 1966 *Design and manufacture of plastic insulated aluminum conductor telephone cable* (Defense Technical Information Center) p 31

[2] Red'ko V V, Leonov A P, Red'ko L A et al 2016 *Journal of Physics: Conference Series* **671** 012049 doi:10.1088/1742-6596/671/1/012049

[3] Fedorov E M, Bortnikov I D 2015 *Technical Physics* **60** 1689–1692 doi: 10.1134/S1063784215110110

[4] GOST (State Standard) 11326.0-78 1978 *Radio-frequency cables. General specifications* (Moscow: Standards publishing house, 2003) (in Russian)

[5] GOST (State Standard) 23286-78 1981 *Cables, wires and cords. Standards for insulation and sheath thickness and voltage tests* (Moscow, Standartinform, 2008) (in Russian)

[6] Cousins K 2000 *Polymers for Wire and Cable - Changes Within an Industry* (Smithers Rapra Technology) p 122

[7] Cadick J 1990 *Cables and Wiring* (Delmar Publishers) p 244

[8] 1969 *Engineering Design Handbook: Electrical Wire and Cable* (National Technical Information Service) p 297

[9] 1956 *Electrical wire and cable handbook* (United States Steel Corporation) p 237

[10] Govorkov V A 1968 *Electric and magnetic fields* (Moscow: Energy) (in Russian)

[11] Gomboeva S G, Shishkin I F, Hamhanova D N et al 2016 *Key Engineering Materials* **685** 463–466 doi: 10.4028/www.scientific.net/KEM.685.463

[12] Blohm W 1999 *Proceedings Intl. IWMA Conf. Economical Processing of Rod to Wire & Cable* 67–75

[13] Prunk H 1999 *Proceedings of the 69th Annual Convention of the Wire Association International* 378–385

[14] Vasendina E, Plotnikova I, Levitskaya A et al. 2016 *IOP Conference Series: Materials Science and Engineering* **110** 012070 doi: 10.1088/1757-899X/110/1/012070

[15] Term and Definition (Custom Wire Harnesses & Specialty Connectors, USA: Fontana, CA) [Electronical resource] URL: http://www.dsmt.com/pdf/resources/glossary.pdf (2017)

[16] GOST (State Standard) 15467-79 1979 *Product-quality control. Basic concepts. Terms and definitions* (Moscow, Standartinform, 2009) (in Russian)

[17] Sclater N 1991 *Wire and cable for electronics: a user's handbook* (McGraw Hill) p 237

[18] CAPAC® /FFT Capacitance measurement systems for cable production lines (Zumbach Electronic AG, SCHWEIZ: Hauptsitz) [Electronical resource] URL: http://www.zumbach.com/pdf/Literature_DE/Catalogs/CAPAC/CAPAC_Familie_CAPA.002.0002.D.pdf (2017)

[19] CAPACITANCE 2000: Capacitance measuring devices for cable production lines (Sikora AG, Germany: Headquarters) [Electronical resource] URL: http://sikora.net/wp-content/uploads/2016/03/CAPACITANCE_2000_GB.pdf (2017)

[20] Fleming P, Coleman L R 2001 *A system for monitoring fluctuations in the thickness of a cable insulating sheath* GB Patent 2358928

[21] Fleming P, Coleman L 2003 *Capacitance monitoring systems* US Patent 20030128038

[22] Goldshtein A E, Vavilova G V, Mazikov S V 2016 *MATEC Web of Conferences* **79** 01009 doi: 10.1051/matecconf/20167901009
[23] 1971 *Cable Engineering* (Defense Technical Information Center) p 54
[24] Drobny J G 2012 *Polymers for Electricity and Electronics: Materials, Properties, and Applications* (John Wiley & Sons) p 352
[25] UL 83-2017 2017 Standard for Safety Thermoplastic-Insulated Wires and Cables (UK)
[26] Nizhegorodov A, Gavrilin A, Moyzes B 2016 *Journal of Physics: Conference Series* **671** 012037 doi: 10.1088/1742-6596/671/1/012037
[27] GOST (State Standard) 54813-2011 2013 *Electric cables, wires and cords. Spark test method.* (Moscow, Standartinform, 2012) (in Russian)
[28] BS 5099:2004 2004 *Electric cables. Voltage levels for spark testing* (UK)
[29] 2004 Advances in high voltage engineering (UK : Cardiff University, Cardiff) p 669
[30] ANSI/UL 2556-2012 2012 UL Standard for Safety for Wire and Cable Test Methods
[31] GOST (State Standard) 27893-88 1989 *Communication Cables. Methods of Tests* (Moscow: Standartinform, 2010) (in Russian)
[32] Nizhegorodov A, Gavrilin A, Moyzes B 2016 *Journal of Vibroengineering* **18** 3734–3742 doi: 10.21595/jve.2016.16994
[33] Donald T.H. 1976 *Physical and Chemical Properties of Water* (IFI Plenum Data Company) p.556
[34] ANSI/UL 1581 2006 *Reference Standard for Electrical Wires, Cables, and Flexible Cords* (USA)

IOP Conf. Series: Materials Science and Engineering **289** (2017) 012018 doi:10.1088/1757-899X/289/1/012018

Stand for monitoring the operational parameters of conjugations "Ball support – body of the tie-rod end" of automotive components

D T Safarov[1], A G Kondrashov[2]
[1]Associate Professor, Naberezhnochelninsky Institute (branch) of Kazan (Volga Region) Federal University, Naberezhnye Chelny, Russia
[2]Associate Professor, Naberezhnochelninsky Institute (branch) of Kazan (Volga Region) Federal University, Naberezhnye Chelny, Russia
E-mail:safarov-dt@mail.ru

Abstract. The method of testing the conjugation "Ball support - body of the tie-rod end" is considered. The design of the test bench for simulating the working conditions of the interface is proposed. The results of testing the conjugation samples in accordance with the test procedure are given. The conformity of test results was assessed.

1. Introduction
Tie rods of various design purposes are responsible suspension components and steering systems. They largely determine the predictability of the behavior of the car when performing maneuvers. Tie rods provide the transfer of forces and provide the prescribed laws of movement between individual connecting levers.

For the performance of its functional purpose, the design of tie rods includes matings having several angular degrees of freedom. Structurally, they are made in the form of a ball bearing, and a nest with a spherical element, made in the body of the tip.

Any defect of this coupling, which manifests itself during the operation of the thrust, affects the safety of the vehicle's movement, therefore special requirements are required for the structural elements of the interface parts for controlling the geometric parameters of the parts, material properties and surface layers contacting in the interfaces. To the interface and the design of the unit as a whole, there are requirements for carrying out tests for operational reliability and resistance to atmospheric influences [1–4].

2. Theoretical part
Limiting backlash in pairing is a key functional requirement. In some cases, it can be determined from the data of measurements of conjugate parts [5, 6]. Its value is practically impossible to predict based on measurement data. Loft is manifested as a result of contact interactions in the process of functioning of the conjugate surfaces in various degrees under the axial and radially acting load.

Therefore, at the manufacturer of thrusts, the task arose to design a special stand simulating the operating conditions of the unit, up to and beyond the limits. It is not possible to manufacture the test stand completely by the forces of the enterprise. The finished stand is produced by special orders by foreign manufacturers. To reduce the total cost of the stand, the power frame of the device is manufactured at the enterprise, and the loading device - servo-hydraulic drive and electronic registration components are ordered in a foreign manufacturing enterprise specialized in manufacturing such stands.

The frame is a sturdy table with T-shaped grooves, the dimensions of which provide reliable fastening of the tested products without destroying the fasteners, with the posts and a movable cross bar into which the test head is mounted (Figure 1). Dimensions of the frame and landing nests in it provide an opportunity to install both horizontal and vertical positions of the entire range of the steering rods produced by the enterprise of different design.

Figure 1. Test bench with installed steering traction during the test.

3. Testing technique

To control the axial and radial play for the stand, a special method for carrying out the control tests has been developed. The tests for determining the magnitude of the axial displacement and the effort of pressing the finger in the tip are carried out according to the scheme shown in Figure 2.

The tip of the rod is installed with a lid on the support sleeve. A static force is applied to the end of the ball pin, aimed at "extruding" it from the tip body, smoothly increasing from zero to a force value of 2…2.5 kN. During the test, the loading diagram is recorded in the coordinates "Moving S, mm – force F, kN.

According to the diagram obtained, the axial movement is determined at a force of 1 kN and the amount of force of the pinching of the finger by the elastic element to the supporting spherical surface of the tip. In the diagram, the moment of overcoming the clamping force is observed in the form of a characteristic sharp curve inflection. In the absence of an inflection, the compression force is assumed to be zero (the liner does not press the finger against the support surface). An example of the diagram is shown in Figure 3.

IOP Conf. Series: Materials Science and Engineering **289** (2017) 012018 doi:10.1088/1757-899X/289/1/012018

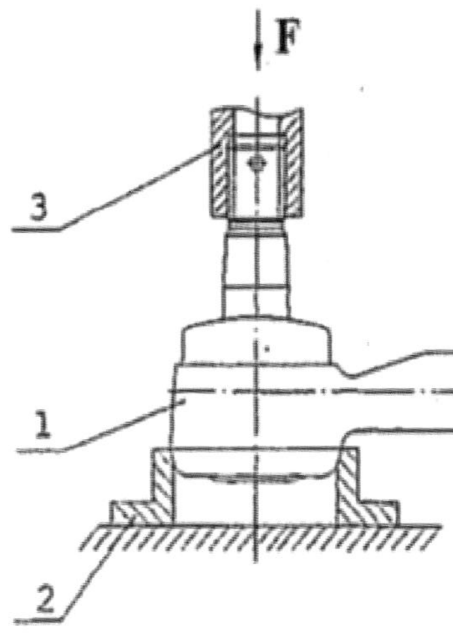

Figure 2. Scheme of fixing and loading the tip of the thrust during the test to determine the axial displacement and the compression force: 1 – test sample, 2 – basic rigging, 3 – transition element.

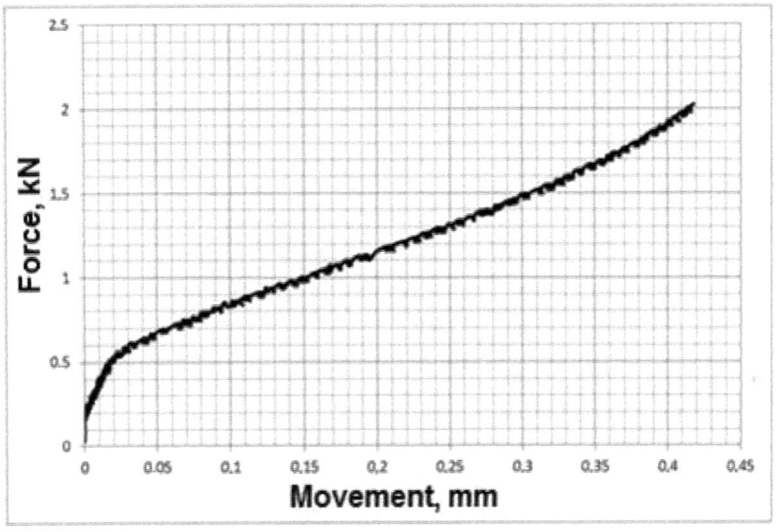

Figure 3. Example of the loading diagram in the test for determining the magnitude of the axial displacement and the force of compression.

The tests are carried out according to the scheme shown in Figure 4 to determine the radial clearance value under the action of cyclically changing forces in the friction pair "ball pin – tip body".

Thrust (or a separate tip) is attached to the ball pin on the stand, a static tension-compression force of ± 20 kN is applied to the other finger. In the course of the tests, the diagram "Movement S, mm – force F, kN is also recorded.

From the diagram obtained, determine the difference between the absolute values of the displacement at the moments of the transition of the curve in opposite directions through the zero value of the force. The value obtained in this way is considered as a gap in the friction pair "ball pin – tip body" (for traction assembly - the total for the two tips). An example of a diagram is shown in Figure 5.

IOP Conf. Series: Materials Science and Engineering **289** (2017) 012018 doi:10.1088/1757-899X/289/1/012018

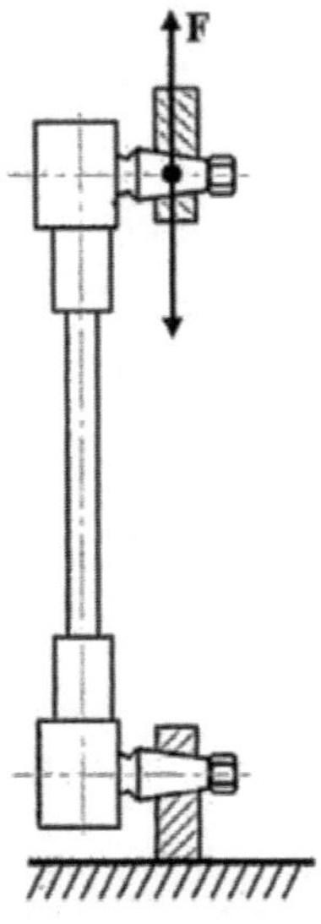

Figure 4. The scheme of fastening and loading of steering drafts in tests for determining the gap in the friction pair "Ball pin - body of the tie-rod end " with the radial cyclic action of the load.

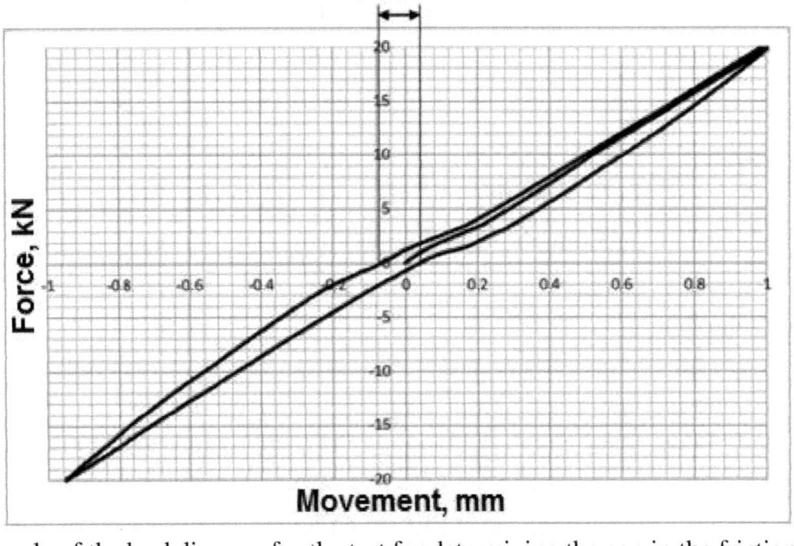

Figure 5. Example of the load diagram for the test for determining the gap in the friction pair "Ball pin – body of the tie-rod end".

In the initial period of the draft, there were no requirements for axial and radial clearance in the conjugation under load, the operation of the stand allowed determining the value of the gap under the action of the testing load.

The tie-rod ends are considered inoperative, in cases where:

- axial movement at a force of 1 kN exceeds 0.6 mm (for tie-rod end with a sphere diameter ø 45 mm);
- the compression force is zero (there is no kink in the diagram of the axial movement, an example is shown in Figure 6);
- on the gap measurement diagram there is a section coinciding with the X axis or close to it (there was a pronounced re-lay at zero force, an example is shown in Figure 7).

IOP Conf. Series: Materials Science and Engineering **289** (2017) 012018 doi:10.1088/1757-899X/289/1/012018

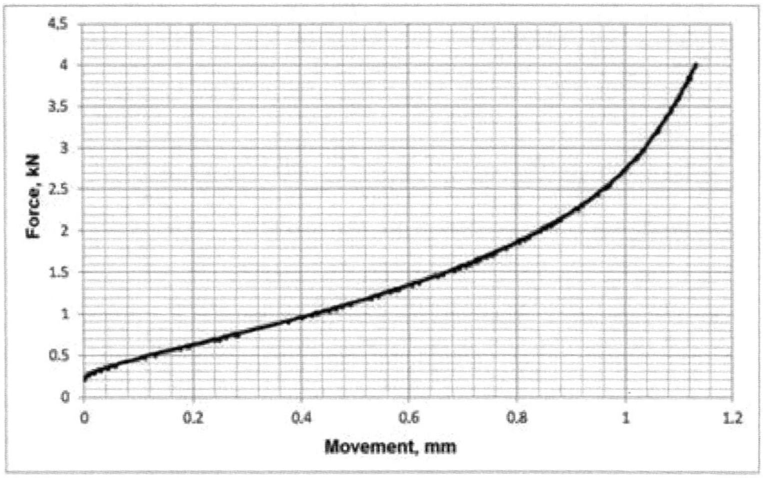

Figure 6. Example of the loading diagram in the test for determining the magnitude of the axial displacement and the force of the pinching: the elastic element is not compressed, the tie-rod end is inoperable.

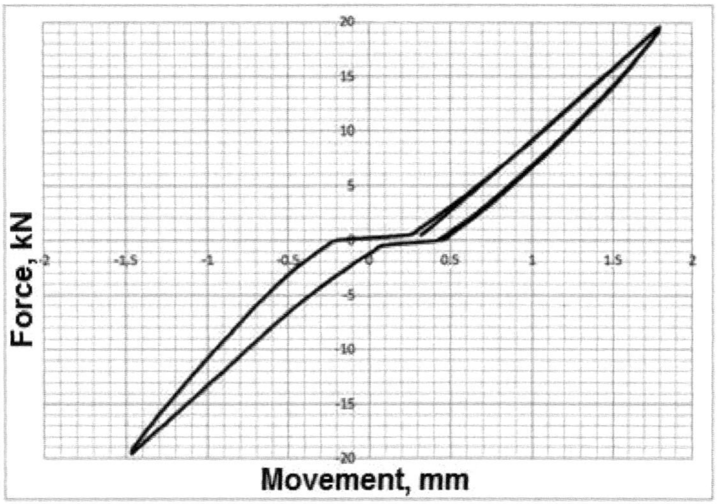

Figure 7. Example of the loading diagram for the gap test: a pronounced rearrangement has occurred, the tie-rod end is inoperable.

4. Results of tests

Let's give an example of tests of drafts of a sod by the resulted technique of a steering for a new product for a heavy dump truck KamAZ 6520, and the last run in 60 thousand in km. in conditions of constant operation at the maximum load when performing road earth during the excavation of soil (Figures 8, 9).

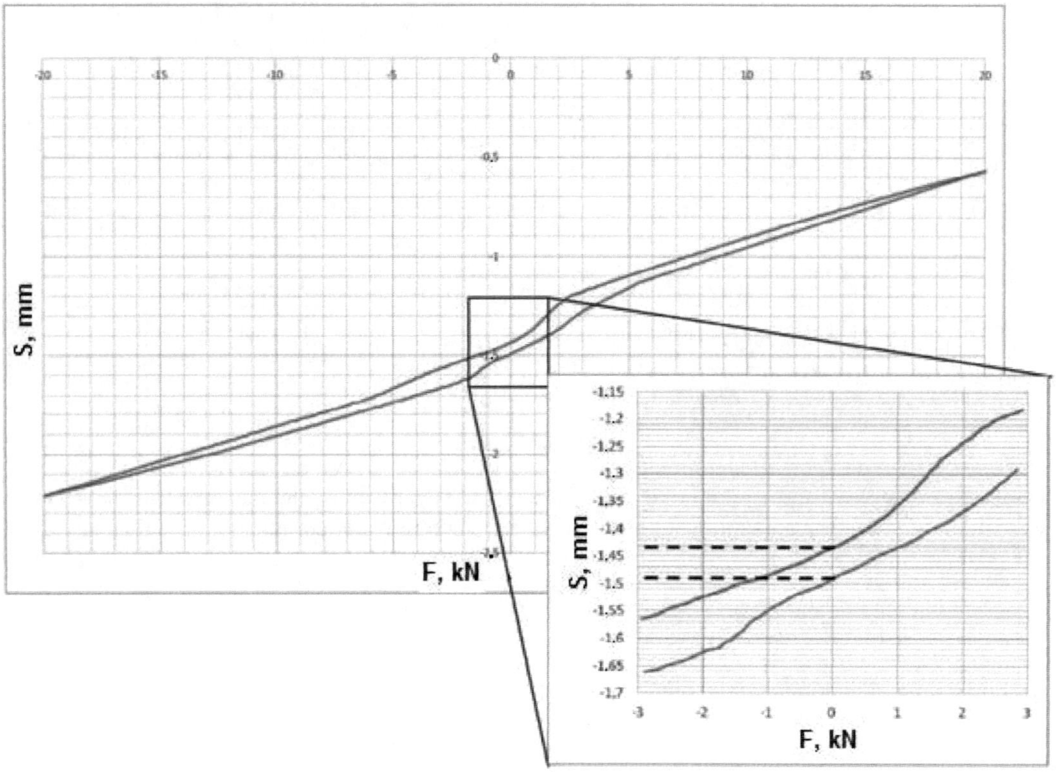

Figure 8. Diagram of tests of new steering link.

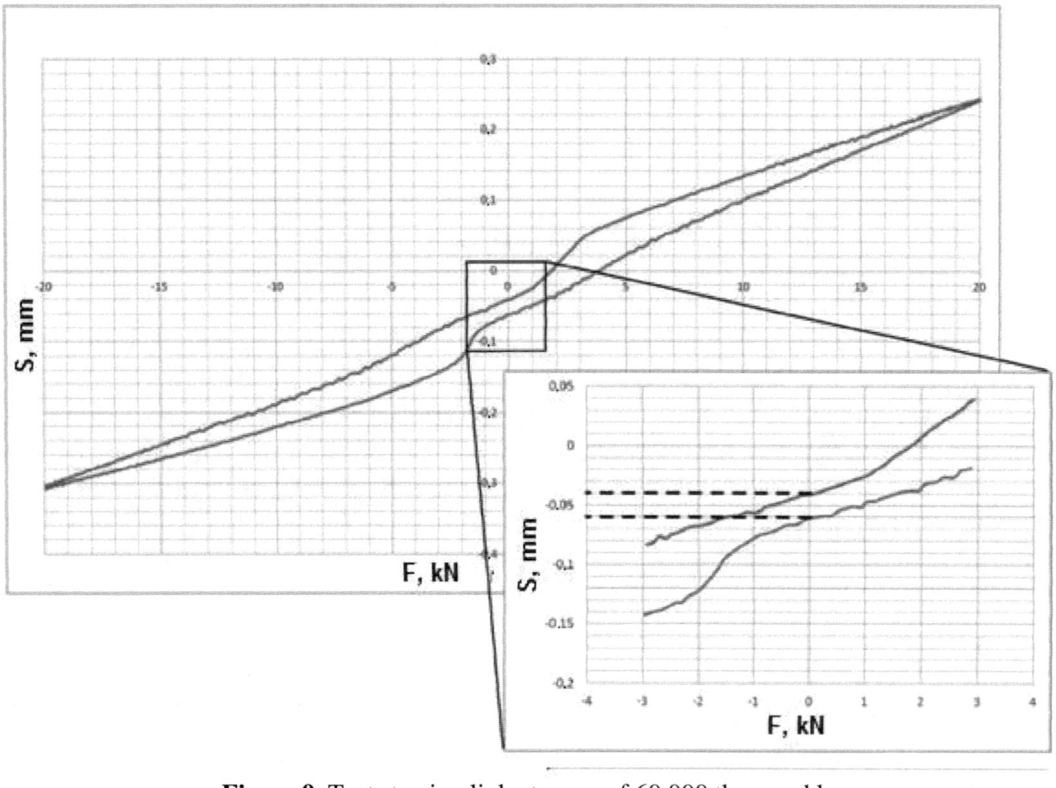

Figure 9. Test steering link at a run of 60,000 thousand km.

The results of changes in the gap in the process of operation from these diagrams are given in Table 1.

Table 1. Results of measurements of a backlash in pairs of friction of tie-rod end.

Sample name	Clearance value, mm
Thrust of bipod of steering control 6520 in assembly, new	0.05–0.055
Thrust of bipod of a steering of 6520 in gathering, run of 60 000 km	0.055–0.06

The control data show a slight change in the clearance in the ball interface, which indicates a high reliability of the interface.

5. Conclusion

Thus, the design and manufacture of the test stand allowed:

- reduce the costs of the company to conduct tests in external laboratories;
- to carry out operational improvement of the design of newly developed traction links for various models of modern trucks;
- to shorten the duration of the APQP process of preparation of production when developing models of new steering link [7].
- reduce the complexity of establishing the causes of the defects of defects without disassembling the interface.

References

[1] GOST R 53835-2010 Vehicles. Elements of the steering actuator and suspension guide vane. Technical requirements and test methods, Moscow, Izdatelstvo Standartov (in Russian)
[2] GOST R 27.403-2009 Dependability in technics. Compliance tests plans for reliability Moscow, Izdatelstvo Standartov (in Russian)
[3] GOST R 51709-2001 Motor vehicles and their trailers. Safety requirements for technical conditions and methods of inspection Moscow, Izdatelstvo Standartov (in Russian)
[4] GOST 16962.1-90 Electrical articles. Test methods as to environment mechanical factors stability Moscow, Izdatelstvo Standartov (in Russian)
[5] Murtazin R M, Safarov D T, Kondrashov A G 2014 *INTERFINISH-SERIA 2014* 62
[6] Kasjanov S V, Kondrashov A G, Safarov D T 2017 *Procedia Engineering* **206** 1508–1514 doi: 10.1016/j.proeng.2017.10.670
[7] Safarov D T, Fedorova K A and Ilyasova A I 2016 *IOP Conf. Series: Materials Science and Engineering* **134** 012036 doi: 10.1088/1757-899X/134/1/012036

Influence of a high-power pulsed ion beam on the mechanical properties of corundum ceramics

V Kostenko[1], S Pavlov[2], S Nikolaeva[3]
[1]Graduate student, Tomsk State University of Control Systems and Radioelectronics, Tomsk, Russia
[2]Engineer, National Research Tomsk Polytechnic University, Tomsk, Russia
[3]Researcher, National Research Tomsk Polytechnic University, Tomsk, Russia
E-mail: kostenko.lerochka@mail.ru

Abstract. The mechanical properties of near-surface layers of corundum ceramics treated by high-power pulsed ion beam of carbon are investigated. The samples for investigation were prepared from corundum substrate, which is usually used in microelectronic. The ion treatment was carried out at the TEMP-4M facility under the following conditions: an accelerating voltage of 160-200 keV, the current density in the pulse varied within 15-85 A/cm^2. It was found that ion irradiation changes the structure and properties of near-surface layers of corundum ceramics. At the same time, melting and erosion of the surface layer takes place. These processes are accompanied by the formation of a network of microcracks. Microcracks are propagated only by the depth of melting layer. The mechanical properties were measured using a NanoTest600 nanohardness testing instrument. It was found that the nanohardness depends of the treatment modes. At a current density of 15A/cm^2, with an increase treatment dose, the nanohardness of the irradiated surface layer increases in comparison with the initial value before irradiation. At higher current densities, the nanohardness of irradiated ceramics decreases relatively to the initial value before irradiation. The dependences of nanohardness off the irradiation dose in this case have the view of a curves with a minimum at irradiation doses of $2.5 \cdot 10^{14}$ and $1.3 \cdot 10^{14}$ cm^{-2}, for current densities of 50 and 85 A/cm^2, respectively.

1. Introduction

At present, ceramic materials are widely used. There are widespread use of oxide ceramics, for example, lithium-ferrite ceramics [1–3], zirconium ceramics [4–7] and corundum ceramics [8–10]. When manufacturing of ceramics for functional and instrumental purposes, both traditional and non-traditional methods of processing finished ceramics are used. It is known that ceramic materials, such as corundum and zirconium, are characterized by high strength. Their mechanical processing, in particular, grinding by traditional methods is a very costly process. In many cases, the properties of the ceramic product as a whole depend on the state of the near-surface layer. The use of ceramics with gradient properties has great prospects, when the surface layer differs in its characteristics from the corresponding parameters of the bulk layers. As shown in works [11–24], surface treatment can be effectively carried out using low energy high current pulsed electron beam (LEHCPEB) [11–13], laser treatment [14,15] and high-power pulsed ion beam (HPPIB) [16–24]. To study thin modified layers of dielectrics, the method of secondary ion mass spectrometry [25–27] and the Rutherford backscattering method [28, 29] have proved to be very useful. The above methods of radiation treatment are used to modify and create structures that can not be obtained by traditional technologies.

By varying the modes of surface treatment, it is possible to impart different properties not only to the surface layers of metals, alloys and semiconductors, but also to dielectric materials, thereby expanding the field of their application. Radiation methods of surface treatment of materials have a whole range of advantages. They allow to carry out high-precision treatment of complex shaped ceramic parts with a short exposure time to the surface, are equipped with simple process automation, etc. Among radiation methods of influence on materials, the method of surface modification by high-power pulsed ion beams (HPPIB) [16–24] widely is used.

At present, studies are underway to modify the surface of metals and alloys by high-power pulsed ion beams [19–22, 30]. At the same time, the physicochemical processes taking place in the near-surface layers of dielectric materials under treatment HPPIB have been little studied. The results of the influence of LEHCPEB and HPPIB on zirconium ceramics are reflected in [11, 24]. It is shown that under the influence of such treatments, the surface layer of the material is rapidly heated to the melting point, and then also rapidly cooled [12, 30]. As a result of such influences, surface modification occurs, which is manifested in a change in the phase and chemical composition of the near-surface layer, structuring the surface in the form of grain recrystallization [9, 15, 24].

Consequently, it is possible to change the optical, mechanical and electrical properties of the near-surface layers of the ceramic, and also significantly change the catalytic and adsorption capacity of the surface.

To study the nature of the radiation effect on ceramic materials, it is of interest to investigate the effect, for example, of HPPIB on a ceramic structure that is resistant to changes in the phase composition.

This study aims of this work is to study the influence of HPPIB on the mechanical properties of near-surface layers of corundum ceramics.

2. Experimental technique

Alumina ceramic (corundum) was researched in the work. The samples measuring 1×1 cm in thickness 0.52 mm was cut from a ceramics plate used in the manufacture of microelectronic devices as substrate material.

Ion irradiation of the samples with accelerated C^{+}, $^{n+}$ and H^{+} carbon ions (85% and 15%, respectively) was carried out with a TEMP-4M pulse accelerator [31, 32] at a residual chamber pressure of $2 \cdot 10^{-2}$ Pa. Irradiation modes are the following: an accelerating voltage of 160 – 200 keV, the pulse duration (at half maximum of the accelerating voltage) of the current was 100 ns, the current density in the pulse varied in the range 15 – 85 $\mathrm{A/cm^2}$. Pulse repetition rate 8 s^{-1}. The energy density in the pulse varied within $W_i = 0.3 - 1.5\ \mathrm{J/cm^2}$.

In the experiment were realized three variants of irradiation. The effect of ion treatment were researched depending on the beam current density and irradiation dose, which was set by the number of pulses. Depending on the density of the ion current, the number of pulses varied from 3 to 300. At current densities of 15, 50 and 85 A/cm^2 the amount of incorporated carbon ions per pulse was f ~ $7.5 \cdot 10^{12}$ cm^{-2}, $2.5 \cdot 10^{13}$ cm^{-2} and $4.25 \cdot 10^{13}$ cm^{-2}, respectively.

The surface of irradiated ceramics was analyzed by SEM using a Hitachi TM-3000 Tabletop Microscope. The static electrical conductivity of the corundum surface was measured in the temperature range T = 25–300°C by the two-probe spreading resistance method [33]. To assess the degree of change in the mechanical properties of the near-surface layers of corundum was obtained of results of nanoindentation with the NanoTest 600 instrument allowing indentation in a wide range of loads. The study was carried out using the method of an unrestored print using the trihedral pyramid of Berkovich [34, 35]. Indentation was carried out in the range of loads from 1 mN to 100 mN. For the statistical analysis, about 12 fingerprints of the indenter was applied to the surface of the samples.

3. Experimental results

In Figure 1*a* is an SEM analysis of the surface of a sample of corundum ceramics prior to exposure to HPPIB. It can be seen that on the surface of samples is observed the block structure of ceramics. The dimensions of the blocks vary between 10 and 60 μm. The image shows pores randomly distributed over the surface of corundum.

It is visually observed that after exposure to HPPIB, the surface of the samples darkens, with increasing current density and treatment dose, the samples become darker. It was shown in paper [24] that such a darkening of the ceramic is caused by the formation of oxygen vacancies due to the violation, under the action of HPPIB, of oxygen stoichiometry. Also, the darkening can be influenced by the pyrolysis process on the surface of samples of carbon-containing oil vapors from the residual atmosphere and contamination of the sample surface by the graphite electrode erosion products generated during HPPIB generation.

The impact of HPPIB on corundum ceramics has a thermal character, which leads to melting, evaporation and subsequent recrystallization of the near-surface layer. SEM analysis showed that after the irradiation of HPPIB, a noticeable change in the microstructure of the material surface (Figure 1*b*). The surface of the ceramics after recrystallization acquires a mosaic structure in the form of separate blocks separated by microcracks. The average block size is 9…11 μm. The process of crack formation is due to the low thermal conductivity of the material and the heat treatment modes. Traces of melting are observed on the surface. Compared to the initial surface of the ceramics, the pores are not observed after HPPIB treatment.

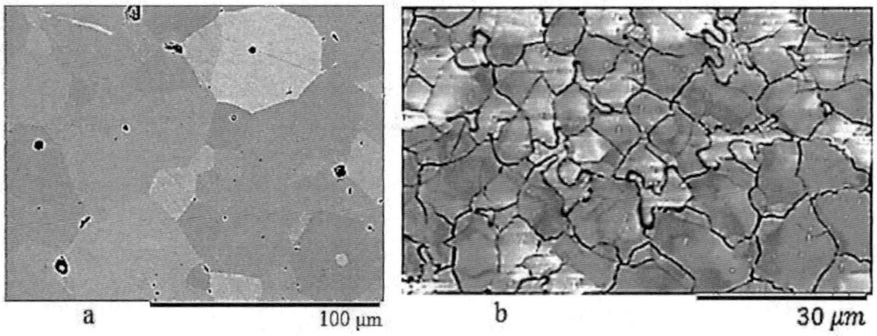

Figure 1. SEM image of the alumina ceramics sample surface: *a* – before HPPIB treatment; *b* – after HPPIB treatment (j = 85 A/cm^2, number of pulses N = 3).

To reduce the intensity of the cracking process, it is possible to choose the heat treatment modes. Such a study is presented in the paper [36], where the optimization of thermal pulse treatment modes is described using the example of the treatment of enamels with a heating light flux. Another possible way to reduce cracking is thermal heating of the sample to a temperature of the order of 0.7 from the sintering temperature and further processing of its HPPIB in the heated state

In Figure 2*a–c* shows the SEM image of the fracture of ceramic samples exposed to HPPIB. Treatment of HPPIB at an ion beam current density of 15 A/cm^2 does not lead to noticeable changes in the surface layers of the ceramic (Figure 2*a*). In Figure 2*b* shows that the modification zone is 6–8 μm. In some areas of this zone a microstructure of the "columnar type" is observed, the grain shapes in the near-surface layers are oriented toward the surface being treated. The process of formation of the "columnar type" structure as a result of the radiation effect was observed in the works [9, 15, 24]. The formed cracks, during the processing of HPPIB, are distributed over the entire depth of the fused layer (Figure 2*c*). With an increase in the ion current density, the microstructure of the "columnar type" is not observed. This may be due to the high heating rates of the material when impact by HPPIB, as a result of which the material does not have time to recrystallize between pulses. The thickness of the melted layer is practically not increased, since simultaneously with the melting the erosion of the surface. The thickness of the melted layer is 9…11 μm.

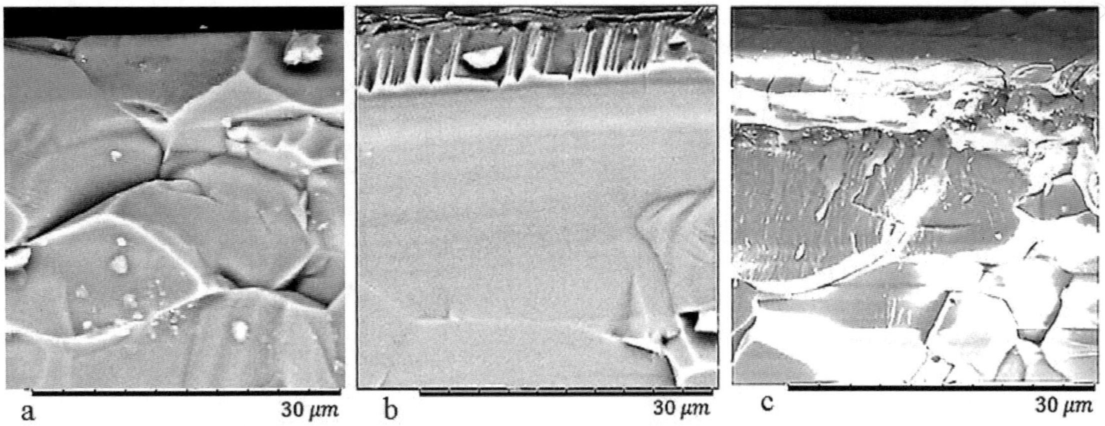

Figure 2. SEM image of the transverse fracture of the ceramics:
a – surface treatment of HPPIB ceramics at current density j = 15 A/cm^2, number of pulses $N = 20$;
b – surface treatment of HPPIB ceramics at current density j = 50 A/cm^2, number of pulses $N = 100$;
c – surface treatment of HPPIB ceramics at current density j = 85 A/cm^2, number of pulses $N = 30$.

The static electrical conductivity of the modified ceramics layers was measured by applying a constant voltage U_{max} = 5 V to the electrodes and the sensitive limit of the universal voltmeter V7-21A 100 nA. When measuring the electrical conductivity of corundum, it was found that when the HPPIB ceramics are treatment, the electrical conductivity of the sample remains practically unchanged compared to its initial state. It follows that after treatment by HPPIB, the oxygen content of the material changes insignificantly.

For different ion current densities, curves for the variation of the nanohardness of the samples are plotted against the HPPIB irradiation dose at the maximum applied load P_{max} to the indenter 10 mN (Figure 3*a*) and 100 mN (Figure 3*b*). As can be seen from the graphs, the choice of the applied load to the indenter does not affect the character of the change in the curves. From Figure 3 that when the current density of the ion beam is 15 A/cm^2, the nanohardness increases continuously with the increase in the dose (fluence) (curve 2). The growth of the nanohardness of the near-surface layers is explained by the decrease in the grain size without cracking during the recrystallization. An increase in nanohardness is observed in comparison with the initial value by 22%. With an increase in the current density of the ion beam to 50 A/cm^2 (curve 3) and 85 A/cm^2 (curve 4), the value of nanohardness at the minimum point on the curve H (Q) decreases. Perhaps this is due to the process of cracks formation. As the dose is increased, partial healing of cracks occurs due to melting, which leads to an increase in nanohardness of surface.

Thus, it has been established that the impact by HPPIB carbon leads to a change in the nanohardness of the surface layers of corundum ceramics. The degree of change depends on the chosen modes of irradiation. In the selection of irradiation modes it is possible to achieve hardening of ceramics. This result is confirmed by the data of paper [23].

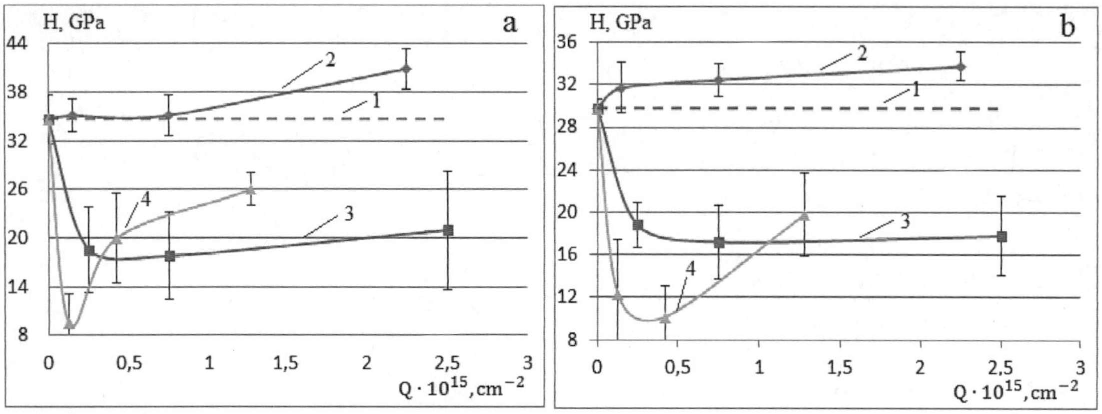

Figure 3. Nanohardness of alumina ceramics as a function of HPPIB treatment dose Q at current densities 15, 50, 85A/cm^2 (curves 2-4, respectively. Curve 1 – the value of nanohardness before treatment by HPPIB):
a – is the maximum applied load to the indenter P_{max} = 10 mN;
b – is the maximum applied load to the indenter P_{max} = 100 mN.

4. Conclusion

The impact of the high-power pulsed ion beam of carbon has been studied on the structure and mechanical properties of corundum ceramics. The following results are obtained.

Under the influence of HPPIB, at a current density of an ion beam of 85A/cm^2, a thin near-surface layer melt to a depth of the order of 9-11 μm. With increasing dose, the thickness of the melting layer remains practically unchanged due to erosion of the surface during ion treatment.

When treatment of HPPIB ceramics, the electrical conductivity of the samples remained practically unchanged from its original value.

The surface treatment of HPPIB corundum with a current density of 50A/cm^2 leads to the formation of a "columnar type" microstructure. When treatment ceramics HPPIB, the near-surface layer acquires a block structure with boundaries in the form of cracks, which propagate to the depth of not more than the thickness of the fused layer.

The nanohardness of the corundum ceramics treated with HPPIB depends on the current density and the dose. At current densities on the order of 15 A/cm^2, the nanohardness of near-surface layers of corundum ceramics increases with increasing radiation dose. The greatest increase in nanohardness is about 20%. At higher ion current densities, nanohardness as a whole decreases and has a minimum at irradiation doses of $2.5 \cdot 10^{14}$ and $1.3 \cdot 10^{14}$ cm^{-2}, for current densities of 50 and 85A/cm^2, respectively.

Acknowledgments
The research was supported by the Russian Science Foundation (Grant № 17-19-01082).

References

[1] Surzhikov A P, Peshev V V, Pritulov A M, Gyngazov S A 1999 *Russian Physics Journal* **42** 490–495
[2] Zatsepin D A, Waistein I C, Cholakh S O 2014 *Ion Modification of Functional Materials* (Ekaterinburg: GOU HPE URFU) (in Russian)
[3] Gyngazov S A, Lisenko E N, Petyukevich M S, Frangulyan T S 2007 *Russian Physics Journal* **50** 134-139 doi:10.1007/s11182-007-0018-3
[4] Ivanova S V 2002 *International Journal of Hydrogen Energy* **27** 819–824 doi:10.1016/S03603199(01)00160-4

[5] Surzhikov A P, Frangulyan T S, Ghyngazov S A 2012 *Russian Physics Journal* **55** 345–352 doi:10.1007/s11182-012-9818-1

[6] Berezneeva V E, Pushilina N, Berezneev V D, Chernov P I, Lider Andrey, Kreoning M 2013 *Applied Mechanics and Materials* **302** 82–85 doi:10.4028/www.scientific.net/AMM.302.82

[7] Surzhikov A P, Frangulyan T S, Ghyngazov S A 2014 *Journal of Thermal Analysis and Calorimetry* **115** 1439–1445 doi:10.1007/s10973-013-3455-y

[8] Kacar E, Mutlu M, Akman E, Demir A, Candan L, Canel T, Gunay V, Sınmazcelik T 2009 *In Journal of Materials Processing Technology* **209** 2008–2014 doi: 10.1016/j.jmatprotec.2008.04.049

[9] Burdovitsin V A, Medovnic A V, Oks E M, Yushkov Y G, Dvilis E S, Khasanov O L 2013 *The Russian Journal of Applied Physics* **58** 111–113 doi:10.1134/S1063784213010039

[10] Zum Gahr K-H, Schneider J 2000 *Ceramics International* **26** 363–370 doi:10.1016/S0272-8842(99)00065-6

[11] Surzhikov A P, Frangulyan T S, Ghyngazov S A, Koval N N 2009 *Nuclear Instruments and Methods in Physics Research, Section B: Beam Interactions with Materials and Atoms* **267** 1072–1076 doi:10.1016/j.nimb.2009.01.144

[12] Rotshtein V, Ivanov Yu, Markov A 2006 *Materials Surface Processing by Directed Energy Techniques* 205–240 doi: 10.1016/B978-008044496-3/50007-1

[13] Leyvi A Ya, Talala K A, Krasnikov V S, Yalovets A P 2016 *Herald South Ural State University* **16** 28–55 DOI:10.14529/engin160103

[14] Šugár P, Frnčík M, Šugárová J, Sahul M 2017 *Solid State Phenomena* **261** 143–150 doi:10.4028/www.scientific.net/SSP.261.143

[15] Savruk E V, Smirnov S V 2011 *Industrial Laboratory* **77** 32-–35 doi:621.315.612:546.62-31

[16] Yotoriyama T, Yoshiaki S, Takaya M, Takeyo T, Masaya I 2005 *Surface & Coatings Technology* **196** 383–388

[17] Sokullu-Urkac E, Oztarhan A, Tihminlioglu F, Nikolaev A, Brown IG 2012 *IEEE Transactions on Plasma Science* **40** 863–869

[18] Renk TJ Provencio, Prasad P V, Shlapakovski Somuri S, Petrov A, Yatsui A, Kiyoshi J, Weihua S H 2004 *Proceedings of the IEEE* **92** 1057–1081 doi:10.1109/JPROC.2004.829024

[19] Schmidt B, Wetzig K 2013 *Ion Beams in Materials Processing and Analysis* (Germany: Springer Verlag)

[20] Ovchinnikov V V, Gushchina N V, Gapontseva T M, Chashchukhina T I, Voronova L M, Pilyugin V P, Degtyarev M V 2015 *High Pressure Research 35* 300–309 doi: 10.1080/08957959.2015.1041522

[21] Davis H A, Remnev G E, Stinnett B W, Yatsui K 1996 *MRS Bulletin 21* 58–62

[22] Elke W, Werner W 2016 *Ion Beam Modification of Solids(Springer Series in Surface Sciences)*

[23] Romanov I G and Tsareva I N 2001 *Technical Physics Letters* **27** 695–697 doi: 10.1134/1.1398972

[24] Ghyngazov S A, Vasil'ev I P, Surzhikov A P, Frangulyan T S and Chernyavskii A V 2015 *Technical Physics* **60** 128–132 DOI: 10.1134/S1063784215010120

[25] Surzhikov A P, Chernyavskii A V, Ghyngazov S A, Frangulyan T S 2002 *Russian Physics Journal* **45** 1190–1194

[26] Benninghoven A 1973 *In Surface Science* **35** 427–457 doi: 10.1016/0039-6028(73)90232-X

[27] Gyngazov S A, Surzhikov A P, Frangul'yan T S, Chernyavskii A V 2002 *Russian Physics Journal* **45** 753–758

[28] Toussaint U V, Krieger K, Fischer R, Dose V 1999 Fundamental Theories of Physics (Dordrecht: Springer)

[29] Li M M, Verma B, Fan X 2008 *Neural Computing and Applications* **17** 391–397 doi:10.1007/s00521-007-0138-2

[30] Grosdidier T, Zou J X, Stein N, Boulanger C, Haoc S Z, Dong C 2008 *Scripta Materialia* **58** 1058–1061 doi: 10.1016/j.scriptamat.2008.01.052

[31] Renk T J, Provencio P, Prasad V S, Shlapakovski S A, Petrov A, Yatsui K, Jiang W, Suematsu H *Proceedings of the IEEE* **92** 1057–1080 doi: 10.1109/JPROC.2004.829024
[32] Remnev G E, Isakov I F, Pushkarev A I 1999 *In Surface and Coatings Technology* **114** 206–212
[33] Miroshkin V P, Panova Ya I and Pasynkov V V 1981 *Physics of the Solid State* **66**
[34] ISO 14577-4:2007 Metallic materials – Instrumented indentation test for hardness and materials parameters – Part 4: Test method for metallic and non-metallic coatings
[35] Golovin Yu I 2008 *Physics of the Solid State* **50** 2205–2236 doi: 10.1134/S1063783408120019
[36] Aleutdinov A D, Ghyngazov S A, Mylnikova T S, Luchnikov P A 2015 *IOP Conference Series: Materials Science and Engineering* **81** 012069 doi:10.1088/1757-899X/81/1/012069

IOP Conf. Series: Materials Science and Engineering **289** (2017) 012020 doi:10.1088/1757-899X/289/1/012020

Development of control system of coating of rod hydraulic cylinders

S Zh Aizhambaeva[1] and A V Maximova[2]
[1]Department director, Karaganda State Technical University, Karaganda, Kazakhstan
[2]Candidate for a master's degree, Karaganda State Technical University, Karaganda, Kazakhstan
E-mail: nice.alina1994@mail.ru

Abstract. In this article, requirements to materials of hydraulic cylinders and methods of eliminating the main factors affecting the quality of the applied coatings rod hydraulic cylinders. The chromium plating process - one of ways of increase of anti-friction properties of coatings rods, stability to the wear and corrosion. The article gives description of differences of the stand-speed chromium plating process from other types of chromium plating that determines a conclusion about cutting time of chromium plating process. Conducting the analysis of technological equipment suggested addressing the modernization of high-speed chromium plating processes by automation and mechanization. Control system developed by design of schematic block diagram of a modernized and stand-speed chromium plating process.

1. Introduction

Presently the hydraulic and pneumatic systems are used in a power supply [1–4] and land-reclamation, on all types of transport [5], in building, robotics [6], management [7], metallurgy and engineering [8]. Large attention at technical progress in an engineering must be spared to upgrading quality of mining machines, their reliability and longevity. Reliability enhancement – one of major tasks of modern metrology, that it is related to the continuous height of intensification of technological processes of mining, productivity enhancement of equipment and increase of the affecting him loading.

In the device of hydraulic cylinder, materials of contacting pairs, besides the required high durability, must possess good anti-friction properties at high enough speeds of recurrently - forward motion, by enhance able stability to the wear at a long friction. The details of power cylinders (a research object is a power hydraulic cylinder) make from corrosion-resisting materials or coated their surface the galvanic coating [9–11].

2. Attributes of anodic stand-speed chromium plating process

The chromium plating process provides this protection by providing of high stainless in such environments, as water, salt water and nitric acid. In order that to increase stainless of contacting surfaces, protect from corrosion and to recover threadbare or defective details the electrolytic chromium plating process of rod hydraulic cylinders is used [11, 12].

Quality of corrosion-resisting chromic surfaces is determined on next criterions:

1) their appearance;
2) coupling durability;
3) thickness of layer.

Considering that properties of chromic surface substantially depend on the mode of chrome-plating and not enough characterized by appearance of surface, it is expedient to carry out in practice control on operations, watching out for operation of the anodic activation process and mode of chrome-plating (current density and temperature of electrolyte). For intensification of chromium plating process of rod hydraulic cylinders, it is necessary to apply the anodic stand-speed chromium plating process. The anodic-stream method of chromium plating process usually carry out in self-regulating electrolyte at a temperature from +55 to +66 °C (as marked higher this parameter must be controlled exactly).

Automation and mechanization of this process are used in heavy and harmful for a man productive terms, to increase the labor productivity and improve quality of electroplating rod hydraulic cylinders, also to remove ineffective hand labor. Therefore in modern terms on enterprises automatic operator galvanic lines (further – AOGL) are used [13–15].

Presently for the further decline of labor content of chromium plating process at providing of the required quality of coating perspective technology is technology of the anodic stand-speed chromium plating process in the stream of electrolyte at the set technological parameters. The distinctive of technological process of the stand-speed chromium plating process is that this process takes place at pumping of electrolyte through a workpiece-to-electrode with certain speed, pressure, temperature, current density and other parameters. Basic advantages:

- Speed-up deposition of chrome - to three μm/mines.
- Irregularity deposition of chrome on a diameter – 10%.
- The healthy terms of labor by complete absence of open bath of electrolyte.

3. Modernization of the speed anodic stand-speed chromium plating process

The technological equipment of the stand-speed chromium plating process of period of 1980–90th years, applied on the enterprises of Republic of Kazakhstan, is characterized by the physical wear of basic equipment, physically and by the morally out-of-date element base of control system, which results operating trouble and low-grade coating. Scientific and technical modernization is therefore required this equipment. Essence of modernization of the speed anodic stand-speed chromium plating process consists in providing of new element base: introduction of programmable logical controllers (further – PLC) [14].

The schematic block diagram of modernized AOGL of the speed chrome-plating (Figure 1) include eight programmable logical controllers, automatic thermostat (further – AT), conveyor, personal computer (further – PC) and block of preparation of electrolyte. Conceptually all element base of control system subdivided into four basic working blocks:

1) A1 – control block of conveyor.
2) A2.1, A2.2, A2.3 – control blocks of current.
3) A3 – control block of pressure.
4) Block of preparation of electrolyte.

Such dividing into certain zones allows more exactly producing control on operations necessary in control system such chromium plating process of rods. By block A1 controls moving by the conveyor of object of chrome-plating (rod of hydraulic cylinder). Control blocks of current (A2.1, A2.2, A2.3) control power steam-lines electric current, the same is provide regularity of coating, removing or decrease fringe effect on the chrome-plated surface. Fringe effect is an unfavorable factor influencing on distributions of chromic coating. In addition, automatic control of power steam-lines is necessary for the exception of interruptions of current feeding in the chromium plating process, inasmuch as at the repeated galvanic plating shelling of chrome layer is possible. The block of preparation of electrolyte is straight related with automatic thermostat and accordingly with control block of pressure (A3). The block of preparation of electrolyte controls the temperature of electrolyte, because a defect (mat precipitates of chrome) will follow from the decline of temperature. This type of defect causes difficulties at polish of details.

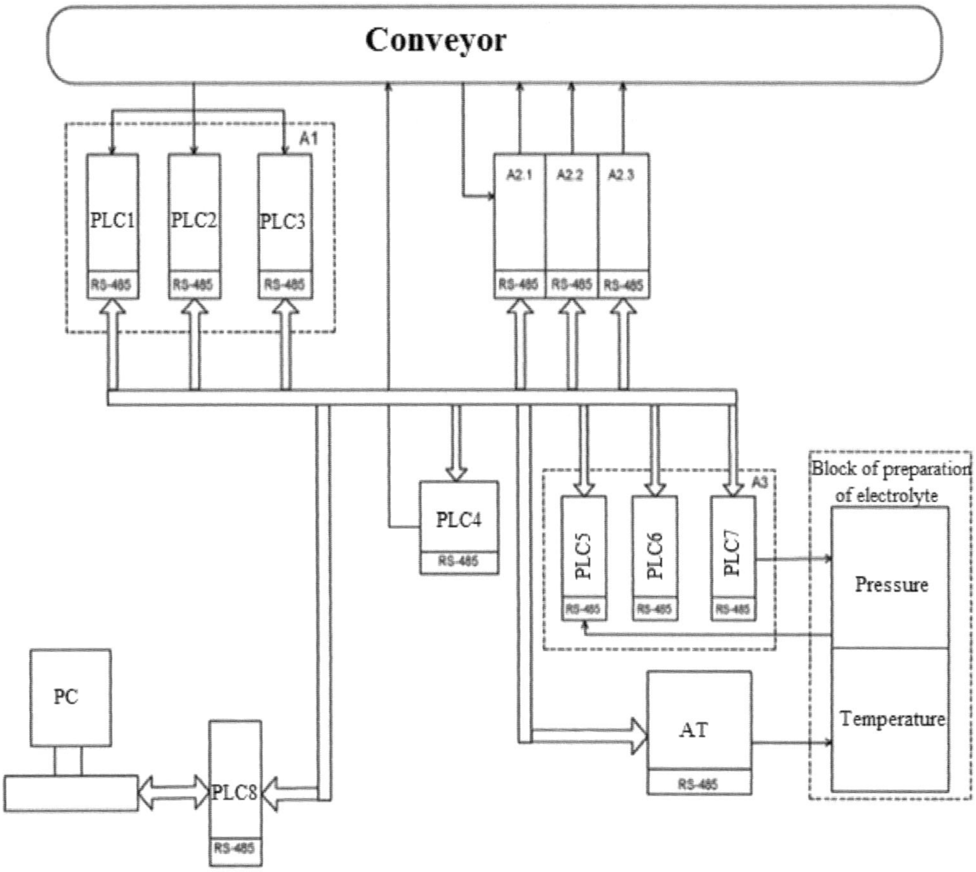

Figure 1. The schematic block control diagram of the chromium plating process of rod hydraulic cylinder.

The block of A3 include the controllers of PLC 5, PLC 6, PLC 7, that responsible for a control pressure and temperature of electrolyte (automatic thermostat maintain the standard temperature in a range from +55 to +66 °C) [10]. Controllers of PLC 1, PLC 2 and PLC 3 are designed to control conveyor. The controller of PLC 4 is necessary for realization of manipulations with the object of chromium plating process. The controller of PLC 8 is designed for connection with the personal computer. Programmable logical controllers are united inter se and with the personal computer by means of interface of RS – 485.

The hookup of the stand-speed chromium plating process of rods (Figure 2) shows what control elements are involved on the certain stage of technological process for control of temperature and amperage. The technological process of chromium plating process of rod hydraulic cylinders contains positions of loading, unloading, depriving of fat, washing and chromium plating process. Control system by programmable logical controllers coordinates performance of separate operations of chromium plating process in time and in space. Intercommunication of this control system and the personal computer allows carrying out monitoring the stand-speed chromium plating process.

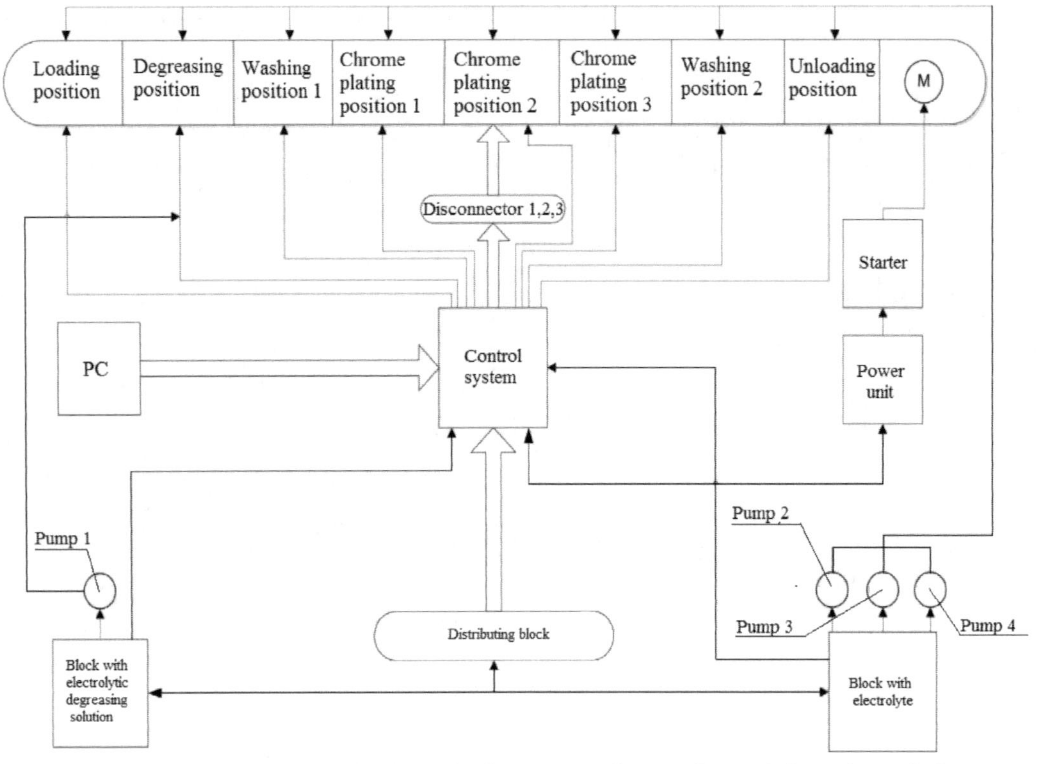

Figure 2. The hookup of the modernized stand of treatment the stand-speed chromium plating process of rods.

The chromium plating process of rods consists of next strict sequence: electrolytic degreasing (for details from high-strength steel possibly only chemical degreasing), washing, 3 positions of chrome plating, washing after chrome plating, drying and heat treatment [16]. By control block of conveyor rod of hydraulic cylinder passes all positions of the stand-speed chromium plating process. Positions of loading and unloading come true due to a programmatic logical controller 4, responsible for manipulations with the object of chromium plating process. Position of electrolytic degreasing is interrelated with a block in that there is electrolytic degreasing solution (depending on the type of steel composition of solution is determined). In addition, at co-operating with a distributing block the exactly that concentration of solution, that is necessary in accordance with technical documentation for rod of hydraulic cylinder, is given. The process of drying and heat treatment are produced not on the modernized stand, and after unloading position rod of hydraulic cylinder heads for other productive area (drying options, stoves) [17].

4. Summary

Thus, it is possible to do a conclusion on modernization of chromium plating process: realization of computer control the chromium plating process will allow to carry out monitoring at work of technological equipment real-time. The planned introduction of line of the stand-speed chromium plating process will allow shortening time of coating on 30…35 % as compared to an existent technological process. Reduction of time feasibly due to property of besieging of chrome (to three μm/mines), that is main advantage of the stand-speed chromium plating process.

Acknowledgments

We thank Galina Vavilova, assistant professor, director of laboratory, Tomsk Polytechnic University for comments that greatly improved the manuscript. We thank our colleagues from Karaganda State

Technical University who provided insight and expertise that greatly assisted the research. We are immensely grateful to Ozhigin Sergey (vice-rector on scientific work, doctor of Engineering Science) and Esenbaev Salim (assistant professor) for their comments on earlier versions of the manuscript, although any errors are our own and should not tarnish the reputations of these esteemed professionals.

References

[1] Obukhov S G *et al* 2017 *IOP Conf. Ser.: Mater. Sci. Eng.* **189** 012008 doi: 10.1088/1757-899X/189/1/012008
[2] Akhmetbayev D S *et al* 2017 *Results in Physics* **7** 1644–1649 doi: 10.1016/j.rinp.2017.03.010
[3] Sarsikeev Y et al 2014 *Advanced Materials Research* **953–954** 529–532 doi**:** 10.4028/www.scientific.net/AMR.953-954.529
[4] Sokol R et al 2011 *Proceedings of the 6th International Scientific Symposium on Electrical Power Engineering (ELEKTROENERGETIKA)* 380–383
[5] Goldshtein A E, Vavilova G V, Mazikov S V *J. Phys.: Conf. Ser.* **671** 012062 doi: 10.1088/1742-6596/671/1/012062
[6] Plotnikova I V, Efremova O N, Tchaikovskaya O N *J. Ponte* **73** 158–162 doi: 10.21506/j.ponte.2017.4.45
[7] Korneva O Yu, Chicherina N V *J. Ponte* **73** 355–360 doi: 10.21506/j.ponte.2017.8.58
[8] Nizhegorodov A I, Gavrilin A N, Moyzes B B 2016 *Key Engineering Materials* **685** 320–324 doi: 10.4028/www.scientific.net/KEM.685.320
[9] Safonov A, Szilevich M 2010 *Hydraulic pneumatic systems of mobile and technological machines* (Minsk: BNTU)
[10] Luchnikov P A *et al* 2017 *IOP Conf. Ser.: Mater. Sci. Eng.* **189** 012027 doi**:** 10.1088/1757-899X/189/1/012027
[11] Vernhes L *et al* 2013 *Materials Chemistry and Physics* **140** 522–528 doi: 10.1016/j.matchemphys.2013.03.065
[12] Vasendina E A, Plotnikova I A, Levitskaya A A, Kvesko S B 2016 *IOP Conf. Ser.: Mater. Sci. Eng* **110** 012070 doi: 10.1088/1757-899X/110/1/012070
[13] Gavrilin A N, Moyzes B B, Cherkasov A I, Mel'nov K V, Zhang X 2016 *MATEC Web of Conferences* **79** 01078 doi: 10.1051/matecconf/20167901078
[14] Istomina N, Sosnovskaya N, Kovaluk E 2010 *Equipment of electrochemical productions: tutorial* (Angarsk: ASTA)
[15] Goldshtein A E, Vavilova G V, Mazikov S V 2016 *MATEC Web of Conferences* **79** 01009 doi: 10.1051/matecconf/20167901009
[16] Novikov A., Darintseva A. 2006 *Flexible automated lines* (Ekaterinburg: USTU)
[17] DeGarmo Paule E., Black J, Kohser Ronald A 2003 *Materials and Processes in Manufacturing* (John Wiley & Sons, Inc)

IOP Conf. Series: Materials Science and Engineering **289** (2017) 012021 doi:10.1088/1757-899X/289/1/012021

Research of aluminum alloys with using eddy-current transducers on the basis of cores of various form

S F Dmitriev[1], A V Ishkov[2], A O Katasonov[3], V N Malikov[4], A M Sagalakov[2]
[1]Assistant professor, Altai State University, Barnaul, Russia
[2]Professor, Altai State Agricultural University, Barnaul, Russia
[3]Student, Altai State University, Barnaul, Russia
[4]Graduate student, Altai State University, Barnaul, Russia
E-mail:osys11@gmail.com

Abstract. The research aims to develop a microminiature eddy current transducer for aluminum alloys. The research topic is considered relevant due to the need for evaluation and forecasting of safe operating life of aluminum. A microminiature transformer-type transducer was designed, which enables to perform local investigations of unferromagnetic materials using eddy-current method based on local studies conductivity. Having the designed transducer as a basis, a hardware-software complex was built to perform experimental studies of aluminium. Cores with different shapes were used in this work. Test results are reported for a flaws in the form of hidden slits and apertures inside the slabs is derived for excitation coil frequencies of 300–700 Hz.

1. Introduction

Monitoring the quality of conducting materials occupies an important place in modern applied physics and industrial production. In the industrially developed countries, the total expenditure on monitoring the quality of work pieces is up to 3% of the production output. In areas of industry with elevated safety and reliability requirements (military, nuclear, aerospace), the cost of quality control can reach 20% [1].

One of the most widespread methods of control of conductive materials is non-destructive eddy current testing. One of the important advantages of the eddy current testing is a possibility to carry out researches without the direct contact between sensor and object that allows achieving of reliable results under high speed of scanning. However, at the same time, removal of noises affects, influencing significantly the value of desired signal is an important task. Limitation of eddy current testing can include small depth of scanning and also difficulties in determining of the burial depth of defect. Another small parameter restricting the method seriously is the area of the control zone. Maximum burial depth of defects and also minimal size of control zone are significantly limited by the constructive features of the transformers and insufficient hardware and program processing of the signal of eddy-current transformer. It's often difficult to access the geometrical parameters of separate defect in the depth of metal. In this connection, the topical task becomes the development of microminiature eddy current measurement systems for the local measurements of electroconductivity in nonuniform materials, search of defects in the alloys.

Construction produced of the aluminum-based alloys became widespread in modern machinery. Use of these alloys in such spheres as aircraft construction, automobile construction, space engineering, imposes high requirements on the structure and mechanical properties.

This task is the subject of a number of papers [2–6]. Flaw detection in cracks between two aluminum plates with a model defect in the center is discussed in [2]. The diameter of the measurement coil in this design for a eddy-current transducer was 7 mm. Scans were made at frequencies of 1 and 5 kHz, and the characteristic penetration depth of the eddy currents into the test plates at these frequencies were 3.82 and 1.71 mm. Experimental methods based on two eddy-current transducers operating in a differential mode are discussed in [3]. These kinds of circuits can greatly reduce the level of parasitic noise produced during continuous scanning in real time.

The tendency is to miniaturize the eddy-current transducer because of the need for local measurements. Recently eddy-current transducers have been developed with dimensions of 5×5 mm and a wire diameter of 0.15 mm in the working region [4]. These devices, however, do not ensure the required penetration depth and localization of the magnetic field required for local measurements in inhomogeneous media. This problem is solved using magnetic field concentrators made of ferrite with a high magnetic permeability. Ferrite concentrators significantly reduce the dispersion of the eddy currents [5, 6].

The localization of the mentoring by eddy-current transducers is improved when conical cores are used [7]. This shape makes it possible to reduce the measurement error and to detect flaws with small linear dimensions.

The worked out conception of the virtualized measuring device (transformer) enables a researcher to make a great number of measurements in different spheres with the help of one sensor only. To these spheres belong fault detection of conducting materials and stratified compos, thickness measuring, profilometry, tension measuring of permanent and variable magnetic fields and conductivity measuring of unferromagnetic materials and some others.

Producing such devices demands making a special technological line, due to that the final price of the device will increase significantly. To make the price lower we came to the conclusion that we should replace the expensive hardware blocks with the software for personal computers (PC). As a result, out device consists of eddy-current transducer (ECT) only, which is connected to the sound card of the PC, managed with the special software.

Software allows carrying out signal injection to the exciting winding of the transformer and also provide registration of the voltage on the measuring winding. Values of this voltage are later converted into values of electroconductivity.

2. Materials and methods

Microminiature eddy current transducer is developed for the control of parameters of items of aluminium alloys on small sites [8, 9]. Definable parameter is electroconductivity of the material and its distribution on the surface and depth of the control object. ECT is connected to the sound board of a personal computer with a special C++ program in the Windows system. The program controls the voltage delivered to the exciter winding of the eddy-current transducer and reads the voltage (in arbitrary units) from the measurement winding. It then uses a prior calibration to convert the received data into values of the electrical conductivity.

In general the scheme of the sensor is represented in Figure 1. In this scheme there are actuating (E) and measuring (M) windings, placed in the ECT.

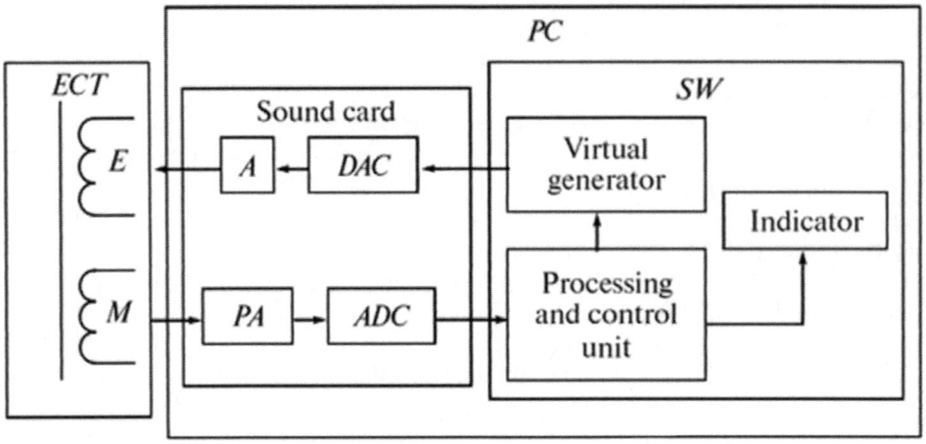

Figure 1. Scheme of the measuring device.

Specially formed signal of the virtual generator is transformed to the digital-to-analog converter. Digital-to-analog converter transforms the signal from digital to analog and transfers it through special intensifier to the exciting winding of eddy current transducer. This signal induces electromagnetic field interacting with control object. Arising field of eddy currents creates the electromotive force in the measuring winding of eddy current test. Electromotive force from the measuring winding of eddy current test allows receiving of data about the structural inhomogeneities of the control object in the active zone of eddy current transducer. This electromotive force appears and is intensified in special microphone amplifier. After filtration and selective amplification, signal is transferred into analog-digital converter, and moves to the input of sound card. The signal transforms into digital and is transferred into the unit of software processing and management. This unit allows processing of the signal and its displaying on the screen of personal computer.

Cores with different shapes were used in this work. Figure 2 shows a trapezoidal core VTP-1 (Figure 2*a*) and a core in the form of a sharpened cone, VTP-2 (Figure 2*b*).

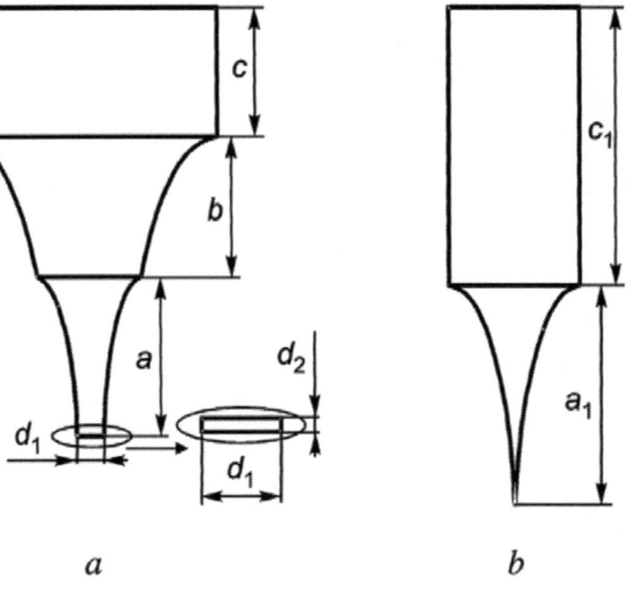

Figure 2. Scheme of the measuring device.

The dimensions of the cores are: a = 2.2 mm, b = 1.8 mm, c = 1.6 mm, d_1 = 0.15 mm, d_2 = 0.03 mm, a_1 = 3.3 mm, and c_1 = 3.6 mm. The cores were made of NMZ 2000 ferrite with a magnetic permeability of 500. They were used to construct eddy-current transducers with exciter, measurement, and compensation windings. The compensation winding attached to a measurement winding is used for subtracting the signal of the exciter winding from the resulting signal. The exciter and measurement windings have 200 turns, and the compensation winding, 160–180 turns. 5-μm-thick copper wire is used for the windings. The diameters of the measurement and exciter coils of VTP-1 are 0.15 and 0.3 mm, respectively, and the core has a trapezoidal shape. The corresponding dimensions of the measurement coil for VTP-2 are 0.03–0.15 mm; the diameter of the exciter winding is 0.3 mm, and the core is in the shape of a sharpened cone.

Developed eddy current transducer (Figure 3) contains measuring (1), exciting (2) and compensating (3) windings. The windings are winded on the core (4), placed inside the special platform (5). To protect the windings, platform is impregnated with compound (6). For localization of magnetic field, ferrite screen is imposed on the construction (7). For the protection of the converter from physical impact, it's placed into the corundum washer (8).

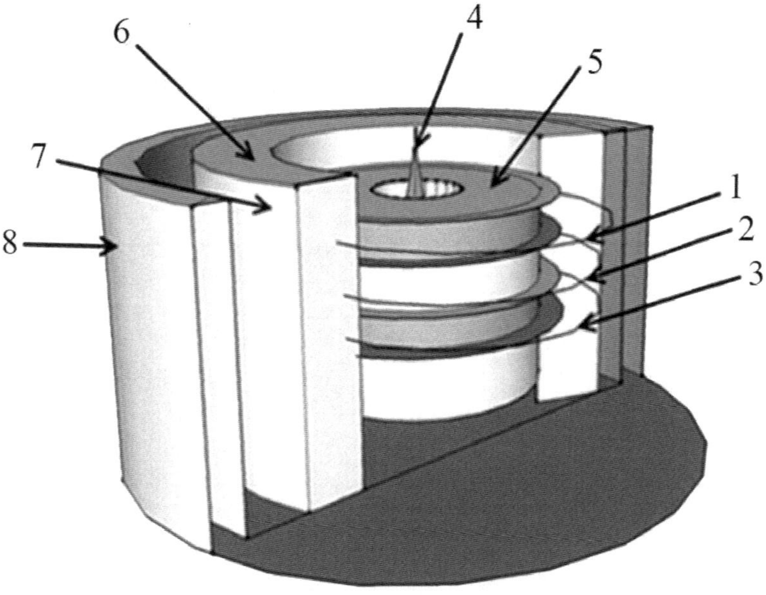

Figure 3. Scheme of the eddy current transducers.

Characteristics of the developed eddy current tests provide localization of magnetic field on the sites of about 2500 μm^2 and provide significant depth of its penetration into the studied object during the work on rather low frequencies.

3. Experimental results

This measurement system was tested using samples with freely accessible model defects. Each sample was scanned by several of the eddy-current transducers with different core designs. The depth of the flaws was estimated indirectly.

Plates of aluminum-magnesium alloy (94% Al, 3% Mg) with thicknesses of 5.5 mm were used as samples. The first plate contained six faults in the form of 0.25-mm-thick slits at depths of 1.0, 2.0, 3.0, 4.0, 5.0, and 5.3 mm. The results of the scans are shown in Figure 4, where U is the voltage induced in the measurement winding of the sensor, and l is the position of the sensor relative to the tip of the object.

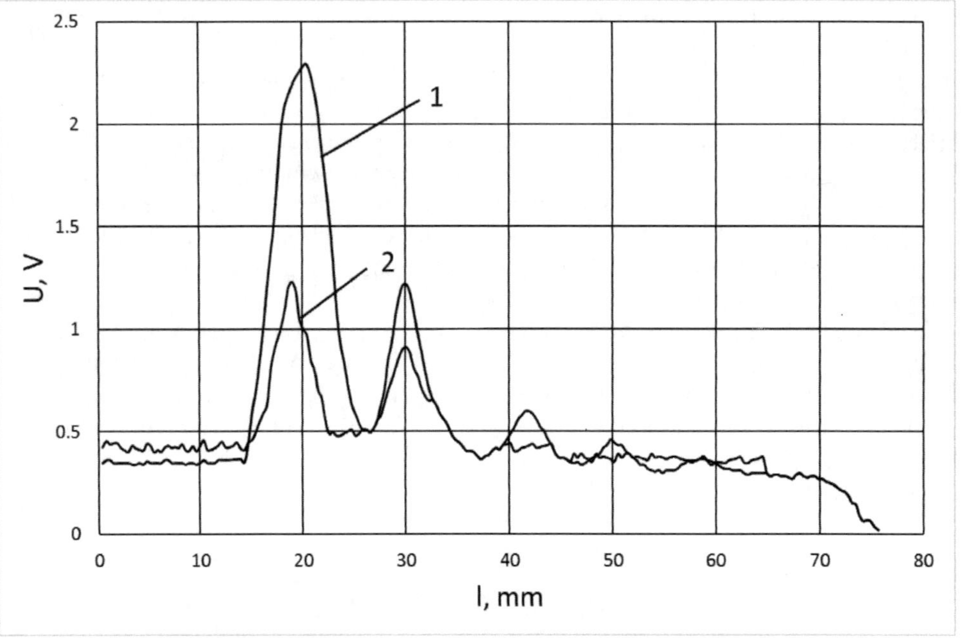

Figure 4. Results of scanning plates No. 1: 1) VTP-1; 2) VTP-2.

Flaw detection with the transducer VTP-1 revealed five of the six flaws. Scans with VTP-2 revealed only two of the flaws.

The second sample was in the form of a plate with six flaws at the same depth. The results of the scans are shown in Figure 5.

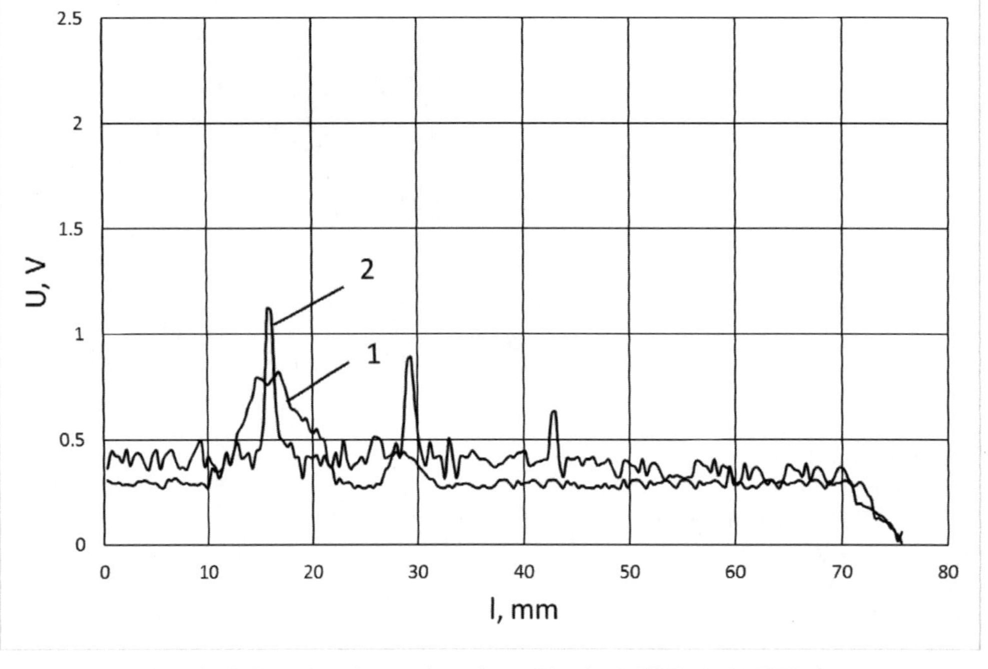

Figure 5. Results of scanning plates No. 2: 1) VTP-1; 2) VTP-2.

Flaw detection at a frequency of 500 Hz with the transducer VTP-2 revealed three of the flaws. Scanning with VTP-1 identified 2 flaws, but it was difficult to localize the position of the flaws with these transducer cores.

4. Conclusion

A measurement system has been developed for studying flaws in plates of aluminum-magnesium alloys with the aid of subminiature eddy-current transducers. The present study demonstrates the capabilities of the eddy-current monitoring technique in searching for small, deep-lying defects in aluminum alloy plates. Subminiature eddy-current transducers based on cores with a trapezoidal shape can efficiently detect long subsurface cracks with widths of 250 μm. For finding flaws in the form of hollows, it is better to use eddy-current transducers with cores in the shape of sharpened cones.

Use of microminiature eddy current transducer with ferrite cores of special shape and modified filters allowed achieving of significantly bigger localization of magnetic field and increasing of the depth of its penetration into the control object.

References

[1] Semenov V, Ryabtsev A, Mudrov A 2003 University *Vestnik Tekhnics State University* **278** 48–54 (in Russian)
[2] Prance R, Clark T, Prance H 2000 *Sensors and Actuators A: Physical* **85** 361–364 doi:10.1016/S0924-4247(00)00375-7
[3] Prance R, Clark T, Prance H 2003 *Review of Scientific Instruments* **74** 3735–3739 doi:10.1063/1.1590745
[4] Kurilyak D 1988 Overhead eddy-current transducers *Patent USSR* 4054140/25-28
[5] Barbato L, Poulakis N, Tamburrino A 2015 *IEEE Transactions on Magnetics* **51** 7047881 doi: 10.1109/TMAG.2015.2406765
[6] Rocha J, Ramos H, Ribeiro L et al. 2015 *Sensors & Actuators: A. Physical* **228** 55–61 doi:10.1016/j.sna.2015.02.004
[7] Dmitriev S, Malikov V, Sagalakov A et al. 2015 *Journal of Physics: Conference Series* **643** 012058 doi: 10.1088/1742-6596/643/1/012058
[8] Dmitriev S, Malikov V, Sagalakov A et al. 2017 *Welding International* **31** 608–611 doi: 10.1080/09507116.2017.1295563
[9] Dmitriev S, Malikov V, Sagalakov A et al. *AIP Conference Proceedings* **1785** 030014 doi: 10.1063/1.4967035

IOP Conf. Series: Materials Science and Engineering **289** (2017) 012022 doi:10.1088/1757-899X/289/1/012022

Modeling of a bimetallic MEMS-based infrared detector

E Ph Pevtsov[1], S V Breev[2], T A Demenkova[1]
[1]Associate professor, Moscow Technological University (MIREA), Moscow, Russia
[2]Engineer, Moscow Technological University (MIREA), Moscow, Russia
E-mail: pevtsov@mirea.ru

Abstract. In work techniques of calculations of the key parameters of the bimetal capacitive IR detectors matrix are presented. Sensitivity characteristics of the bimetal detector were calculated and prototype model of detector in the form of bimetallic console was produced. The possibility of use of capacitive MEMS-receivers on basis of bimetallic effect as an alternative to pyroelectric and microbolometer IR detectors is shown.

1.Introduction

Currently, the market for relatively inexpensive infrared cameras is dominated by devices based on microbolometer matrices. In our opinion, the improvement of characteristics and the expansion of their functionality is based on the use of new materials in these devices. An example is the work on creating matrices with ferroelectric detectors [1, 2], as well as structures in which the matrix pixels are made of bimetallic materials. A principle of operation of MEMS bimetallic infrared detectors is based on change of bimetallic plate capacitance when the plate is heated with radiation. Theoretical evaluation of the detector parameters and results of the prototype models experimental studies [3, 4] showed that this kind of detector demonstrates low noise, high sensitivity, and therefore lower for detectors noise equivalent delta temperature (NEDT ~ 5…10 mK) in the wavelengths range 8…14 μm. Compatibility with standard silicon technology defines relatively low cost of such detectors production for a monolithic array configuration while simultaneously achieving high reliability characteristic for all MEMS. This work is dedicated to the development of methods for such device modeling using specialized CAD software.

2. Design of Bimetallic IR Detector

Change of z coordinates of the bimetallic plate center in the course of heating (temperature sensitivity) is calculated by Equation [1]:

$$\frac{dz}{dT} = \frac{3L^2}{8t_1}(\alpha_1 - \alpha_2)\frac{8(1+x)}{4 + 6x + 4x^2 + nx^3 + \dfrac{1}{nx}}, \quad (1)$$

where α_1 and α_2 – thermal expansion coefficients of the respective layers; t_1 and t_2 – layers thickness, L – length of bimetallic console, $x = t_2/t_1$ – ratio of the bimetal layers thicknesses, $n = E_2 / E_1$ – ratio of Young's modules for the correspondent materials.

When developing detector design, the detector pixel parameters were defined taking into account the following considerations:

- difference between thermal expansion coefficients of the bimetallic layer materials should be as large as possible;

– the material thicknesses should meet requirements of standard silicon technology for production of integrated circuits;
– ratio of the bimetal material thicknesses should correspond to the maximum sensitivity;
– the material of the thermal insulating elements should have low thermal conductivity;
– length of the bimetal element should be maximized within the selected topology design;
– high absorption should be provided in the range of far infrared atmospheric transmission band (8…14 μm);
– pixel design should provide high mechanical strength;
– films of the materials used should have a low internal stresses and good adhesion to each other;
– materials used should be fully applicable in standard technological processes of integrated circuits production.

Based on these criteria, in this work, a bimetal pair Al/Si_3N_4 was selected as a sensitive element of the detector. Other parameters of the bimetallic pixel: the area of the pixel – 50x50 μm^2; the length of the bimetallic element – 52 μm; the length of the heat-insulating element (supporting console) – 47 μm; the thickness of the Al layer – 0.5 μm; the thickness of the SiO_2 layer varied within 0.2...0.5 μm; the height of the initial gap of the sensing capacitor – 0.5 μm.

3. Results of device modeling

CAD software [5], which is a versatile tool for analysis and design of MEMS devices, was used for modeling of bimetallic IR detector. The modeling purpose was to study mechanical behavior of the device exposed to external flux of radiation, and also study of the temperature distribution in the model volume and corresponding change of electrical capacitance of the sensing capacitor based on this device, for different geometrical characteristics.

When the 3D model was synthesized, CAD used both standard procedures from the built-in libraries of technological operations of microsystem technology and prospective procedures for the formation of sacrificial layers from organic materials [6, 7].To improve expression (1) and define the detector characteristics, temperature distribution in the detector volume and temperature distribution change in time in the course of heating with radiation were calculated of this CAD, for given boundary conditions, using a mesh method for thermal problems. The results are illustrated in Figure 1 and Figure 2, which show the results of calculating the transition function of the mean temperature T_{av} of a bimetallic element when a constant heat flux of 1 W/m^2 is switched on and the results of simulation of the temperature distribution of the receiver in a stationary mode.

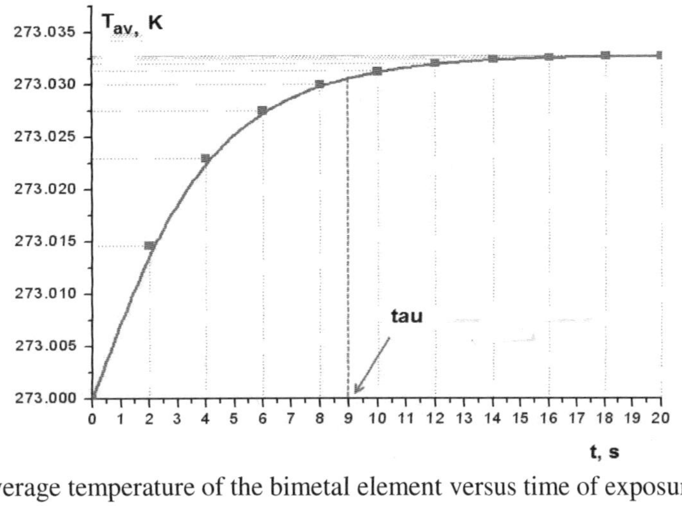

Figure 1. Average temperature of the bimetal element versus time of exposure to heat flux.

The characteristic heating time (thermal constant, tau), determined from the simulation results, is 8 ms for this design.

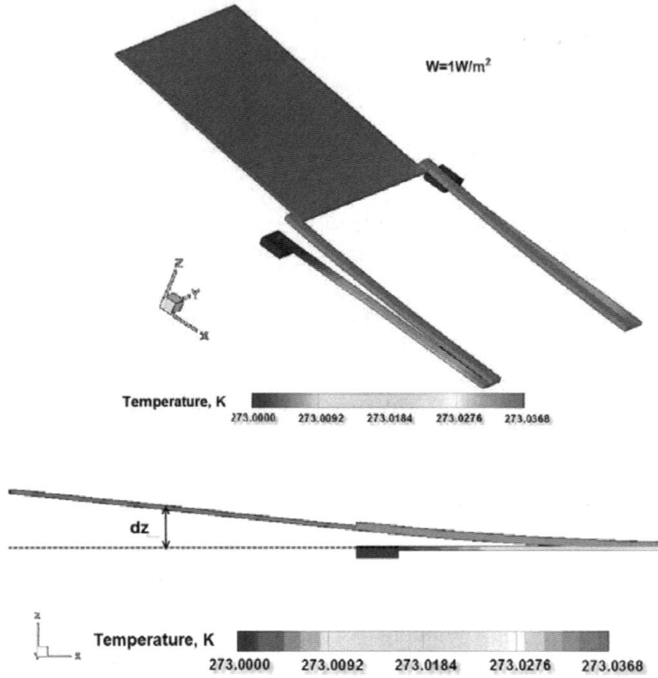

Figure 2. Distribution of the temperature in a bimetallic receiver when heated by radiation in a stationary mode.

The detector sensitivity to radiation was calculated in the next stages of modeling:

$$S_C = \frac{1}{C_0}\left(\frac{dC}{dT}\right) \approx \frac{1}{z_0}\left(\frac{dz}{dT}\right) \qquad (2)$$

where z_0 – initial value of the distance between the plates of the sensing capacitor (in this model, $z_0 = 0.5$ µm); C_0 – corresponding value of the bimetallic plate electrical capacitance relative to the substrate (see Figure 3). It is shown that the relative change of applicate of the detector pixel center is 0.072 µm/W, and the radiation converter sensitivity S_C for this design variant is 14.4 %.

In the modeling, analysis of the device frequency characteristics was performed as well. Frequency parameters of the detector being a mechanical oscillator are very important for selection of the sensing voltage frequency and evaluation of thermomechanical noise. A model used the boundary condition in the form of variable electrostatic forces between the plates of the sensing capacitor that occurs when vibrations due to sinusoidal electric voltage with amplitude of 5 V are excited was used for calculation of the resonant frequency. Modeling results are given in Figure 3.

The resonant frequency of this device is about 23 kHz. Subsequent evaluations showed that at frequency of the sensing voltage more than 170 kHz, and in presence in the sensing circuit of additional RC damping filter [4], thermodynamic mechanical noise has virtually no effect on the output characteristics of the device.

The analysis of the limiting parameters of sensitivity was carried out according to the classical technique [8, 9], in which, with known characteristics of the optical path for a given transfer function of the device, a scene temperature equivalent to the noise source noise density (NETD) is determined. The spectral densities of the noise components due to thermodynamic fluctuations during heat transfer, thermomechanical oscillations, noise of readout circuit were calculated for the given characteristics of the system (optical force F:1, surface absorption coefficient 0.9, spectral range 8...14 µm, operating voltage 5 V, etc.).

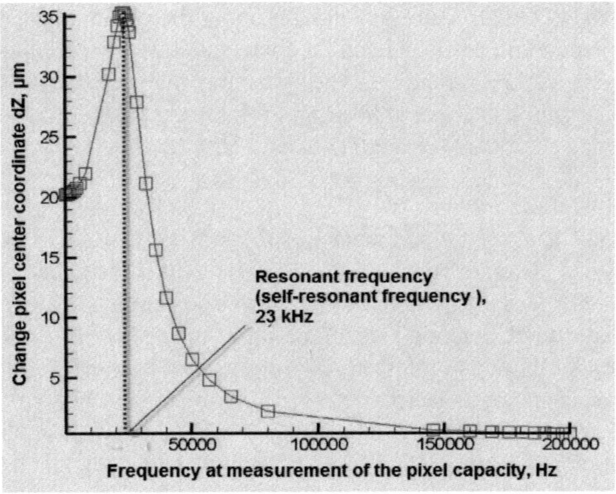

Figure 3. Frequency characteristic of the bimetal detector.

The results of calculations of the components of NETD, depending on the ratio of the thicknesses t_1 and t_2 of the bimetallic pixel layers ($m = t_1 / t_2$) are summarized in Figure 4.

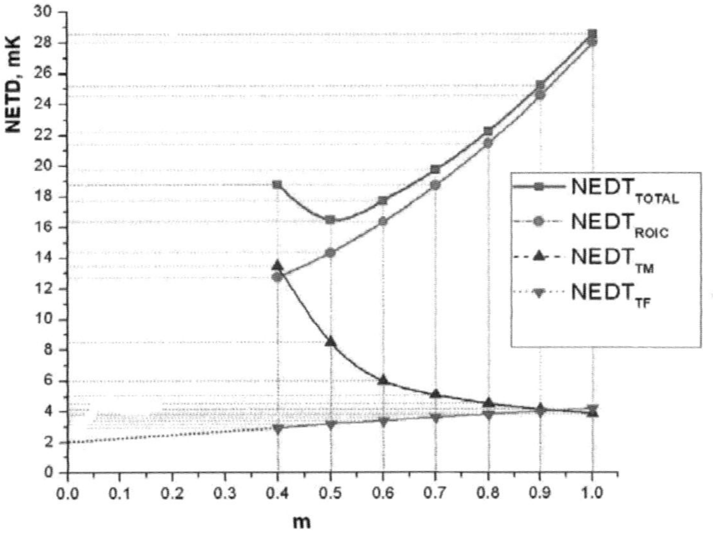

Figure 4. Dependences of the NEDT components of a bimetallic receiver on the design parameters: $NEDT_{TF}$ – the contribution of thermal fluctuation noise; $NEDT_{TM}$ – the contribution of mechanical noise; $NEDT_{ROIC}$ – the contribution of the noise of the electrical circuit of reading; $NEDT_{TOTAL}$ – total value of temperature difference equivalent to noise.

4. Prototype model of bimetal infrared detector pixel

The model of the prototype MEMS bimetallic console IR detector was made by ion etching technology using Quanta 200 3D equipment. The equipment combines a scanning electron microscope and a thermionic cathode intended for formation of a focused ion beam that enables applying and removing of the materials, and it also contains system for energy dispersive microanalysis and analysis of crystalline materials texture.

To obtain the cantilever structure, substrate of low resistance silicon coated with a layer of silicon dioxide with a thickness of about 1 micron was applied. A layer of platinum in the form of a pad with dimensions of 20×20 μm^2 and outgoing stripes with width of 1 μm and length of 20 μm was deposited on oxide using Quanta 200 3D equipment. Ion etching under the platinum layer was carried out at the angle of 52° that was preconditioned by technical characteristics of the equipment. As a result, the structure shown in Figure 4 was produced. The platinum layer and substrate form a bimetallic capacitor with capacitance which changes at heating. Preliminary results of the temperature coefficient measurements for capacitance of such bimetallic detector structure prove the above modeling results.

5. Conclusion

Technique for calculations of the basic parameters of the bimetal capacitive IR detectors matrix pixel was developed. Sensitivity characteristics of the bimetal Al/SiO_2 detector with the pixel size of 50x50 μm^2 were calculated. It was shown that the thermal constant is 7.3 ms for this design variant of detector matrix pixel that allows obtaining of high-quality image with a standard frame rate. In this case, calculated value of the bimetallic detector sensitivity is 14 % that is much larger than similar characteristics of bolometric and pyroelectric detectors for which the sensitivity is less than 2 %/K.

The bimetallic IR detectors elements producibility was studied with application of electron and focused ion beam equipment Quanta 3D. A prototype model of detector in the form of bimetallic console with length of 20 μm and width of 1 μm, consisting of Pt (1 nm) and SiO_2 (1 μm) layers, was produced.

The results of the characteristics analysis for the model of bimetallic IR detector matrix pixel show the prospect of development of such type of thermal detectors. Simplicity of the MEMS bimetallic IR detector design and also application of materials which are fully compatible with standard silicon manufacturing techniques for integrated circuits production should provide lower cost of such devices in comparison with pyroelectric and microbolometer IR detectors. Higher sensitivity and availability of manufacturing processes for capacitive MEMS sensors based on bimetallic effect enables their use as an alternative for pyroelectric and microbolometer IR detectors.

Acknowledgement

This work was supported by the Ministry of Education and Science of Russian Federation (project 2017/8.5098).

References

[1] Sigov A S et al 1999 *Ferroelectrics* **226.** 183–190 doi: 10.1080/00150199908230299
[2] Petrovsky V I et al 1994 *Journal: MRS Online Proceedings Library Archive* **361** 135 doi:10.1557/PROC-361-135
[3] Steffanson M, Rangelow I W 2014 *Opto-Electronics Revue* **22** 1–15 doi: 10.2478/s11772-014-0176-0
[4] Amantea R, Lawrence A 1998 *Infrared Technology and Applications XXIV* **3436** 647–659 doi: 10.1117/12.328065
[5] Pevtsov E Ph, Demenkova T A, Shnyakin A A, Kravchenko V I 2013 *Humanities&Science University Journal* 164-174 (in Russian)
[6] Luchnikov P A et al 2017 *IOP Conf. Ser.: Mater. Sci. Eng.* **168** 012092 doi: 10.1088/1757-899X/168/1/012092
[7] Luchnikov P A et al 2017 *IOP Conf. Ser.: Mater. Sci. Eng.* **189** 012014 doi: 10.1088/1757-899X/189/1/012014
[8] Rogalski A 2003 *Progress in Quantum Electronics* **27** 59–210 doi: 10.1016/S0079-6727(02)00024-1
[9] Pevtsov E P et al 2017 *IOP Conf. Ser.: Mater. Sci. Eng.* **168** 012095 doi: 10.1088/1757-899X/168/1/012095

Modeling of liquid flow in surface discontinuities

I S Lobanova[1], V A Meshcheryakov[2] and A N Kalinichenko[3]
[1]Teacher, National Research Tomsk Polytechnic University, Tomsk, Russia
[2]Associate professor National Research Tomsk State University, Tomsk, Russia
[3]Associate professor, National Research Tomsk Polytechnic University, Tomsk, Russia
E-mail: konarevai007@tpu.ru

Abstract. Polymer composite and metallic materials have found wide application in various industries such as aviation, rocket, car manufacturing, ship manufacturing, etc. Many design elements need permanent quality control. Ensuring high quality and reliability of products is impossible without effective nondestructive testing methods. One of these methods is penetrant testing using penetrating substances based on liquid penetration into defect cavities. In this paper, we propose a model of liquid flow to determine the rates of filling the defect cavities with various materials and, based on this, to choose optimal control modes.

1. Theory

High quality and reliability of products made of metal and non-metallic materials cannot be provided without effective modern methods of nondestructive testing used at all stages of the production cycle: design, manufacture, operation and repair. The defects in structures made of different materials can be detected by means of acoustic, radiation (X-ray), thermal, radio wave, optical, electrical and other methods of nondestructive testing, including liquid penetrant test [1, 2].

Liquid penetrant testing techniques are based on the ability of liquids to penetrate into the cavities of defects. These techniques can be employed to detect all types of surface closed-end and open-end discontinuities, such as cracks, delaminations and leaks in products of any non-porous material, including ferrous and non-ferrous metals, glass, ceramics, plastics and other materials. The hydrodynamics of the liquid flow in capillaries obeys the equations for a Newtonian liquid with a reasonable degree of accuracy. A colored or fluorescent liquid used in penetrant testing to ensure the quality of products is referred to as penetrants. These liquids are typically based on various alcohols and kerosine [3–5].

When liquids fill open-end capillaries, capillary absorption occurs, i.e. wetting forces rise the liquid to a certain height h above the surface. The filling of the closed-end capillary with liquid differs from that of the open-end capillary since the entrapped air-vapor mixture limits the penetration depth. Thus, the process involves two stages that are fundamentally different in physical nature. At the beginning, when the liquid gets into contact with the surface discontinuity (microscopic rupture of the surface), it penetrates rapidly deep into the defect to a certain depth h under the impact of capillary forces. However, the process then slows down substantially as the capillary pressure increases. The first stage ends when the capillary pressure of the liquid approaches the pressure of the air-vapor mixture. Then the second stage begins. The capillary stage of filling changes into the diffusion stage, when the gas compressed in the defect cavity gradually dissolves in the liquid and diffuses towards the defect mouth [6,7].

A problem of practical interest is to reveal the patterns of the filling of closed-end and open-end capillaries by various liquids and to determine the limiting depth of capillary filling and the time of liquid penetration to this depth.

This paper provides the result of the modeling of liquid flow in open-end and closed-end capillaries with flat and parallel walls. A differential equation of liquid flow in the capillary is obtained with respect to its slope, surface roughness, regions with different wetting parameters, an analytical change in the size of the gap in different regions under the assumption of the liquid flow close to a laminar one:

$$h_1''+\frac{1}{h_1}(h_1')^2+\frac{8\mu(r,r_a)}{r^2(h_1)\rho}h_1'-\frac{2\sigma(h_1)a(h_1)cos(\theta_s(h_1))}{r(h_1)\rho h_1}+\frac{p_a}{\rho(h_0-h_1)}+g\,sin(\alpha)=0,$$

where $h_{0,1}$ is the total and instantaneous capillary depth, θs is contact angle, a is roughness coefficient, σ is surface tension, is atmospheric pressure, μ is dynamic viscosity, r_a is the radius of the penetrant macromolecule (if dimensional effects are taken into account), α is the angle of inclination of the capillary, *g* is gravitational acceleration, and ρ is liquid density. This equation takes into account the dependence of the dynamic shear-viscosity coefficient of the penetrant during reduction or expansion of the capillary radius (allows for dimensional effects through the procedure). The equation enables modeling of absorption with allowance for a continual change in all the parameters of the problem.

To simulate liquid flow in closed-end and open-end capillaries, a software package is developed to perform the numerical solution of the differential equation obtained. The interface of the package for computer simulation of the liquid flow is shown in Figure 1.

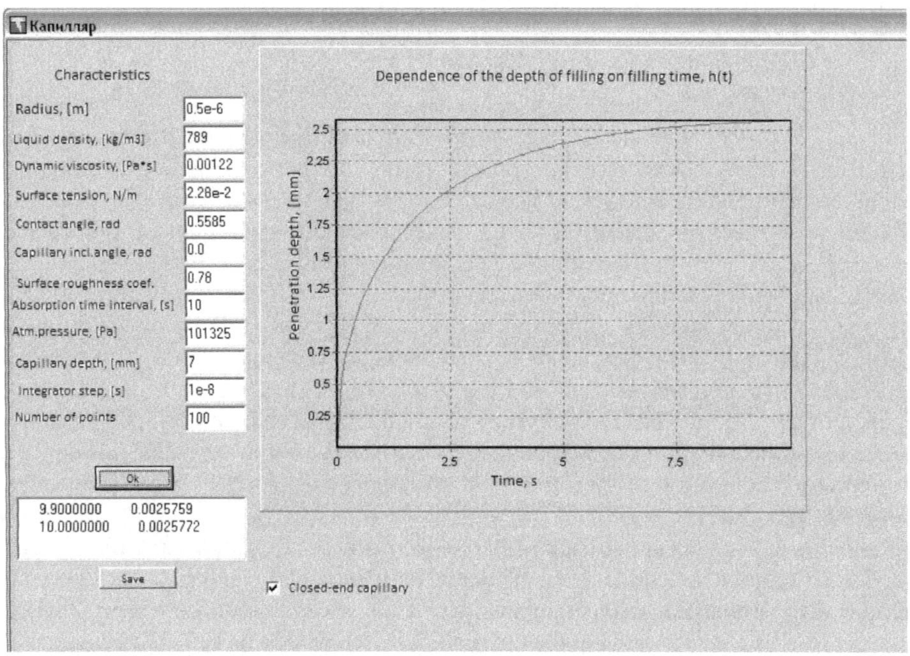

Figure 1. Interface of the package for computer simulation of liquid flows.

The windows to enter the parameters are displayed on the left side. These parameters fall into two groups. The first group is related to the geometric and material parameters of the model. The second group includes the calculation parameters and graphical interpretation of the results.

The first group involves the following parameters:

- liquid density (kg/m^3);
- dynamic viscosity (Pa s);
- surface tension(N/m);
- contact angle (rad);
- capillary inclination angle relative to the horizontal (rad);
- surface roughness coefficient (dimensionless value determined by the ratio of rough and smooth surface areas);
- absorption time interval (s);
- atmospheric pressure (Pa);
- capillary depth (mm). It should be noted that this parameter comes into operation if the mode "Closed-end capillary" is chosen. Otherwise, the program will perform the calculations for an open-end capillary.

The second group includes the following parameters:

- integrator step (s). Despite the fact that the method automatically selects the integration step, it is desirable to set its initial value at the start. This is due to the fact that the differential equation refers to the category of stiff equations and has a singularity for the point h=0;
- number of points on the graph. This parameter is required to reduce the modeling results on a common graph during variation of the problem parameters. Otherwise, automatic selection of the integration step generates the resulting files of different lengths.

An output field for intermediate results can be found in the lower left corner. This field is a simplified text editor with the ability to view all intermediate values and operate them if required. This editor comprises two columns – the current time value and the sorption depth.

2. Model experiments:

To construct a mathematical model of the liquid flow in open-end and closed-end slot-like capillaries. The capillary dimensions of the chosen samples made of glass and aluminum were as follows: widths of 10, 7, 5 and 3 µm, and depths of 10.0, 7.0, 5.0, 3.0, 1.0 and 0.3 mm. Kerosine and alcohol described in [3–5] as the main common components, as shown in Table 1, of flaw detection materials were used as penetrants.

The contact angles measured by the droplet detachment method (A.O. Osipov, O.P. Osipov, I.I. Gogonin) are presented in [8].

Table 1. Summarizes the values of the basic equation parameters.

Liquid	Dynamic viscosity coefficient, Pa×s, 10^{-3}	Surface tension coefficient, N/m, 10^{-3}	Contact angle, deg		Liquid density kg/m^3	Surface roughness	
			Aluminium	Glass		Aluminium	Glass
Kerosine	1.85	24.00	37	25	820	0.78	1.0
Alcohol	1.22	22.80	32	17	789		

The observation time for all the experiments was chosen to be equal. During that time, the liquid filled the capillary to a certain depth.

The results of modeling the flow of the liquids (alcohol and kerosine) in the slot-like capillaries were interpreted graphically. Figure 1 shows that the rate of filling the open-end capillaries with liquids is inversely proportional to their opening width, that is the narrower the capillary, the longer it takes for the liquid to travel to the same depth. Figures 2 and 3 illustrate the dependencies of the liquid penetration depth on the filling time for an aluminum-alcohol pair with a capillary width of 5 µm. For all other material-liquid pairs, the graphs appeared to be similar.

When filling closed-end capillaries with liquids, the filling depth is limiting. The graph presented in Figure 2 shows that the limiting depth of filling closed-end capillaries is about 10% of the total capillary depth.

IOP Conf. Series: Materials Science and Engineering **289** (2017) 012023 doi:10.1088/1757-899X/289/1/012023

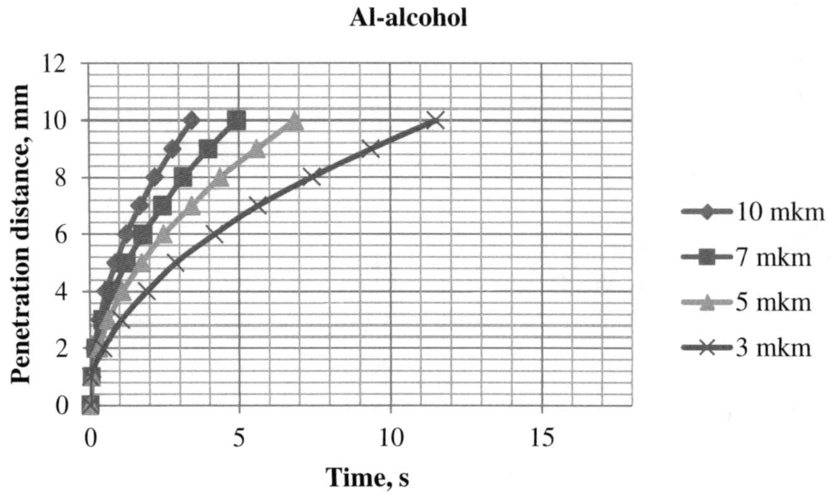

Figure 2. Dependence of the depth of filling a slit-like capillary on filling time (open-end capillary).

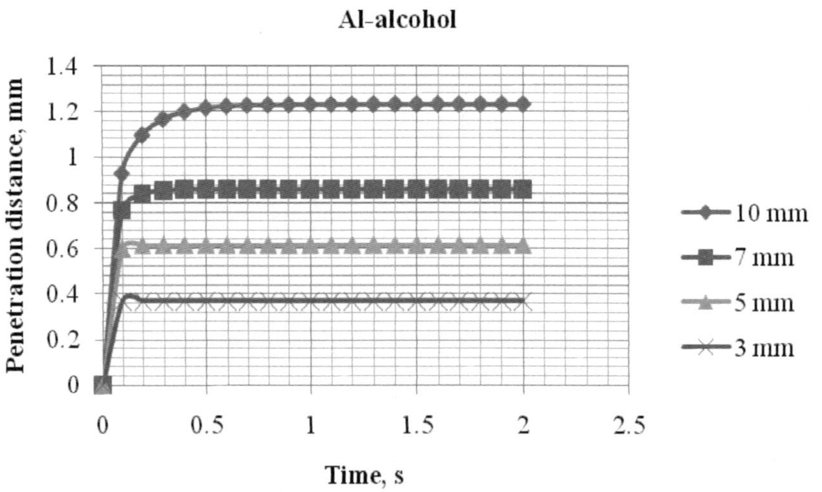

Figure 3. Dependence of the depth of filling a slit-like capillary on filling time (closed-end capillary).

Figures 4 and 5 show the results of modeling the flow of liquids (alcohol and kerosine) in open - and closed-end capillaries with a 5 μm opening made of aluminum and glass on filling time. The speed of flow of alcohol in the slit-like capillary is higher than that of kerosine. The speed of liquid flow in products made of non-metal is higher than that in metal materials.

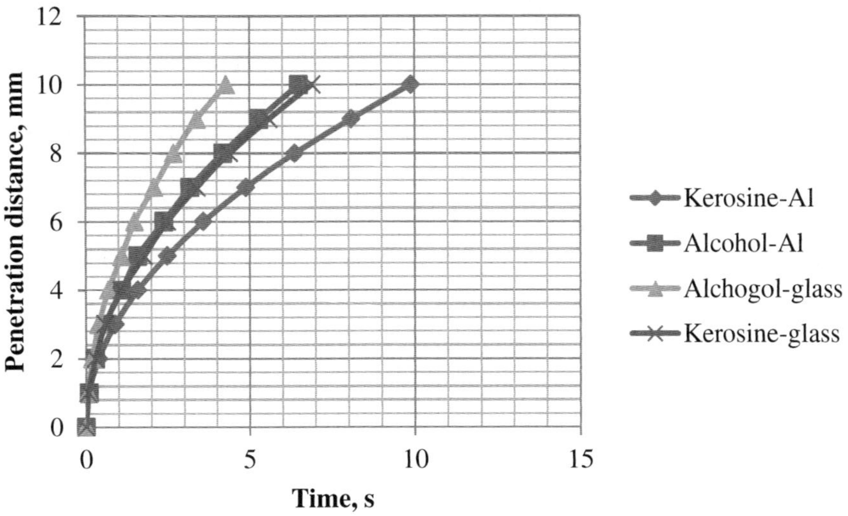

Figure 4. Dependence of the depth of filling a slit-like capillary on filling time (open-end capillary) for material-liquid pairs.

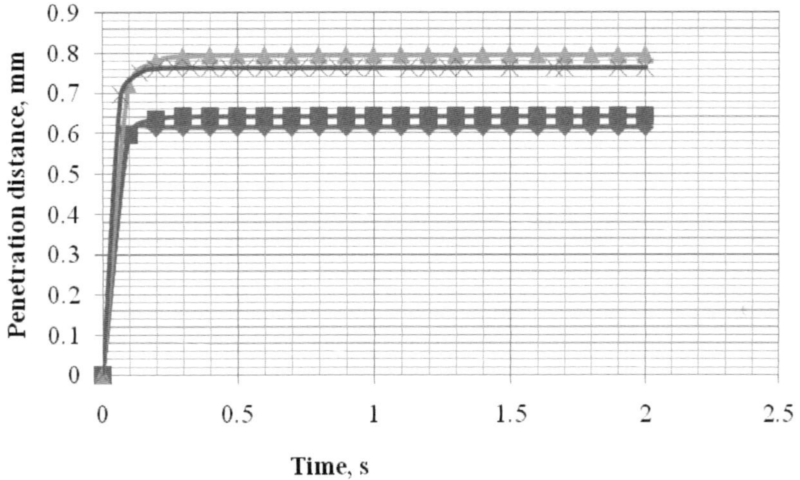

Figure 5. Dependence of the depth of filling a slit-like capillary on filling time (closed-end capillary) for material-liquid pairs.

3. Conclusion

Thus, the graphs show that the rate of filling glass capillaries with liquids is higher than that of filling aluminum capillaries. Numerous model experiments on other types of materials proved the necessity to take into account all the model parameters to construct the quality control procedures for products. However, it should be noted that the time required for sorption of capillaries in a number of polymeric and nonmetallic materials is slightly less than that for metals. The reduced time of capillary sorption accelerates penetrant testing.

References

[1] Kerber M, Vinogradov V, Golovkin G et al. 2008 *Polymer materials. Property. Structure. Technology* (S. Peterburg: Professia)

[2] Nizhegorodov A, Gavrilin A, Moyzes B 2016 *Journal of Vibroengineering* **18** 3734–3742 doi: 10.21595/jve.2016.16994
[3] Morozov G, Kablov E , Sokolova L 2006 Penetrant for penetrant testing (RU 2278372) (in Russian)
[4] Denel A, Sokolova L , Kondrashov E 2004 Penetrant for penetrant testing (RU 2238543) (in Russian)
[5] Sherwin A 1972 Fluorescent penetrant composition and method (US 3735131 A)
[6] NDT resource centre [Electronic resourse] URL: https://www.nde-ed.org/EducationResources/CommunityCollege/PenetrantTest/cc_pt_index.htm
[7] Migun N, Gnusin N 2011 *Heat Exposure at the Capillary Non-destructive Testing* (Minsk: Nauka)
[8] Osipov A , Osipov O 2011 *Bulleten of Moscow state technical university of civil aviation* **173** 134–136 (in Russian)

IOP Conf. Series: Materials Science and Engineering **289** (2017) 012024 doi:10.1088/1757-899X/289/1/012024

The effect of the DSSC photoanode area based on TiO_2/Ag on the conversion efficiency of solar energy into electrical energy

N Ibrayev[1], T Serikov[2], A Zavgorodniy[3] and A Sadykova[4]
[1]Professor, Institute of molecular nanophotonics, Karaganda State University named after E.A. Buketov, Karaganda, Kazakhstan
[2]Senior teacher, Institute of molecular nanophotonics, Karaganda State University named after E.A. Buketov, Karaganda, Kazakhstan
[3]PhD student, Institute of Applied Mathematics, Karaganda, Kazakhstan
[4]Graduate student, Institute of molecular nanophotonics, Karaganda State University named after E.A. Buketov, Karaganda, Kazakhstan
E-mail: niazibrayev@mail.ru

Abstract. A module based on dye-sensitized solar cells with Ag/TiO_2 structure was developed. It is shown that the addition of the core-shell structure to the semiconductor film of titanium dioxide, where the nanoparticle Ag serves as the core, and the TiO_2 is shell, increases the coefficient of solar energy conversion into electrical energy. The effect of the photoanode area on the efficiency of conversion of solar energy into electrical energy is studied. It is shown that the density of the photocurrent decreases with increasing of the photoanode area, which leads to a drop in the efficiency of solar cells.

1. Introduction

Dye-sensitized solar cells (DSSC) are promising photovoltaic devices using inexpensive materials whose production technology is quite cheap [1, 2]. The disadvantage of DSSC cells is their low efficiency of conversion of solar energy into electrical energy compared to existing photoconverters based on single-crystal silicon. One of the promising methods for increasing the efficiency of conversion of light energy into electrical energy in DSSC cells is the use of surface plasmon resonance, which Ag nanoparticles (NP) possess [3]. Due to its high reactivity of silver, direct use of NP Ag in DSSC cells is limited due to the possibility of their direct contact with the electrolyte, which will lead to their oxidation. The use of "core-shell" structures, where the nucleus is a metal nanoparticles with plasmon resonance, the shell – a semiconductor metal oxide (TiO_2) will protect the metal NP from oxidation. In addition, the presence of a semiconductor shell will avoid recombination processes between the metal and the dye [4, 5]. Research in this direction is quite active, therefore, it is expected that a large-scale production of organic solar cells using roll-to-roll technology will soon begin. The characteristic area of the photoanode used in the DSSC is 0.5 cm^2. With this geometry, the efficiency of one cell is about 11% [6]. In this case, to produce a module of a solar cell with a total area of 100 cm^2, dozens of consecutively and in parallel connected such solar cells must be made. Such a technology is not promising, since losses at the contacts between the cells will lead to loss of photocurrent density and lowering of the voltage, and the materials used to connect these cells will lead to an increase in its total cost. Many works are mainly focused on the fabrication of the DSSC module of a cell based on TiO_2 NP [7], and the manufacturing of a plasmon-enhanced solar cell module is not practical. In addition, the effect of the area of the DSSC photoactive layer on the basis

IOP Conf. Series: Materials Science and Engineering **289** (2017) 012024 doi:10.1088/1757-899X/289/1/012024

of the Ag/TiO_2 core-shell structure on its efficiency is unknown. In this connection, this work is devoted to the study of the influence of metal NP and the area of the photoactive layer on the efficiency of the solar cell module.

2. Experiment

A film of TiO_2 NP was prepared as follows: 50 mg of colloidal TiO_2 (Sigma-Aldrich, >21 nm), triturated in a porcelain mortar with the addition of 2 ml of deionized water and 0.2 ml of acetone. To prevent cracking of the resulting films, polyethylene glycol-sebacate was added to the main paste formulation. The polymer was added to the paste together with ethylene glycol in an amount of 25% by weight with respect to TiO_2. Stirring was carried out for 48 hours on a magnetic stirrer to form a homogeneous paste. A colloidal solution of silver NP was synthesized by chemical reduction of metal ions from a silver nitrate salt ($AgNO_3$) in an ethanol solution. Sodium borohydride ($NaBH_4$) was used as the reducing agent. $NaBH_4$ mass 0.001 g was dissolved in 25 ml of ethyl alcohol. A solution of $AgNO_3$ in a ratio of 0.0045 g per 10 ml was dissolved in ethyl alcohol. A solution of $NaBH_4$ cooled to -2 °C was added to the resulting $AgNO_3$ solution with constant stirring in several stages of 50 μl each. The average particle size of silver was 3.6 nm. To synthesize core-shell structures, the following procedure was used: in 10 ml of the solution of silver NP, 1 ml of a solution of titanium tetraisopropoxide (TIPT ($Ti(OCH(CH_3)_2)_4$) was added. To synthesize the TiO_2 coating, ethanol solutions of TIPT and silver NP were thorough mixing. The resulting mixture was stirred for 12 hours on a multifunctional rotator PTR-35. The particle size of the core-shell is 25 nm, the particle size of the titanium dioxide is 22 nm. Ag/TiO_2 NP was added during the preparation of the paste. Their concentration in the films was equal to 1 mass% with respect to the mass of the TiO_2 NP. The choice of this concentration of Ag/TiO_2 NP is associated with the optimal concentration of Ag/TiO_2 NP in the manufacture of plasmon-enhanced solar cells sensitized with dye [4, 8]. To obtain porous films based on plasmon core-shell NP and titanium dioxide, the screen printing method was used using the automatic installation Print Screen ATMA AT-45PAB. The films were deposited on FTO substrates (Sigma Aldrich, 8 Ohm/cm^2). The surface morphology of the films was studied by scanning electron microscopy using the MIRA 3LMU microscope (Tescan, Czech Republic). Geometric dimensions of the deposited films were determined using mesh screens (43 mesh) As a comparison, we obtained films with the area of the photoactive layer: 0.25, 2.25, 4, and 6.25 cm^2. After drying, the porous films were annealed at the following temperatures and time: 325 °C - 5 min; 375 °C - 5 min; 450 °C - 15 minutes; 500 °C for 15 min. Later the samples were immersed in an ethanol solution of ruthenium dye N719 (Di-tetrabutylammoniumcis-bis (isothiocyanato)bis(2.2'-bipyridyl-4.4'-dicarboxylato) ruthenium (II), SigmaAldrich) for 24 hours. Platinum, which was also deposited by screen printing, was used as the counter electrode. Platisol T/SP (Solaronix, Switzerland) was used as a starting material for the production of platinum layer. To form a platinum layer on a conducting electrode, it is necessary to treat it at 500 °C for 1 hour. A 60 μm thick polymer film of Meltonix (Solaronix, Switzerland) serves as a spacer between the working electrode and the platinum electrode in the solar cell. Iodolyte H30 (Solaronix, Switzerland) was used as an electrolyte in a solar cell, which was poured between two glued electrodes through a special hole. Ultrasound soldering station UltrasonicCS55-X151 (CHEERSONIC) was used to obtain the contacts. Solder material is Ecosolder RMA-98. The volt-ampere characteristics of the solar cells were measured by illuminating the cells with the light of a Xe lamp, with a light output of 100 mW/cm^2 (AirMass 1.5) on the CT50AAA measuring complex (PET PHOTO Emission TECH. INC., USA).

3. Results

Figure 1 shows a photomicrograph of the surface of the Ag TiO_2 film obtained on a scanning electron microscope.

Figure 1. SEM image of a film based on NP Ag/TiO_2.

It can be seen from the Figure that the surface of the film has a granular structure, there are no visible defects. The use of a screen for the application of films allows the production of films with the same thickness. The study of the transverse cleavage of the sample showed that the film thickness is of the order of 10 μm with a minimum thickness spread of 200 to 500 nm along the entire perimeter.

The influence of the area of the photoactive layer of the solar cell on its efficiency was studied. For this purpose, DSSC was obtained on the basis of the Ag/TiO_2 "core-shell" structure with the area of the photoactive layer of 0.25, 2.25, 4, and 6.25 cm^2. The results of measurements of the current-voltage characteristics of solar cells are shown in Figure 2.

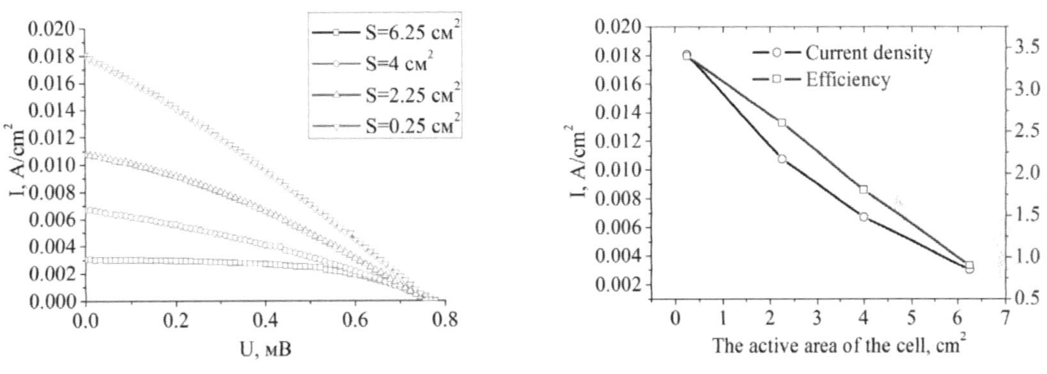

Figure 2. I-V characteristics (*a*) and the effect of the active layer area on the efficiency of a solar cell based on Ag/TiO_2 (*b*)

From the data presented, it can be seen that the photocurrent density decreases linearly with increasing photonode area. Thus, for a photoanode area of 0.25 cm^2, the photocurrent density is equal to 0.018 A/cm^2. When the area is increased to 6.25 cm^2, its value decreases to 0.003 A/cm^2. Perhaps this is due to the fact that as the area of the photoactive layer grows, the density of surface defects increases, which are primarily formed as a result of unformed bonds between NPs that form a film. Unformed connections will lead to a sharp increase in recombination processes in the film and will impede the transport of electrons. The reason for the lowering of the current density can also be the low rate of removal of electrons from the surface of titanium dioxide by the platinum counter electrode.

Next, the influence of the Ag/TiO_2 "core-shell" structure composition on the efficiency of the solar cell module was investigated. As it is known, the open curcuit voltage of DSSC depends on the position of the quasi Fermi level and the level of the oxidation-vestibular potential of the electrolyte used. When using the electrolyte Iodolyte H30, the maximum open curcuit voltage in the DSSC is about of 0.8 V. But for most electrical appliances, it is necessary to obtain a voltage of the order of 5 V. Therefore, to obtain the required voltage, several cells must be connected in series. With this in mind, we were able to place 9 separate elements connected in series with a total area of 48.6 cm^2 on a FTO conductive glass substrate with the area of 100×100 mm. The geometrical arrangement of the electrodes is shown in Figure 3.

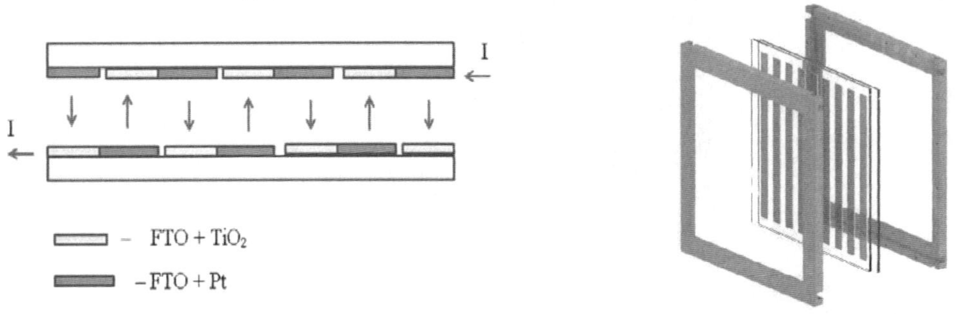

Figure 3. Connection scheme for electrodes (the arrow indicates the direction of the electric current) (*a*) and external view of solar cell module (*b*).

The modules of solar cells were assembled without and with the addition of Ag/TiO_2 nanoparticles. Under standard conditions, their current-voltage characteristics were measured and their efficiency was determined. The results of the measurement are shown in Table 1.

Table 1. IV-characteristics solar modules without and with the addition of NP Ag/TiO_2.

Sample	U_{oc}, V	J_{sc}, mA/cm^2	FF	R_{ser}, Ом	R_{sh}, Ом	Efficiency, %
Ag/TiO_2	4.8	0.089	0.06	43	731	1.26
TiO_2	4.2	0.081	0.06	124	1059	0.93

From the data obtained, it can be seen that the addition of Ag/TiO_2 NP at the concentration of 1% by weight in the semiconductor film of titanium dioxide leads to an increase in the efficiency of the solar cell that results in improving of the basic characteristics of the module. Thus, with the addition of the Ag/TiO_2 "core-shell" structure, the successive R_{ser} and shunting R_{sh} resistances decrease.

4. Summary

The effect of the area of the DSSC photoactive layer on the basis of the Ag/TiO_2 "core-shell" structure was investigated. It was found that with an increase in the area of the photoactive film from 0.25 cm^2 to 6.25 cm^2, the photocurrent density decreases and the cell efficiency decreases. A laboratory module of a solar cell with a total area of 48.6 cm^2 was developed. I-V characteristics were measured and the efficiency of DSSC cells was determined without and with the addition of Ag NP. It is established that the addition of 1 mass% of silver NP to the structure of the titanium dioxide film leads to an increase in the efficiency of the module.

Acknowledgments

This work was carried out with the financial support of the Ministry of Education and Science of the Republic of Kazakhstan, grant No. 0088/PCF.

References

[1] Regan O B, Grötzel M. 1991 *Nature* **353** 737–740
[2] Khalil E J 2011 *Solar Cells - Dye-Sensitized Devices* (Rijeka, InTech Europe) doi:10.5772/1757
[3] Catchpole K R, Polman A 2008 *Optics Express* **16** 21793–21800 doi: 10.1364/OE.16.021793

[4] Afanasyev D A, Ibrayev N Kh, Serikov T M, Zeinidenov A K 2017 *Russian Journal of Physical Chemistry A* **90** 833–837 doi: 10.1134/S0036024416040026

[5] Qi J, Dang X, Hammond P T, Belcher A M 2011 *ACS Nano* **5** 7108–7116 doi: 10.1021/nn201808g

[6] Chiba Y, Islam A, Watanabe Y, Komiya R, Koide N, Han L 2006 *Japanese Journal of Applied Physics* **45** L638–L640 doi: 10.1143/JJAP.45.L638

[7] Vendra K, Absher J, Ellis S R, Amos D A, Druffel T, Sunkara M K 2012 Photoanode Area Dependent Efficiency and Recombination Effects in Dye-Sensitized Solar Cells *Journal of The Electrochemical Society* **159** H728–H733 doi: 10.1149/2.067208jes

[8] Afanasyev D A, Ibrayev N Kh, Serikov T M, Epp A A 2015 *Proceedings Topical problems of Biophotonics-2015* 111

IOP Conf. Series: Materials Science and Engineering **289** (2017) 012025 doi:10.1088/1757-899X/289/1/012025

Experimental research of the influence of the strength of ore samples on the parameters of an electromagnetic signal during acoustic excitation in the process of uniaxial compression

L V Yavorovich[1], A A Bespal`ko[2], P I Fedotov[3]
[1]Senior researcher, National Research Tomsk Polytechnic University, Tomsk, Russia
[2]Leading researcher, National Research Tomsk Polytechnic University, Tomsk, Russia
[3]Reseacher, National Research Tomsk Polytechnic University, Tomsk, Russia
E- mail.ru: Lusi@tpu.ru

Abstract. Parameters of electromagnetic responses (EMRe) generated during uniaxial compression of rock samples under excitation by deterministic acoustic pulses are presented and discussed. Such physical modeling in the laboratory allows to reveal the main regularities of electromagnetic signals (EMS) generation in rock massive. The influence of the samples mechanical properties on the parameters of the EMRe excited by an acoustic signal in the process of uniaxial compression is considered. It has been established that sulfides and quartz in the rocks of the Tashtagol iron ore deposit (Western Siberia, Russia) contribute to the conversion of mechanical energy into the energy of the electromagnetic field, which is expressed in an increase in the EMS amplitude. The decrease in the EMS amplitude when the stress-strain state of the sample changes during the uniaxial compression is observed when the amount of conductive magnetite contained in the rock is increased. The obtained results are important for the physical substantiation of testing methods and monitoring of changes in the stress-strain state of the rock massive by the parameters of electromagnetic signals and the characteristics of electromagnetic emission.

1. Introduction

Extraction of minerals by mining method with increasing depth inevitably leads to the growth and redistribution of the stress-strain state (SSS) of the rock massif. With explosive methods of mining, the rate of SSS change increases. Probability of damage to the mine field, equipment and maintenance personnel increases. Electrometric, seismic and acoustic methods of monitoring are used to monitor the state of the massive. One of the most promising methods for continuous monitoring of the development of geodynamic events is the method based on mechanoelectric transformations in rocks with simultaneous generation of electromagnetic signals.

Laboratory investigations of changes in the electromagnetic signals parameters of various types of rocks subjected to uniaxial compression are important for understanding the preparation and development of destruction processes in mines when the massive SSS is changing. In this regard, electromagnetic radiation, arising from the change in the mechanical effect on rock, is the subject of study in many countries of the world in order to control and monitor geodynamic manifestations. Registered signals carry valuable information on the mechanisms and sources of mechanoelectric transformation in a variety of materials such as crystals [1], composite materials [2], plastics [3–5], rocks [6–9], ice [10, 11] and metal [12]. The study of EMS parameters allows to know the dynamics of formation and propagation of cracks in real time. EMS is recorded in a frequency range from a few Hertz to a few megahertz.

Measurements, made on rock samples, show that EMS is a good precursor for determining the maximum strength of samples [13]. This can be used to monitor changes in the stressed state of the rock mass [13–17]. The basis of these studies is an increase in electromagnetic activity during the pre-destruction stage experimentally proven on rock samples [18]. However, the process of mechanoelectric transformations occurs at all stages of deformation, from the moment of the initiation of microcracks to failure. Depending on the stage of deformation, EMS characteristics have their own peculiarities [19, 20].

The main attention of researchers is drawn to EMS sources. It was found in [21, 22] that the origin of EMS corresponds to the creation of cracks in time. Simultaneous recording of EMS and acoustic signal (AC) was performed during the monotonous loading of the granite samples. It is established that EMS is associated with the creation or growth of microcracks. It was also observed that the EMS appears several microseconds before the generation of the AS. The same lag of the AS from the EMS was noticed in [5]. This delay, according to the authors, is due to the fact that the EMS propagates at the speed of electromagnetic waves. Also, this effect can be used to determine the locations of cracks.

Creation of cracks is accompanied by a redistribution of electric charge due to weakened chemical bonds, noted in [23, 24]. Electric charges create dipole moments. The vibrations of the beads of the cracks change the configuration of the dipoles, and, consequently, create electric and magnetic moments. This confirms the dipole mechanism of EMS generation in the process of mechanical action on rocks.

The study of model transparent samples made it possible to determine the sequence of crack initiation, their growth and interactions. Investigations of EMS in [25, 26] were carried out during compression of rock samples and model samples. It was established that, just as in [18], the EMS duration changes at certain stages of the load.

EMS measurements of hard coal, gray dolomite, sandstone and magnesite samples, subjected to uniaxial compression presented in works [27]. Measurements show that EMS is a good precursor for determining the maximum strength of materials and is suitable for determining the state of tension in a rock massif. This was demonstrated in laboratory tests on coal samples, where EMS and AS were observed immediately before the sample failure.

In work [15] it was shown that the generation of EMS in rocks is caused both by the effects associated with the formation of microcracks, and by the passage of acoustic waves through them, which cause vibrations of charged defects. In this case, as will be shown below, the influence of the material properties on excitation of EMS by acoustic waves is similar to the effect of material properties on the parameters of EMS, which is excited by mechanical action. In the presented work, excitation of EMS was created during the passage of AS at various stages of the samples stress-strain state under uniaxial compression.

In connection with this, the aim of this work is to determine the influence of the petrographic properties of rock samples on the parameters of the EMS excited by a deterministic acoustic pulse during uniaxial compression.

2. Objects of research

The test samples of rock, represented by magnetite ore, were selected at the Tashtagol iron ore deposit (Russia, Western Siberia). Samples were cut from the core material and had the shape of a cylinder 42 mm in diameter and 80 mm high. Before carrying out the research, the end faces of the sample were ground to a plane parallelism. This ensured the alignment of the sample and the press plates and the uniformity of the load distribution on the end surface of the sample. Table 1 shows the petrographic data of the investigated samples of iron ore.

Table 1. Petrographic data of the test samples.

Magnetite ore samples							
Sample number	Sample mass, g m_o	Sample density, g/cm^3 ρ_o	Sample volume, cm^3 V_o	Magnetite volume in the sample, cm^3 V_{mag}	Magnetite mass in the sample, g m_{mag}	Magnetite content in the sample,%	Ultimate compressive strength, kN P_{ult}
4	456	4.15	110	59	302	66	167
5	399	3.66	109	33	172	43	261
7	384	3.52	109	27	137	36	120
10	401	3.68	109	34	177	44	273
13	434	3.98	109	50	256	59	131
16	463	5.45	104	70	362	78	213

The mass of magnetite in the samples was calculated by the equation:

$$m_{\text{mag}} = \rho_{\text{mag}} \cdot V_0 \cdot \left(1 - \frac{\left(\rho_0 - \rho_{\text{mag}}\right)}{\left(\rho_{encl} - \rho_{\text{mag}}\right)}\right)$$

where m_{mag} – mass of magnetite; ρ_{mag} – density of magnetite; V_0 – volume of the sample; ρ_{mag} – the sample density; ρ_{encl} – density of the enclosing rock.

3. Research Methodology

Experimental studies were carried out on a stand, the block diagram of which is shown in Figure 1. Uniaxial compression was carried out on a press with a movable (1) and supporting (2) plates, developing a force of up to 50 tons and an integrated force measuring system (8). The analogue signal of the load-measuring system and the values of the longitudinal deformation arising during the compression of the sample were recorded in the computer (PC) memory (9) using a special program. A deterministic acoustic signal was introduced to the sample through an acoustic contact (mineral oil) using a piezoelectric transducer (PET) (3). The acoustic signal, passing through the sample, was recorded with a piezoelectric sensor (PES) (7). A differential capacitive electric field sensor with an integrated power amplifier (EFS) (6) received the electrical component of the EMS generated by the sample during the passage of the acoustic signal and then recorded the signal using a multifunction digital input / output board BNC-2120 (5) for further analysis of its parameters.A subsubsection. The paragraph text follows on from the subsubsection heading but should not be in italic.

3.1. Acoustic stimulation

In experiments, the source of the deterministic pulsed acoustic signal was PET made on the basis of piezoceramics ZTS-19 (lead zirconate-titanate). For such a piezoceramics, the coefficient of electromechanical coupling is 0.523 [28]. The transmitter was supplied with rectangular pulses, which were obtained using a high-voltage generator (HVPG). The duration of high-voltage pulses can vary within 10-6-10-4s. In this case, the value of the high-voltage pulses can vary from 100V to 800V. A wide-band PES, also based on the ZTC-19 [29], was used to record the acoustic signal transmitted through the sample. The signal, generated by the similar in construction and characteristics receiving piezoelectric sensor, was applied to the Tektronix TDS2024B oscilloscope. In the experiments described in this paper, the duration and voltage of the HVPG pulses were 5 μs at the excitation voltage of the 800 V.

3.2. Recording of an electromagnetic signal

The acoustic signal from a piezoelectric transducer passing through a sample generates EMS, which is recorded using an electric field sensor, mounted 2 mm from the sample. The sensor has two copper plates that convert the electrical component of the electromagnetic field into a voltage supplied to the differential input. The high-impedance differential input of the sensor is made on precision amplifiers. The signal is differentiated to exclude the noise signal from the surrounding space, then the purified signal is amplified 100 times. Enhanced by the EFS, it is fed to the second input of the Tektronix TDS2024B oscilloscope. From the oscilloscope, the EMS is recorded in the PC.

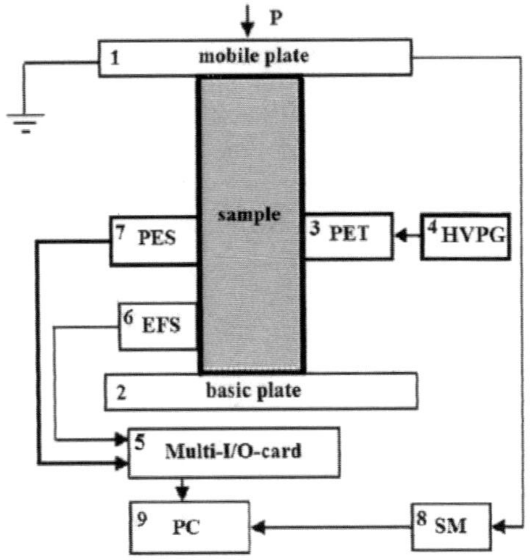

Figure 1. Block diagram of an experimental setup for recording electromagnetic signals under uniaxial compression of samples and deterministic acoustic action.

4. Experimental studies

Uniaxial compression of samples on the press was carried out stepwise. The loading rate was 0.3 kN/s. From the beginning of loading, after every 30–40 kN, the loading was stopped and the force was maintained at the reached level, thus it was possible to produce an acoustic impact. And so on until the sample failure. Depending on the samples strength, the number of steps were from 4 to 10, i.e. the stronger the sample, the more steps. At these levels, deterministic acoustic excitation was carried out. Figure 2 shows a typical loading curve showing stepped uniaxial compression of sample before failure.

Figure 3 shows the summary sequence of oscillograms of the change in the amplitude of the electromagnetic response (EMRe) under acoustic action at various stages of the sample loading. The numbering steps of compression are also shown in Figure 3.

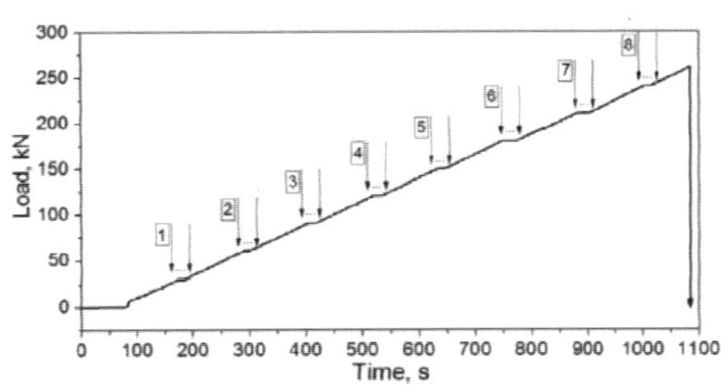

Figure 2. Typical loading curve for samples.

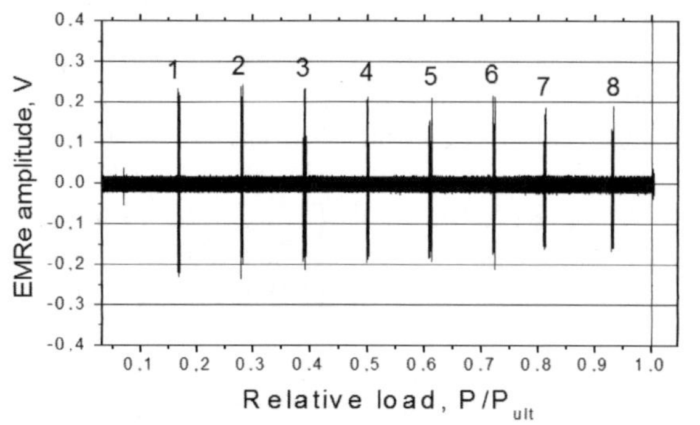

Figure 3. Dependence of the EMRe amplitude by the acoustic effect on the loading stages by uniaxial compression for sample No. 5. The stages are marked with numbers.

It can be seen that the amplitude of the electromagnetic response to the acoustic effect changes at all loading stages. To determine the amplitude-frequency characteristics of the EMS at each stage of loading, a calculation was made using a fast Fourier transform (FFT). In Figure 4 shows analogue EMRe and their amplitude-frequency characteristics for sample No. 16, obtained during the experiment. For this sample, excitation was carried out at seven stages with a load interval of 30 kN. The ultimate strength P_{ult} was 213 kN.

Similar results were obtained for the remaining samples. The Figure 4 shows the results obtained for the five stages (1, 3, 4, 5 and 7) of keeping under load. These stages corresponded to the relative load of P / P_{ult} – 0.14, 0.28, 0.56, 0.7 and 0.98. The load is indicated in the upper right corner on the analog signal pictures. Analyzing the results presented in Figure 4, it can be said, that the amplitude and type of EMRe are changing with the increasing load. The spectral components of the EMRe are also changing their amplitude as the load increases.

5. Discussion

It is known, that rocks have pores filled with gas and mineralized liquid, and existing cracks increase in the process of loading and new ones appear with different orientations and sizes [30]. These violations of the samples homogeneity lead to the formation of double electrical layers (DEL). Compression and expansion of a rock, containing DEL, lead to a transformation of a part of the acoustic energy to electromagnetic energy. In addition, rocks consist of different minerals, which have different dielectric permeability and electrical conductivity. In addition, rocks consist of different minerals, which have different dielectric permeability and electrical conductivity.

Minerals are differently oriented in space in the process of their formation and growth. In [31], when studying the reaction of DEL to ultrasound, it was found that if ultrasound exposure to rocks with an intensity of 10^{-3} W/m^2 and above, than the DEL will generate EMS. In this case, the change in the DES parameters leads to an adequate change in the EMS spectral composition.

The test samples had differences in the texture structure, cracks of different orientation and sizes were observed, healed and unhealed with calcite. As a result, the heterogeneity of composition, inclusion, defectiveness lead to an increase in the number and extent of double electrical layers, and hence to an increase in the number of dipoles emitting electromagnetic pulses. When this sample is acoustically influenced, this circumstance facilitates the active transfer of mechanical energy into the energy of the electromagnetic field. The dependences of the peak-to-peak value on the relative strength calculated as P / P_{ult} were constructed to determine the regularities of the change in the amplitude of the EMRe. Here P is the current load value on which the acoustic effect was carried out, and P_{ult} – the load of the sample failure. Figure 5 shows the dependences obtained for sample No. 16 and Figure 6 for sample No. 5

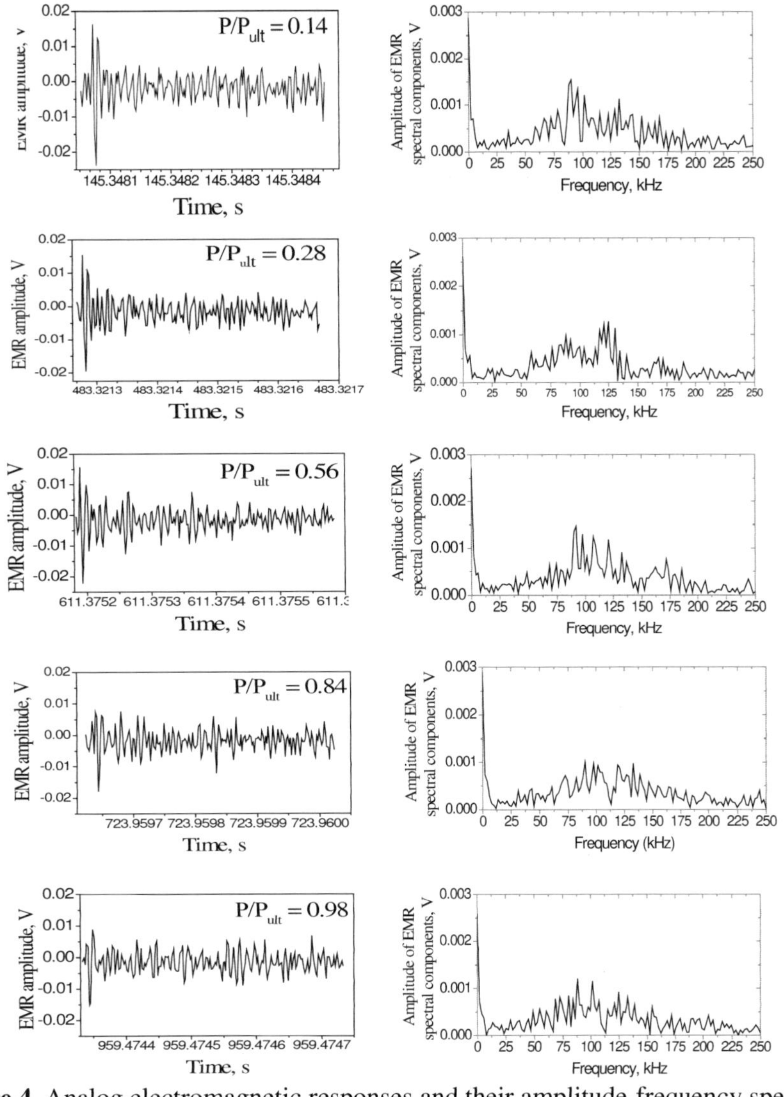

Figure 4. Analog electromagnetic responses and their amplitude-frequency spectra at the five excitation stages of sample No. 16.

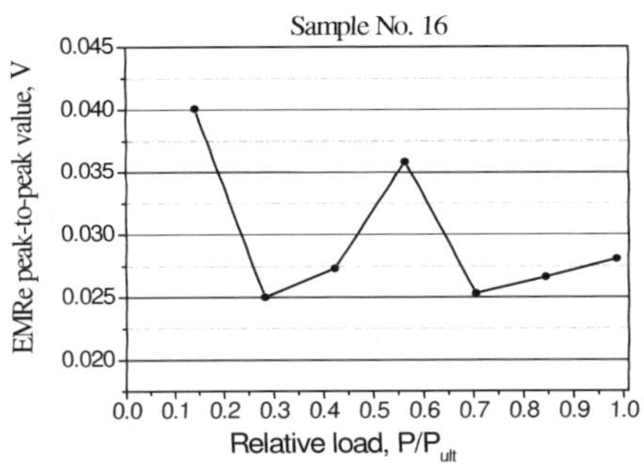

Figure 5. Dependence of the EMRe amplitude from uniaxial load for sample No. 16.

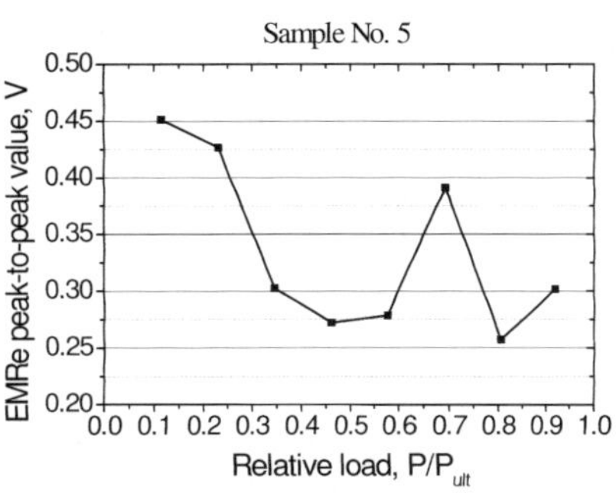

Figure 6. Dependence of the EMRe amplitude from uniaxial load for sample No. 5.

Analyzing the results, it should be noted that no matter how many times acoustic excitations are made, the trend of the EMO amplitude variation remains. The amplitude decreases from the beginning of the loading for the dependences shown in Figures 5 and 6. Since the strength of these samples was different, the decrease for sample No. 16 was noted at a relative strength of 0.28, and for sample No. 5 the reduction continued to a relative strength of 0.5-0.6. The increase in the amplitude of the EMO amplitude is observed for sample No. 16 at a relative load of 0.56, and for sample No. 5 at a relative load of 0.68. This relative load for the samples corresponds to the formation of a failure focus [32]. Then the amplitude decreases and, starting at a relative strength of 0.7–0.8, the amplitude of the EMO begins to grow until the samples reach ultimate strength and fracture.

Some difference in the EMRe amplitude of samples with different contents of magnetite can be explained by the fact that the ultimate strength of these samples is different. It was 213 kN for sample No. 16, and 261 kN for sample No. 5. These samples differ by 50 kN for ultimate strength. It is known that the strength of real rock samples is greatly influenced by factors such as structural and texture structure, the presence of inclusions with different strength, pores, healed and untreated cracks. And the more inhomogeneities, the lower the ultimate strength, and hence the more sources of mechanoelectric transformations. However, it is seen from the experimental results shown in Figure 5 and 6 that sample No. 16, with a lower ultimate strength of 213 kN, has the less EMRe amplitude by an order than the sample No. 5, with the ultimate strength of 261 kN. This became the basis for analyzing the effect of the sample composition on the EMRe characteristics. It is known that the presence of piezoelectric quartz [33,34] and electrically conductive minerals such as pyrite, chalcopyrite, galena and graphite exerts a significant influence on the EMS amplitude of dielectric samples, including samples of magnetite ore. Quartz clusters in magnetite ore samples under the influence of mechanical forces (acoustic wave and loading) are capable of inducing an electric charge, which causes a large amplitude of EMRe. The higher the content of quartz in the sample, the higher the EMRe amplitude. The polarization effect is important for electrically conductive minerals of pyrite and chalcopyrite.

At the initial stage of loading, the sample becomes compacted due to presented cracks and pores. This causes the displacement of the gravity centers of unlike charges and as a result, electric dipoles arise. Their formation and relaxation are associated with the motion of charged particles, i.e. with the current. At the stage of consolidation of failure focus and formation of a main rupture ($P / P_{ult} \geq 0.7$), the current decreases, since the process of crack growth prevails at this stage. The interval 0.5–0.7 of P/P_{ult} is distinguished by the fact that the charged defects directed toward the newly formed microcracks, the contacts of grains and the fragments of inclusions of different minerals, directed by the applied static stresses. New DELs form or existing ones recharge. As a result of oscillation of these layers by any method, the intensity of EME and the amplitude of EMS will increase, which we observe in the experiment. The enlargement of failure focus and the formation of cracks in the separation lead to a change in the distance between their shores and to the partial destruction of the double electrical layers, which leads to a decrease in the amplitude of the EMS [32]. This is what we observe in our experiments at a load of 0.7–0.8 for both samples.

The dependences in Figure 7 were plotted for three samples of magnetite ore to determine the dependence of the peak-to-peak value on the relative strength for samples of different petrographic composition. The percentage of quartz (SiO2), pyrite (FeS2) and magnetite (Fe3O4) for each sample was indicated. The percentages of these minerals in the samples were obtained using an X-ray powder diffractometer ARLX`TRA [35]. The results of measuring the EMRe parameters of sample No. 13, whose ultimate strength was 113 kN, were added for comparison to the results of samples No. 5 and No. 16 (Figures 5 and 6).

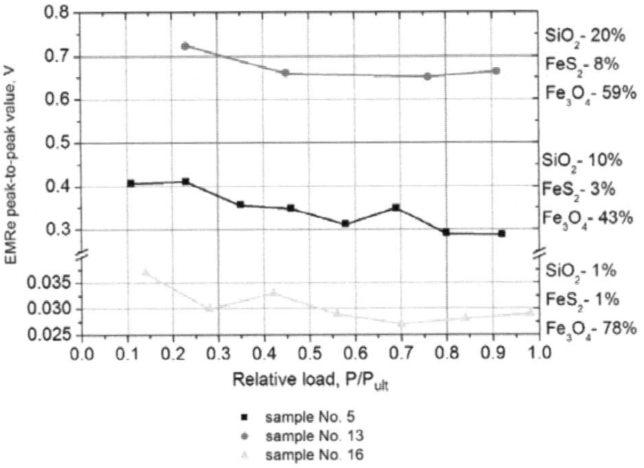

Figure 7. Dependence of the EMRe peak-to-peak value on the relative load for samples with different content of quartz and conducting minerals.

Analyzing Figure 7, it should be noted that a high content of magnetite (Fe_3O_4) 78% and 1% quartz (SiO_2) were detected for sample No. 16, whose ultimate strength was 213 kN, and the EMRe amplitude is an order of magnitude lower than for sample No. 5, which has almost half the amount of magnetite – 43%, but a higher content of quartz (SiO_2) – 10%, and sample No. 13, which has 20% quartz and 59% magnetite in its composition. For a low-strength sample No. 13, the amplitude of the EMRe is the highest. And this is due to the fact that he has a large content of quartz.

Thus, it can be concluded that the EMRe amplitude for samples with different contents of magnetite ore will be determined by the content of mineral inclusions of quartz and pyrite. When the quartz is close in quantitative content in the samples of magnetite ore, the amplitude of the EMRe will be determined by the amount of magnetite. One possible explanation for the observed effect is the loss of EMS energy in a conducting medium. Magnetite, as is well known, has an appreciable electrical conductivity and its EMS absorption coefficient of can be written as $\beta \approx \sqrt{\frac{\pi\mu}{\rho} f}$, where μ – is the magnetic permeability; ρ – is the resistivity; f – is the frequency. For magnetite at a frequency of 1 kHz, $\rho = 10^3$ Ω·m and $\mu = 10^5$. Then the absorption coefficient for a frequency of 1 kHz is $\beta = 500$ m^{-1}. At a frequency f = 10 kHz, $\beta = 1700$ m^{-1}. The change in the amplitude of the harmonic at a frequency of 1 kHz is: $\frac{E}{E_0} = \sqrt{e^{\beta x}} \approx 0.1$, where E – is the electric field strength.

For a frequency of 10 kHz, this ratio is equal to $\frac{E}{E_0} = \sqrt{e^{\beta x}} \approx 2 \cdot 10^{-4}$, i.e. the amplitude of the harmonic more than 10 kHz will decrease by 1000 times than the 1 kHz harmonic.

There is another possible explanation for the correlation of the EMS amplitude and their conductivity in the case of acoustic excitation of rocks. When an acoustic pulse is applied and the acoustic wave propagates the total effective charge of the sample changes as a result of mechanoelectric and acoustoelectric transformations. The magnitude of this charge will depend on the conductivity of the rock and will basically determine the drain time or charge compensation. As a result, the effective charge will decrease, and the electric field strength also decreases. Oscillations in the resulting acoustic field lead to a change in the electric field, and, as a consequence, to the occurrence of displacement currents. As a result of these oscillations, EMS is recorded with an amplitude that reflects the conductivity of the rock samples. The assumption is confirmed by the results of studies with acoustic effects on samples with different contents of magnetite. The amplitude of the EMRe from sample No. 16 is about of 30-35 mV, and for sample No. 5 – 3–4 mV.

6. Conclusions

As a result of the conducted studies, it was established that the sensitivity of electromagnetic signals to defectiveness with multiple passage of an acoustic excitation wave through a monitoring object makes it possible to track the evolution of accumulation of defects in conditions of a stress-strain state.

With an increase in the amount of a high-conductivity mineral contained in rocks, such as magnetite, the amplitude of the analog electromagnetic signal and the amplitude of the spectral components of the electromagnetic response are reduced in the uniaxial compression process, and, consequently, when the stress-strain state of the rock samples changes. The amplitude-frequency spectrum of electromagnetic responses of samples with a large magnetite content is characterized by ruggedness and wide limits.

The presence of quartz in samples of magnetite ore leads to an increase in the amplitude of the EMRe under acoustic excitation during uniaxial compression.

Thus, the physical modeling of the influence of the ore samples strength and petrophysical composition on the parameters of the electromagnetic respons, generated during acoustic excitation in the uniaxial compression process, promotes nondestructive testing and monitoring methods of the change in the stress-strain state of rock samples before their failure by parameters of the electromagnetic signals.

Acknowledgement

The work was supported by the Ministry of Education and Science of the Russian Federation within the framework of the State task in the field of scientific activity, project No. 11.980.2017 / 4.6.

References

[1] Hadjicontis V, Mavromatou C, Antsygina T N, Chishko K A 2007 *Phys. Rev.* B **76** 024106 doi: 10.1103/PhysRevB.76.024106
[2] Koktavy P 2009 *Meas. Sci. Technol.* **20** 015704 doi: 10.1088/0957-0233/20/1/015704
[3] Gade S O et al 2014 *J. Nondestruct. Eval.* **33** 711 doi: 10.1007/s10921-014-0265-5
[4] Sause M G R 2016 In-Situ Monitoring of Fiber-Reinforced Composites doi: 10.1007/978-3-319-30954-5
[5] Gade S O, Sause M G R 2009 *J Nondestruct Eval* **36** 9 doi: 10.1007/s10921-016-0386-0
[6] Tsutsumi A, Shirai N 2008 *Tectonophysics* **450** 79 doi: 10.1016/j.tecto.2008.01.001
[7] Koktavy P, Pavelka J, Sikula J 2004 *Meas. Sci. Technol.* **15** 973 doi: 10.1088/0957-0233/15/5/028
[8] Ogawa T, Oike K, Miura T 1985 *J. Geophys. Res.* **90** 6245 doi: 10.1029/JD090iD04p06245
[9] Goldbaum J, Frid V, Bahat D, Rabinovitch A 2003 *Meas. Sci. Technol.* **14** 1839 doi: 10.1088/0957-0233/14/10/314
[10] Petrenko V F 1996 Electromechanical phenomena in ice CRREL Special Report 96-2
[11] Vorotilov K A, Sigov A S 2012 *Physics of the Solid State* **54** 894 doi: 10.1134/S1063783412050460
[12] Misra A *et al* 2007 *Int. J. Fract.* **145** 99 doi: 10.1007/s10704-007-9107-0
[13] Yavorovich L V, Bespal`ko A A, Fedotov P I, Baksht R B 2016 *Acta Geophys.* **64** 1446 doi: 10.1515/acgeo-2016-0081
[14] Frid V, Vozoff K 2005 *Inter. J. Coal Geol.* **64** 57 doi: 10.1016/j.coal.2005.03.005
[15] Koktavy P 2009 *Meas. Sci. Technol.* **20** 015704 doi: 10.1088/0957-0233/20/1/015704
[16] Bespal'ko A A et al 2010 *J. Min. Sci.* **46** 136 doi: 10.1007/s10913-010-0018-5
[17] Sigov A, Podgorny Y, Vorotilov K, Vishnevskiy A 2013 *Phase Transitions* **86** 1141 doi: 10.1080/01411594.2013.790033
[18] Bespal'ko A A, Yavorovich L V, Fedotov P I 2011 *Russian Journal of Nondestructive Testing* **47** 680 doi: 10.1134/S1061830911100068
[19] Vorotilov K et al 2013 *Phase Transitions* **86** 1152 doi: 10.1080/01411594.2013.794276

[20] Bespal'ko A A, Surzhikov A P, Yavorovich, L V, Fedotov P I 2012 *Russian Journal of Nondestructive Testing* **48** 221 doi: 10.1134/S1061830912040043
[21] Mori Y, Obata Y, Sikula J 2009 *J. Acoust. Emiss.* **27** 157
[22] Yamada I, Masuda K, Mizutani H 1989 *Phys. Earth Planet. Int.* **57** 157 doi: 10.1016/0031-9201(89)90225-2
[23] Sedlak P A *et al* 2008 *Measurement Science and Technology* **19** 045701 doi: 10.1088/0957-0233/19/4/045701
[24] Koktavy P. 2009 *Meas. Sci. Technol.* **20** 015704 doi: 10.1088/0957-0233/20/1/015704
[25] Rabinovitch A, Frid V, Bahat D 2007 *Tectonophysics* **431** 15 doi: 10.1016/j.tecto.2006.05.027
[26] Rabinovitch A, Frid V, Bahat D, Goldbaum J 2000 *Int. J. Rock Mech. Min. Sci.* **37** 1149 doi: 10.1016/S1365-1609(00)00042-3
[27] Pralat A, Wojtowicz S 2004 *Acta Geodyn. Geomater. Experimental measurements.* **1** 111
[28] URL: http://www.elpapiezo.ru/porous.shtml
[29] Korolev M V 1973 *Defektoskopiya* **4** 12 (in Russia)
[30] Mavko G, Mukerji T, Dvorkin J 2009 *The Rock Physics Handbook* 2nd ed. Cambridge University Press
[31] Perel`man M E, Khatiashvili N G 1983 *Dokl. Akad. Nauk SSSR* **271** 80 (in Russia)
[32] Bespal`ko A A, Yavorovich L V, Fedotov P I 2011 *Russian Journal of Nondestructive Testing* **47** 680 doi: 10.1134/S1061830911100068
[33] Bespal'ko A A, Yavorovich L V, Fedotov P I 2007 *Journal of Mining Science* **43** 472 doi: 10.1007/s10913-007-0049-8
[34] Kobayashi H, Horikawa K, Ogawa K, Watanabe K 2014 *Phil. Trans. Roy. Soc. A* **372** 20130292 doi: 10.1098/rsta.2013.0292
[35] ОФС.1.2.1.1.0011.15 X-ray powder diffractometry (in Russian)

Method for selecting parameters and assessing efficiency of wind-diesel power plants for autonomous electrical supply systems

S G Obukhov[1], I A Plotnikov[1], V G Masolov[2]
[1]Associate Professor, National Research Tomsk Polytechnic University, Tomsk, Russia
[2]General manager, VDM-tekhnika LLC, Moscow, Russia
E-mail: serob99@mail.ru

Abstract. The article presents an original technique for selecting parameters and evaluating the efficiency of wind-diesel power plants for isolated power supply systems. The initial data to perform energy calculations are simulation models of electric load and wind speed. The load is simulated using typical schedules of electric loads of a decentralized consumer, taking into account a random component for each hour of the day. To create a simulation model of the wind, a typical climatic series of wind speeds at a prospective site of the power plant has been constructed according to the data of long-term meteorological observations. The proposed technique was verified through the example of choosing a wind-diesel power plant for the village of Ust-Olenyok of the Republic of Sakha (Yakutia).

1. Introduction

The problem of power supply to the Northern territories has been, for many years, crucial to Russia. Electricity cannot be delivered to consumers living in the Northern territories through the central electric network due to huge distances and low population density, and the only source of electricity available for these territories is diesel power plants (DPP). High depreciation of the main generation equipment and serious problems in the fuel delivery result in a less reliable electricity supply to consumers and a higher cost of electricity produced. The so-called *Northern Deliveries* program involves annual transportation of up to 6–8 million tons of combustive-lubricating materials, and the share of transport component goes up to about 70% of the total cost of fuel [1]. As a result, the power supply to the Northern territories requires significant financial expenses from the federal and regional budgets.

Meanwhile, the Northern territories of Russia have a huge renewable energy potential that can be used to improve the reliability of power supply and to reduce the cost of electricity. For example, the annual average wind speed along the entire Arctic coast of Russia is 6–7 m/s, which enables using wind-diesel power plants (WDPP).

The study aimed to develop the method for selecting parameters and assessing efficiency of wind-diesel power plants for autonomous power supply systems. The proposed method was verified through the selection of a suitable WDPP for the village of Ust-Olenyok, the Sakha Republic (Yakutia).

2. The object of the study

The The scheme with generation units connected to the DC bus was chosen as a basis for the WDPP construction, since it provides the efficient control of operating modes of autonomous power supply systems with a high replacement level [2, 3]. Figure 1 presents a generalized structural diagram of the WDPP adopted for the study.

In general, WDPP comprises two main generation sources: diesel generator sets (DGS) and wind turbines (WT). Rechargeable batteries (RB) are responsible for supplying electricity to consumers during a shortage of electricity generated by wind turbines. All generation sources are connected to the DC bus via semiconductor power converters controlled by the operating system (OS). The consumer is connected to the DC bus through the autonomous inverter (AI), which provides the conversion of DC to AC with standard parameters. To discharge surplus power generated by wind turbines in strong wind and light load, WDPP provides ballast load (BL), which is the heating elements, and ballast regulator (BR), which changes power consumption.

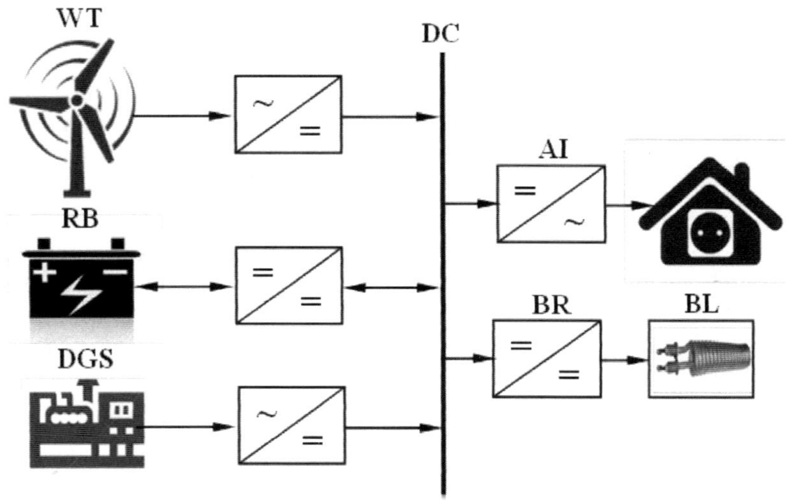

Figure 1. A generalized structural diagram of WDPP.

In addition, RB is equipped with a residual capacity monitoring device. In accordance with the limit set points of the monitoring device, the OS generates control signals to ensure charge/discharge modes for the battery converter.

3. Modeling of the object of the study

The power efficiency of WDPP largely depends on the parameters of the main generation equipment, which should be selected with regard to the characteristics of wind regime at the power plant site. Mathematical modeling was used for studying the power system due to the nonlinear parameters of its main elements and the complexity of physical processes. Simulated models of power load and wind speed are the initial data for the power calculations.

The load is modeled on the standard load schedules for a decentralized consumer with regard to a random component for every hour of the day [4].

A standard wind speed sequence at the considered power plant site is constructed to create the wind model. The data on long-term meteorological observations of the wind speed at the nearest meteorological station are used for the construction of a standard climate sequence [5, 6]. Since the data on the wind speed available on the free-access meteorological sites are averaged over 3- or 6-hour time periods, the original number series is transformed into the annual series with hourly sampling, taking into account a random component.

The method for calculating the fuel consumption at the DGS under different load conditions is based on specific fuel consumption rates for diesel power plants. The operational characteristics of the wind turbine are used to determine its output power [7].

The simulated model enables a comprehensive assessment of the power efficiency of WDPP and optimization of its main generation and storage equipment for a particular power plant.

4. Results of the study

The WDPP located in the village of Ust-Olenyok, the Sakha Republic (Yakutia), was used to verify and improve the simulated model. To date, power supply to the village is provided by the DPP. According to available statistics, in 2015 specific fuel consumption was 560 g/kWh. The village is situated on the Arctic coast, in the zone of strong winds and, according to preliminary estimates, this creates good prospects for using wind turbines. Consider three possible options for the construction of autonomous power supply systems:

- based on DPP;
- based on DPP and WT;
- based on WDPP with a battery storage.

The standard daily load schedules were modeled for Ust-Olenyok in compliance with the seasons of the year to choose DPP that comprises two DGS with the nominal power of 12 kW and specific fuel consumption of 314 g/kWh [8]. The results of calculating the energy characteristics of WDPP were analyzed to determine a rational number of the required batteries, which are based on gel batteries Haze HZY 12 230 [9].

The WT models by FD [10] were used for the comparative analysis of power and technical and economic characteristics of WDPP, and WT of different power classes. The choice of the brand is due to the fact that its inexpensive and reliable models with a horizontal axis of rotation are offered for sale by numerous sales representatives in Russia. The FD brand offers models of the entire power range required for a low-power WDPP. A detailed specification for the supplied power equipment is provided on the manufacturer's website.

A mathematical model of WDPP is a simulated model, therefore the energy characteristics were calculated in series of 6 experiments, with unchanged initial data and subsequent averaging.

For the considered options of the construction of autonomous power supply systems, the annual average electricity consumption was 28345 kWh, and the winter maximum load was equal to 10.5 kW.

For the power supply systems based on DPP, the annual fuel consumption was 12.030 kg, with the average specific consumption of 425 g/kWh. In this case, the annual number of operation hours for DGS will be 8.760 hours. Wind turbines operating together with diesel power plants will enable a partial replacement of diesel fuel used to generate electricity. The number of operating hours for DGS will be reduced as well. This will prolong the operational lifetime of diesel generator sets.

The power efficiency of wind turbines is estimated by the capacity factor *CF*, which is defined as a ratio of the actual (calculated) energy produced by WT W_{actual} over a certain time interval to the energy that can be produced by WT running at the rated power P_{WT} over the same time interval. For a time interval of 1 year, *CF* values are determined by the Equation:

$$CF = \frac{W_{\text{actual}}}{8760 \cdot P_{WT}} \tag{1}$$

The capacity factor depends mainly on the WT quality and the characteristics of wind regime at the WT site. The energy efficiency factor *EEF* provides a more precise assessment of the WT power efficiency. *EEF* is defined as a ratio of the energy consumed W_{load} to the total amount of energy produced by WT W_{actual}:

$$EEF = \frac{W_{\text{load}}}{W_{\text{actual}}} \tag{2}$$

EEF values enable estimation of the quantitative contribution of WT to the power supply. Figure 2 shows the calculated dependences of the energy characteristics of WDPP without RB depending on the installed power of WT.

IOP Conf. Series: Materials Science and Engineering **289** (2017) 012026 doi:10.1088/1757-899X/289/1/012026

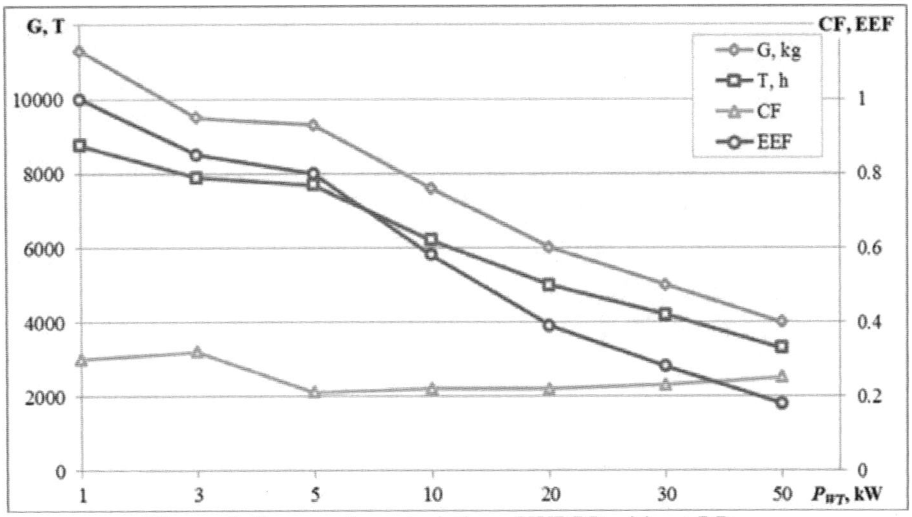

Figure 2. Energy characteristics of WDPP without RB.

The calculated dependencies show that the WT operating as a part of the power plant contributes to the improvement of the plant energy characteristics. An increase in the installed power of wind turbines leads to a reduction in both the total fuel consumption *G* and the operating hours of the diesel generator *T*. However, the energy efficiency factor *EEF* decreases as the WT power grows. *EEF* for the WT with a rated power of 1 kW equals 1.0, which indicates that the energy generated by WT is usable; *EEF* for the WT with a rated power of 50 kW is equal to 0.16, which indicates that more than 80% of the energy produced by WT will be dissipated on ballast resistances.

Meanwhile, the cost of WT with a power of 50 kW is much higher than that of 1kW. Therefore, it is necessary to find the optimal ratio of the installed power for DPP and WT, which will ensure a rational *EEF* for wind turbines at acceptable costs. In this case, the criterion for the optimal solution is the following economic indicators: annual costs, net present value, net cost of kWh of electricity, etc.

The disadvantage of the WDPP without RB is the increased operating time of the diesel generator for a low-power load. When the diesel generator is loaded less than 25% of the nominal load, the specific fuel consumption significantly increases, and the effect of carbonization caused by the accumulation of unburned fuel fractions in the cylinders can be observed. All these factors reduce the engine's life. The calculation results show that the joint operation of the WT and the DPP reduces the total operating time of the DGS, but it leads to an increase in its relative operating time under heavy operating conditions with a low load. For the considered example, the maximum relative operating time of the DGS under partial low-load conditions reaches 57% (for WDPP with a 20 kW WT).

The WDPP with RB enables an increase in *EEF* of WT due to the energy stored in the battery at the output power exceeding the consumed load, and significantly improves the operating conditions for diesel engines. The problem of the optimal battery capacity is crucial to be solved when designing the WDPP with RB.

Figure 3 presents the calculated dependences of the energy characteristics of the WDPP with RB depending on the number of batteries *N* in the WT with the installed power of 30 kW. The comparison of the characteristics of the WDPP with RB and those of the power supply system based on DPP and WT of the same power (see Figure 2) shows a significant improvement in the energy efficiency of the system.

If RB comprises 60 batteries (the total capacity is 144 kWh), there will be a decrease in the annual fuel consumption from 5.185 to 1.650 kg, and the annual number of operating hours of the DGS will drop from 4.325 to 490 hours. The positive effect is due to the increased *EEF* value from 28 to 38%. In addition, the number of operating hours of the DGS under a low load condition (with a high specific fuel consumption) is significantly reduced.

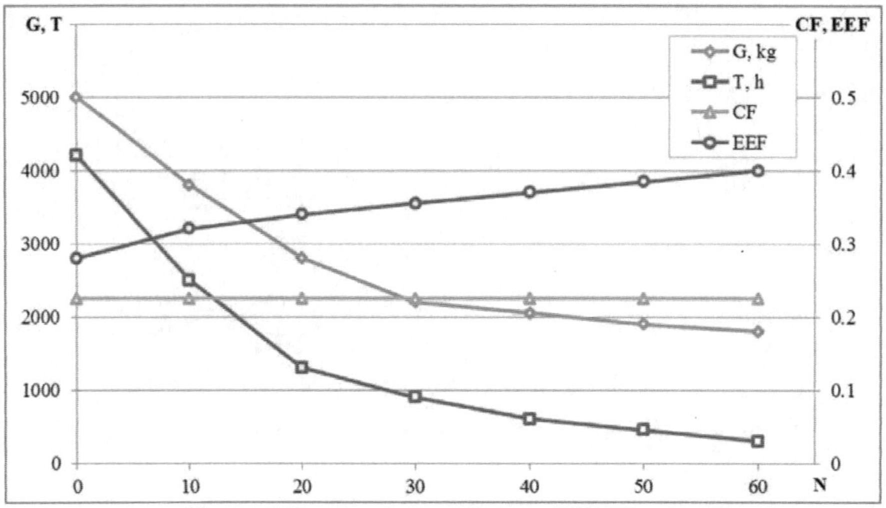

Figure 3. Energy characteristics of WDPP with RB

The results of the study show that the WDPP with RB allows minimizing not only the total number of operating hours of the DGS, but also ensures its optimal loading. The number of operating hours of DGS with a load of less than 25% will be only 40 hours for the WDPP with RB containing 60 batteries, which is less than 8% of the total operating time.

Figure 4 illustrates the daily operating mode of the WDPP based on DGS with a rated power of 12 kW, a 30 kW WT and RB containing 40 batteries with a total capacity of 96 kWh. The energy flows in WDPP are controlled by the OS.

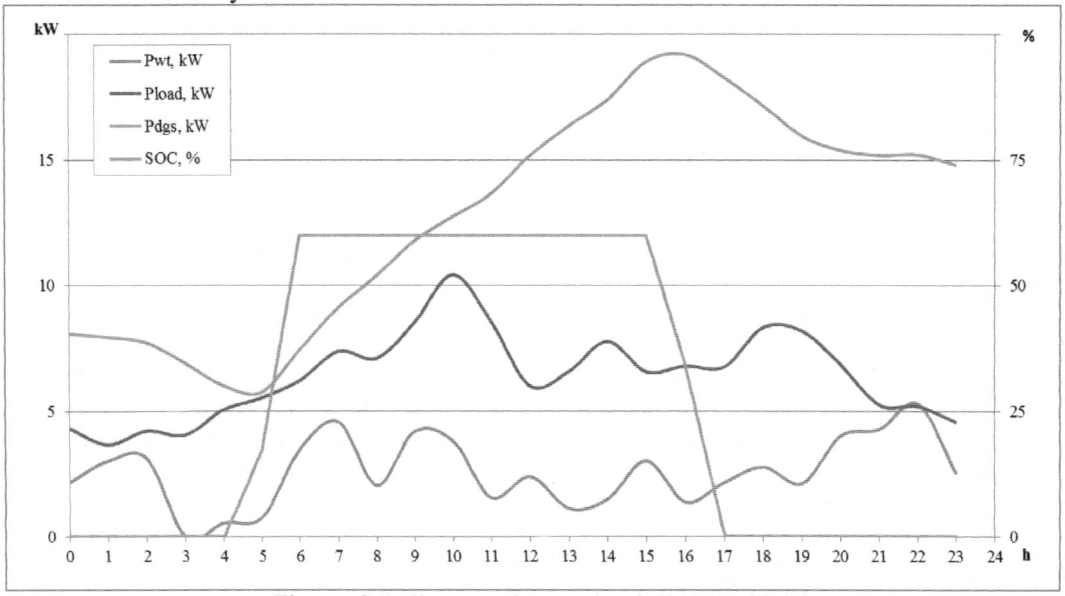

Figure 4. Daily operating mode of WDPP with RB.

Figure 4 shows a decrease in the RB residual energy down to 30% of the total capacity within the time interval 0–4 hour. The OS generates a signal to start the DGS, which, together with the WT, provides both the power supply to consumers and a full charge of the batteries within 5 to 16 hours. At the same time, the DGS operates in the optimal mode with a minimal specific fuel consumption. After 16 hours, the batteries are fully charged, the DGS switches off, and the load is supplied by the WT with RB within 16 to 24 hours.

5. Conclusions

The theoretical studies of the operating modes of the WDPP proved the possibility of improving the energy characteristics of autonomous power supply systems based on DPP: reducing the diesel specific fuel consumption, extending the life of diesel engines, and increasing the reliability of power supply to consumers.

The improved energy characteristics of the WDPP are due to a rational choice of the installed power of generation and accumulation sources, which are determined by the parameters of wind regime at the power plant site and the nature of electric load, and optimal control of energy flows in the closed energy system provided by a unified system for controlling operating modes. The rational ratio of the installed power for DPP, WT and RB is not standard, it is determined for each WDPP with regard to the specific conditions of its location and operation. Thus, technical and economic characteristics are appropriate to be used when choosing the main power equipment.

Acknowledgments

The research was funded by the Ministry of Education and Science of the Russian Federation (gov. contract no. 14.576.21.0098).

References

[1] Berdin V H, Kokorin A O, Yulkin G M, Yulkin M A 2017 *Renewable energy sources in isolated settlements of the Russian Arctic* (Moscow: Worldwide Fund for Nature (WWF)) 80 p. (in Russian)

[2] Chauhan A, Saini R P 2014 *Renewable and Sustainable Energy Reviews* **38** 99–120 doi: 10.1016/j.rser.2014.05.079

[3] Shivarama K K, Sathish K K 2015 *Renewable and Sustainable Energy Reviews* **52** 907–916 doi: 10.1016/j.rser.2015.07.187

[4] Obukhov S G , Hoshnau Z P 2012*Electrical Stations* **11** 43–47 (in Russian)

[5] Russian Weather Server: Weather archive [Electronic resource] URL:http://meteo.infospace.ru (in Russian)

[6] Russian Weather Site :Weather Schedule [Electronic resource] URL: http://www.rp5.ru (in Russian)

[7] Obukhov S.G. 2013 Increase of efficiency of hybrid autonomous power supply systems with renewable energy sources: thesis for the degree of Doctor of Technical Sciences (Tomsk) 386 p. (in Russian)

[8] OOO Disel-Systems Co [Electronic resource] URL: http://www.d-system.ru/ (in Russian)

[9] Realsolar Co [Electronic resource] URL: http://realsolar.ru/ (in Russian)

[10] Mitsuko LTD RF [Electronic resource] URL: http://mitsuko-ltd.kazprom.net/ (in Russian)

IOP Conf. Series: Materials Science and Engineering **289** (2017) 012027 doi:10.1088/1757-899X/289/1/012027

Coulunometry of oxygen, hydrogen and moisture molecular impurities in inert gases

I A Kondratyev[1], A A Shakhmatov[2], P V Melnikov[3], A V Ragutkin[3], G Yu Dalskaya[3], V P Pirog[4], A A Sukhov[4], L F Nosenko[4]
[1]Chief engineer, Experimental design bureau of automation, Angarsk, Russia
[2]Chief of department, Experimental design bureau of automation, Angarsk, Russia
[3]Associate professor, Moscow Technological University (MIREA), Moscow, Russia
[4]Engineer, Experimental design bureau of automation, Angarsk, Russia
E-mail: fisika@mail.ru

Abstract. This paper studies the method of simultaneous measurement of oxygen, hydrogen and water impurities that is based on combining moisture coulometric and solid electrolyte. The basic reduced measurement error for the molecular impurity in inert gases for measurements ranging from 0 to 5…50 and …500 mln^{-1} of oxygen totals 6, 3.8 and 2.8 % respectively. Similarly, measurement error totals respectively 8, 3 and 3.8 % for hydrogen, and respectively 6, 3 and 3 % for water molecules.

1. Introduction

Production of modern hardware components for micro- and nanoelectronics involves a wide range of high-efficiency electron-ion manufacturing processes, such as molecular beam epitaxy, ion implantation doping, laser annealing, plasma and ion beam etching, and applying vacuum coatings. These processes require particular purity of the used operating inert gas environments [1–4]. In this context it is extremely important to control such molecular components of operating gas impurities as humidity, oxygen and hydrogen to ensure the quality of the produced microelectronic devices. There are several methods of measuring the volume ration of oxygen and hydrogen in inert gases, e.g. using coulometric electrolysis and solid electrolyte cells [5–7]. A practical way to measure partial oxygen pressure in gases at high temperatures (above 600 °C) is by using a solid electrolyte cell [6, 7] designed as a tube made of solid electrolyte based on zirconium dioxide stabilized with oxides of calcium or yttrium. The most efficient way is to apply platinum electrodes in the form of spirals or pastes onto the external and internal surfaces of the tube [7]. Continuous simultaneous measurement of water, oxygen and hydrogen content is achieved by placing in the flow circuit of the studied gas a set of active electrolysis cells that simultaneously register the concentration of each impurity component.

2. Coulometric electrolysis cell

If this method is applied, the humidity is measured using a coulometric cell based on phosphoric anhydride film continuously absorbing moisture from a precisely regulated flow of the analyzed gas, with simultaneous electrolytic decomposition of the absorbed moisture into oxygen and hydrogen that are carried away by the flow of the analyzed gas. The amount of electricity spent on the electrolysis of the absorbed moisture (electrolysis current) indicates the concentration of moisture in the gas.

In accordance with Faraday's law, the moisture content is calculated as:

$$C_{H_2O} = \frac{I \cdot M_{H_2O}}{Q \cdot m \cdot F}, \quad (1)$$

where C_{H_2O} – moisture content, kg/m^3; I – electrolysis current, A; M_{H_2O} – water molar mass, kg/mole; Q – gas flow through the coulometric cell, m^3/sec; m – number of electron charges needed to decompose one water molecule; F – Faraday constant, C/mole.

3. Coulometric solid electrolyte cell

Oxygen can be measured with coulometric solid electrolyte cells (CSEC) that use direct current electrodes [8, 9]. CSEC is designed as a test tube made of high-temperature ceramics based on zirconium dioxide stabilized with yttrium dioxide that has selective oxygen-ionic conductivity. The cell has external and internal electrodes made of finely dispersed platinum [7]. Externally the cell is washed with atmospheric air due to the natural convection, and the analyzed gas is passed through the internal cavity of the cell. CSEC electrodes are powered (with specific polarity) from a direct current power source, under the effect of which oxygen is transferred through the solid electrolyte from the analyzed gas into the atmosphere.

If the analyzed gas flow is constant, the current through the CSEC can be used to measure the oxygen concentration in the environment (C_{O_2}), that according to Faraday's law is calculated as [7]:

$$C_{O_2} = \frac{I \cdot M_{O_2}}{Q \cdot n \cdot F}, \quad (2)$$

where C_{O_2} – oxygen concentration, kg/m^3; I – current through the solid electrolyte, A; M_{O_2} – oxygen molar mass, kg/mole; Q – gas flow through CSEC, m^3/sec; n – number of electron charges in the ionized oxygen molecule.

Physical and chemical properties of the solid electrolyte make it possible to use CSEC also for measuring hydrogen content in gases. The measurement is performed by means of electrolytic transfer of the amount of oxygen needed for hydrogen titration through the cell into the flow of the analyzed gas that contains hydrogen. In order to feed the titrant (oxygen) into the flow of the analyzed gas voltage is applied to the CSEC electrodes with polarity opposite to that of oxygen measurement. The signal indicating that the titration process is over is the steady transfer current and a sharp change of the cell electromotive force (EMF). EMF dependence on the titrant amount is typically for the titration curve characterized by the sharp voltage surge close to the equivalence point, i.e. the point in which the composition of the reactants corresponds with the stoichiometric proportion of the reaction products. At the moment of crossing the equivalence point the cell's EMF is changed by hundreds of millivolts. This not only indicates the end of the titration, but also makes it possible to use the cell in the automatic self-titration mode.

Figure 1 shows the CSEC flow chart for operation in the automatic self-titration mode.

CSEC consists of two chambers 1 and 2 divided by the solid electrolyte 3. Electrodes 4 are applied to the surface of the solid electrolyte. The electrodes are connected to the direct current source 5 via a current meter (milliammeter) 6. Chambers 1 and 2 correspondingly have an inlet and an outlet for the analyzed and the reference gas. The temperature of the cell is provided by an external electric heater and is sustained at a specified level by a temperature regulator.

It is known that in the stable conditions the cell's EMF depends on the oxygen partial pressure in chambers 1, 2, and is calculated by the Nernst formula [10]:

$$E = \frac{RT}{nF} \ln \frac{P'_{O_2}}{P_{O_2}}, \quad (3)$$

where E – EMF of the cell, B; R = 8,314, J/mole^{-1}·K^{-1}– molar gas constant; T – absolute temperature, K; n – coefficient; F = 96484, A·c·mole^{-1} – quantity of electricity needed for the electrochemical transfer of one mole of oxygen; P'_{O_2}, P_{O_2} – partial oxygen pressure in the reference and analyzed environments respectively, Pa.

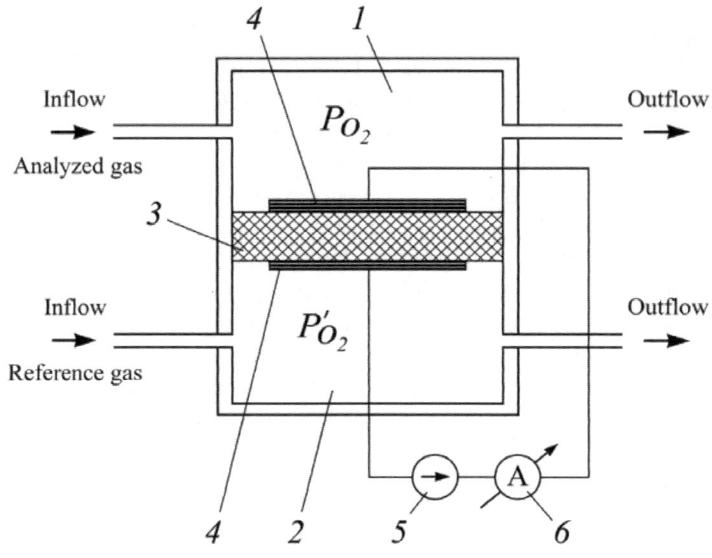

Figure 1. Coulometric solid electrolyte cell.

If the voltage of the direct current power supply is set to compensate the cell's EMF, the electrical equilibrium of the cell will not be changed. If some amount of hydrogen is put into the working chamber 1, it will react with oxygen and reduce the partial oxygen pressure and shift the equilibrium of the electrochemical reaction. The difference in the cell's EMF and the voltage of the direct current power supply will result in electric current in the external electric circuit, and oxygen will be transited from the reference environment into the working chamber through the solid electrolyte. When the ratio of the oxygen partial pressures in both chambers returns to the previous value, the system will return to equilibrium. The speed of equilibration is defined by the diffusion rate in the gas environment, the chemical reaction rate, the electric circuit resistance, and electrode polarization. Increase in cell's temperature, usage of low-polarizable electrodes and a power source with low internal resistance can significantly speed up the equilibration process.

4. Measuring method theory

In order to evaluate the dependence of hydrogen measurement error on the cell's operating conditions in the self-titration mode let's assume that all processes are reversible except for the process of mass-transfer through the solid electrolyte. Therefore, at any given moment the following Equation is true:

$$U + I \cdot r = \frac{RT}{nF} \ln \frac{P'_{O_2}}{P_{O_2}}, \tag{4}$$

where U – voltage of the direct current power supply, V; r – ohmic resistance of the solid electrolyte, Om.

If the inert gas with unknown hydrogen content is constantly passed through the working chamber, and the reference gas is characterized by constant oxygen partial pressure (i.e. atmospheric air), a certain electric current will be formed in the cell's electric circuit. According to the Faraday's law, this current will transfer oxygen from the reference environment into the working chamber. The relative amount of oxygen G_{O_2} that is transferred by the current per one volume of the analyzed gas will be calculated as follows:

$$G_{O_2} = \frac{Vm \cdot I \cdot Э}{F \cdot M_{O_2} \cdot Q}, \tag{5}$$

where Vm – molar volume, m^3/mole; Э – oxygen electrochemical equivalent, kg/C.

Oxygen partial pressure in the cell's working chamber will be calculated according to equation (4). On the other hand the partial pressures of the mixture components in the cell can be calculated in accordance with the reversible reaction:

$$H_2 + {}^1\!/_2 O_2 \rightleftarrows H_2O \tag{6}$$

The reversible reaction proceeds in the working chamber taking into account the equilibrium constant K_p of this reaction:

$$K_p = \frac{P_{H_2O}}{P_{H_2} \cdot P_{O_2}^{1/2}}. \tag{7}$$

Partial pressure of the reacting components (with the total pressure of one) relates to the hydrogen concentration C_{H_2} in the original analyzed as:

$$C_{H_2} = \frac{P_{H_2} + P_{H_2O}}{1 - P_{O_2}}, \tag{8}$$

Partial pressure of the reacting components relates to the relative amount of oxygen transferred through the solid electrolyte according to the Equation:

$$G_{O_2} = \frac{P_{O_2} + {}^1\!/_2 P_{H_2O}}{1 - P_{O_2}}. \tag{9}$$

Simultaneous solution of the Equations (5), (7), (8) и (9) can be recorded as:

$$G_{O_2} = \frac{1}{2} C_{H_2} \cdot (1 - \delta), \text{ where} \tag{10}$$

$$\delta = \frac{2P_{O_2} \cdot (1 + K_p \cdot P_{O_2}^{1/2}) - C_{H_2} \cdot (1 - P_{O_2})}{2P_{O_2} \cdot (1 + K_p \cdot P_{O_2}^{1/2}) + C_{H_2} \cdot K_p \cdot P_{O_2}^{1/2} \cdot (1 - P_{O_2})}. \tag{11}$$

Equation (10) shows that the amount of oxygen transferred through the solid electrolyte per volume of the analyzed gas is stoichiometrically related in terms of reaction (6) to the amount of hydrogen in the analyzed gas with the inaccuracy of δ.

After reducing (11) provided that $P_{O_2} << 1$ и $K_p P_{O_2} >> 1$:

$$\delta = \frac{2P_{O_2}}{C_{H_2}} - \frac{1}{K_p \cdot P_{O_2}^{1/2}}. \tag{12}$$

Simultaneous solution of the Equations (4), (10) and (12) give the following result:

$$\delta = \frac{2P'_{O_2}}{C_{H_2}} \exp\left[\left(U + \frac{C_{H_2} \cdot F \cdot M_{O_2} \cdot Q \cdot r}{44800Э}\right) \frac{nF}{RT}\right] - \frac{1}{K_p \cdot P_{O_2}^{1/2}} \exp\left[\left(U + \frac{C_{H_2} \cdot F \cdot M_{O_2} \cdot Q \cdot r}{44800Э}\right) \cdot \frac{nF}{RT}\right]. \tag{13}$$

Analysis of the Equation (13) shows that at certain conditions (gas flow Q, temperature T, ohmic resistance of the solid electrolyte r), it is possible to find such voltage U, when inaccuracy δ is changed relatively little within a wide range of measured concentrations. Calculations show that under realistic conditions δ does not exceed 0.2 % within the range of measured concentrations from 1 to 500 mln^{-1}. This inaccuracy is significantly lower than inaccuracies connected with measuring and maintaining the flow of the analyzed inert gas. If inaccuracy δ is neglected, Equations (5) and (10) can be used to derive the following Equation for establishing the hydrogen concentration:

$$C_{H_2} = \frac{2Vm \cdot I \cdot Э}{F \cdot M_{O_2} \cdot Q}. \tag{14}$$

If the analyzed gas flow Q is constant, the hydrogen concentration can be calculated using the measured cell current. The described CSEC allows to measure the oxygen and hydrogen concentration with certain accuracy is the analyzed gas contains only one of these two components. If they are contained in the gas in commeasurable quantities, they react with each other and consequently do not participate in the electrode reaction, which makes measurements indeterminate. This drawback is increasingly important taking into account that the inert gases used in industrial and laboratory practice usually contain both oxygen and hydrogen.

5. Experimental test

In order to implement the instrumental gas analysis system it is possible to use the principle of simultaneous measurement of oxygen and hydrogen microconcentrations that is demonstrated in Figure 2.

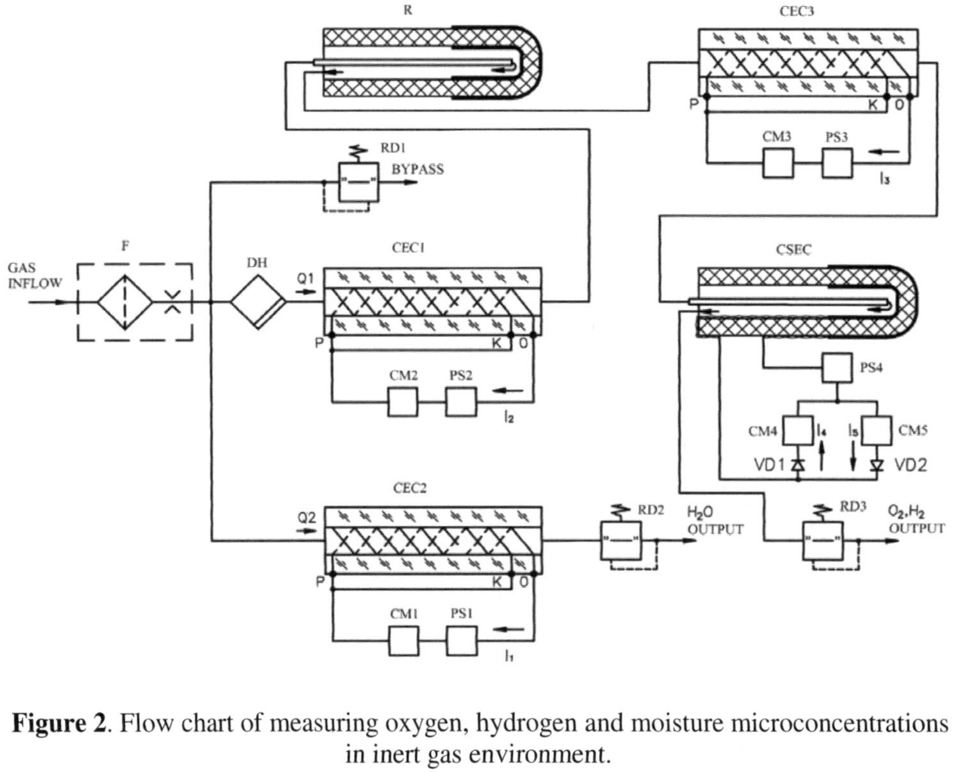

Figure 2. Flow chart of measuring oxygen, hydrogen and moisture microconcentrations in inert gas environment.

In this design the analyzed gas with constant flow Q_1 is successively passed through filter F, dehumidifier DH, coulometric cell CEC1, reactor *R*, coulometric cell CEC3 and further on through CSEC and reducer RD3 is dumped into the atmosphere. Introduction of the dehumidifier DH into the design is conditioned by the need to reduce the error of measuring oxygen and hydrogen microconcentations that occurs because of moisture presence in the gas. Phosphorous pentoxide is used as moisture absorber in the humidifier. In order to control the quality of moisture absorption in the humidifier and simultaneously to correct the results of measuring residue moisture it is reasonable to introduce a coulometric cell CEC1 into the design.

The amount of moisture generated in the reactor R is measured with the coulometric cell CEC3. All cells including the CSEC are galvanically connected with each other, are powered by separate power sources PS1, PS 2, PS3 and are connected to the measurement circuit represented by diodes VD1, VD2 and current meter CM4, CM5 in such a way to ensure algebraic addition of the cells' signals to for the required functional changes of the latter. Basically, the measurement design is a bridge with diagonal connection of CSEC. If the analyzed gas contains excessive hydrogen (scenario $C_{H_2} > 0{,}5\,C_{O_2}$), CSEC generates EMF that exceeds the compensating voltage from power source PS4. In this case CSEC circuit generates the oxygen ions transfer current I'_{O_2} that is proportional to the value ΔC_{H_2}. At the same time current flows through the current meter CM4; this current equals the sum of oxygen ions transfer current I'_{O_2} and the current of moisture electrolysis of the coulometric cell CSEC I'_{H_2O}. Simultaneously current flows through the current meter CM5; this current equals the difference of currents of CEC1 and CSEC (respectively I'_{H_2O} и I_{H_2O}).

In another case, when the analyzed gas contains excessive oxygen, (scenario $C_{O_2} > 2C_{H_2}$) CSEC generates EMF that is lower than the the compensating voltage. In this case the oxygen ions transfer current I_{O_2} that is proportional to the value ΔC_{O_2} will change its direction and will be one of the components of the current that flows through the current meter CM5.

Thus, the microconcentrations of oxygen and hydrogen in the analyzed gas can be analytically measured according to the following eq uations:

$$B_{O_2} = \frac{q}{2}\,\frac{(I_{O_2} + I_{H_2O} - I'_{H_2O})\cdot(t + 273.15)}{Q_1 \cdot P}; \tag{15}$$

$$B_{H_2} = q\,\frac{(I'_{O_2} + I_{H_2O} - I'_{H_2O})\cdot(t + 273.15)}{Q_1 \cdot P}, \tag{16}$$

where B_{O_2} – oxygen microconcentration in the analyzed gas, mln^{-1}; B_{H_2} – hydrogen microconcentration in the analyzed gas, mln^{-1}; I_{O_2} – oxygen ions transfer current if the analyzed gas contains excessive oxygen, μA; I'_{O_2} – oxygen ions transfer current if the analyzed gas contains excessive hydrogen, μA; I_{H_2O} and I'_{H_2O} – currents of moisture electrolysis in the coulometric cells installed in the gas flow beyond respectively reactor and dehumidifier, μA; t – atmosphere air temperature, °C; Q_1 – analyzed gas flow in the oxygen/hydrogen measurement channel (for measurement conditions), cm^3/min; P – atmosphere pressure, kPa; q – coefficient determined by the choice of measurement units, totals 2,582 $\frac{m\ln^{-1}\cdot cm^3\cdot kPa}{\min\cdot\mu A\cdot K}$.

In the process of moisture microconcentrations measurement the analyzed gas is continuously passed through the coulometric cell CEC2 in the direction indicated by arrows. The analyzed gas flow Q_2 is maintained on a strictly constant level.

The cell's electrodes are connected to the direct current power supply PS1 with the applied voltage exceeding the moisture decomposition potential. This ensures the electrolysis of moisture simultaneously with its adsorption, which provides the conditions for uninterrupted electrolytic regeneration of phosphorus pentoxide.

The electrolysis current is measured by the microammeter AM1 installed into the current circuit. Following reactions occur as the result of the moisture sorption from the analyzed gas flow and its electrolysis in the sorbent foil:

$$P_2O_5 + x \cdot H_2O \rightarrow P_2O_5 \cdot x \cdot H_2O. \tag{17}$$

$$P_2O_5 + x \cdot H_2O \rightarrow P_2O_5 \cdot x \cdot H_2 + \frac{x}{2} \cdot O_2, \tag{18}$$

where x – coefficient proportional to the amount of the abstracted moisture.

If the moisture microconcentrations in the analyzed gas are constant in stable conditions, balance is established between the processes (17) and (18). The Faraday's law describes the relation between the amount of substance generated on the electrodes in the electrolysis process and the quantity of electricity spend for it; according to this law, the current measured with the microammeter AM1 makes it possible to establish the moisture concentration in the analyzed gas. The following Equation describes the relation between the current and the moisture concentration:

$$B_{H_2O} = \mathrm{q}\frac{I \cdot (t + 273{,}15)}{Q_2 \cdot P}, \tag{19}$$

where B_{H_2O} – moisture concentration in the analyzed gas, mln^{-1}; I – electrolysis current, μA; Q_2 – analyzed gas flow in the measurement channel (for measurement conditions), cm^3/min.

6. Measurement error

The basic reduced measurement error of the coulometric analysis method described in Figure 2 is defined by the volume ratio of the molecular impurity (oxygen, hydrogen or water) $X_П$ and is calculated according to the following formula using control gas mixtures (CGM):

$$\gamma_{0\,X_П} = \frac{B_{Г\,X_П} - B_{ПГС\,X_П}}{П_{X_П}} \times 100, \tag{20}$$

where $\gamma_{0\,X_П}$ – basic reduced measurement error for the molecular impurity $X_П$, % ; $B_{Г\,X_П}$ – gas analyzer readings $X_П$, mln^{-1}; $B_{ПГС\,X_П}$ – volume ratio of the molecular impurity $X_П$ in the CGM, mln^{-1}; $П_{X_П}$ – measurement range for the molecular impurity type $X_П$, mln^{-1}.

Experiments show that the basic reduced measurement error for the molecular impurity $X_П$ in inert gases for measurements ranging from 0 to 5, …50 and …500 mln^{-1} of oxygen totals 6, 3.8 and 2.8 % respectively. Similarly, measurement error totals respectively 8, 3 and 3.8 % for hydrogen, and respectively 6, 3 and 3 % for water.

7. Conclusion

Study of the coulometric method of measuring concentrations of oxygen, hydrogen and moisture molecular impurities in inert gas environment by passing the analyzed gas through several electrolyte cells demonstrates the technical possibility and high efficiency of registering the impurities. Technological improvement in the manufacture of the coulometric solid electrolyte cell results in wider measurement range of oxygen and hydrogen microconcentrations. The basic reduced measurement error of molecular impurities in inert gases for measurements ranging 0–5–50–500 mln^{-1} does not exceed 8–6–3 % respectively.

Acknowledgments

The research was carried out with financial support of the Russian Science Foundation (project No. 17-79-10439).

References

[1] Piliptsou D G et al 2017 *IOP Conf. Ser.: Mater. Sci. Eng.* **168** 012103 doi:10.1088/1757-899X/168/1/012103

[2] Luchnikov P A et al 2017 *IOP Conf. Ser.: Mater. Sci. Eng.* **168** 012092 doi:10.1088/1757-899X/168/1/012092
[3] Luchnikov P A et al 2017 *IOP Conf. Ser.: Mater. Sci. Eng.* **189** 012014 doi:10.1088/1757-899X/189/1/012014
[4] Luchnikov P A et al 2017 *IOP Conf. Ser.: Mater. Sci. Eng.* **189** 012027 doi:10.1088/1757-899X/189/1/012027
[5] Monreal F J G, Vitter G J 1983 *Pliys. E. Sci. Lustrum* **16** 361–364
[6] Semchevskij A K et al *Arrangement for measuring of the concentration of oxygen in* Patent RF 2234696. Int. Cl. G01N 27/27 gases (in Russia)
[7] Pirog V P et al 2009 *Paste compound for electrodes of solid electrolyte cell* Patent RF 2343471. Int. Cl. G01N 27/30 (in Russia)
[8] Pirog V P et al 2007 *Device for measuring oxygen concentrations* Utility model patent RF 63534, Int. Cl. G01N 15/00 (in Russia)
[9] Lipnin Y A et al 2002 *Instruments* 54–57 (in Russia)
[10] Chebotin V N, Perfilyev M V 1978 *Electrochemistry of solid electrolytes* (Moscow) (in Russia)

IOP Conf. Series: Materials Science and Engineering **289** (2017) 012028 doi:10.1088/1757-899X/289/1/012028

Detection of affectings on fensing surfaces in systems of protection and life-support of buildings

L I Suchkova[1], A G Yakunin[2]
[1]Professor, Altai State Technical University named after I.I.Polzunov, Barnaul, Russia
[2]Head of chair, Altai State Technical University named after I.I.Polzunov, Barnaul, Russia
E-mail: lara8370@yandex.ru

Abstract. Typical features of oscillograms of signals from the piezoelectric transducer are installed at supernumerary affectings on barring surface. Possibility of use for identification of affectings of patterns of behaviour of group of the time rows realising the hybrid approach to the analysis of signals is shown.

1. Introduction

By consideration of functional features of a wide spectrum of security systems for buildings and premises it is possible to note, that early identification of attempts of penetration through the guarded perimeter is a perspective direction of perfection of their software and hardware. For small guarded objects access inside usually is carried out through doors [1, 2].

For anticipation of attempt of penetration through the protected perimeter of object probably to supervise following actions of the malefactor:

- approach to object on distance, smaller the admissible;
- destruction of obstacles;
- attempt of not authorised opening door and window locking mechanisms.

The most vulnerable link of protection after window apertures is the door cloth, locking which mechanisms probably to destroy or open a skeleton key. Therefore for increase of efficiency of hardware-software security systems are actual perfection existing and working out of new algorithms for analysis of the measuring information from gauges and the devices supervising influence on a door cloth. Existing vibrating and capacitor devices do not provide identification of blows, drilling process of locking mechanisms, process of unlocking of the lock by a skeleton key [3–5].

The purpose of researches is studying of features of the signals registered by piezoelectric transducers at oscillations of a door leaf in cases of regular and supernumerary opening of locking mechanisms, and development of algorithms for identification of a condition of controllable object.

2. Theoretical and experimental research methods

Theoretical and experimental methods were applied to research of oscillations of a door leaf. Revealing of a range of an oscillation frequency was carried out by mathematical modeling, namely the equation of oscillations of a plate of the set sizes dared [6]. Modeling was carried out with application of package MathCad.

Results of modeling of oscillations of a surface of a door leaf at an impulse excitation have allowed to establish numerical dependence of an oscillation frequency on density of a material of a door, thickness h, a Poisson's ratio and rigidity E (Figure 1).

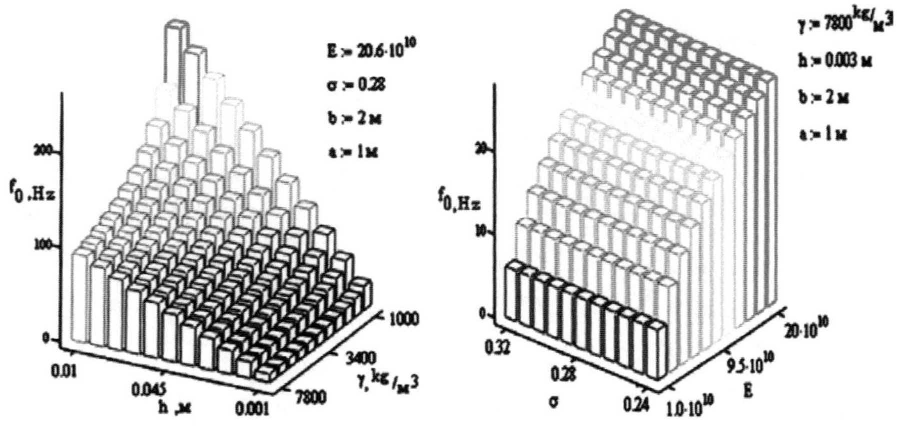

Figure 1. Results of research of an oscilation frequency of a door leaf.

It is established, that at various constructive parameters of a door leaf for identification of type of external affecting it is necessary to analyse area of the low frequencies which are not exceeding 15 кHz. Most considerably the door oscillation frequency is influenced by sizes of a linen and the characteristic of the material used at its manufacturing.

For natural experiment the arrangement which circuit design is shown in Figure 2 and containing, besides actually door leaf and piezoelectric transducers, amplifiers formed on their outputs signals, the personal computer with sound card Creative SB-128 was used [7]. All software for conducting of experimental researches has been realised with application of a package for engineering mathematical calculations MathCAD.

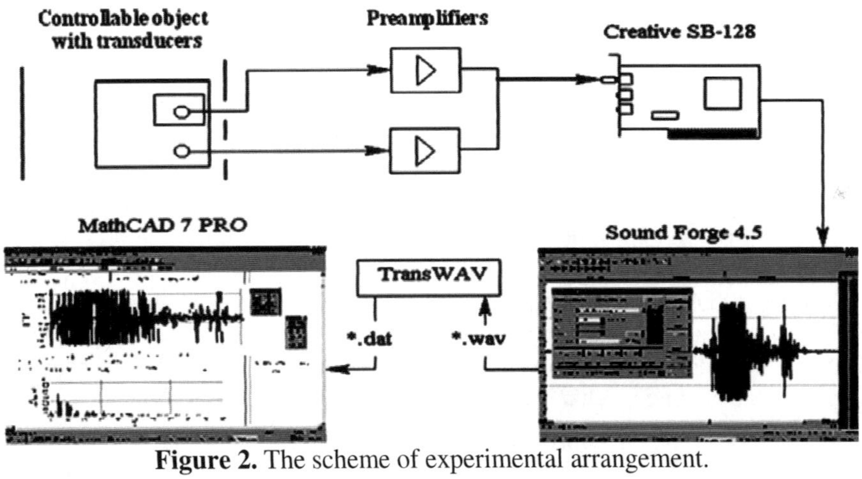

Figure 2. The scheme of experimental arrangement.

Experimental arrangement carries out removal of a signal from the piezoelectric transducer fixed on the fencing surface. Its lateral oscillations change a potential difference between the metal basis of the piezoelectic transducer and a segnetoelectric plate with a dusting from the metal, playing a role of electrodes. The potential difference depends on affecting on controllable object. The signal from the piezoelectric transducer after passage through the broadband preamplifier arrives on an entry of analogue-digitizer sound card which forms the numeric code arriving in the computer for the further complex analysis of behaviour of a signal during the chosen time gap for the purpose of identification of type of affecting [6].

By the spent researches it is established, that the oscillogram of a signal and its amplitude responses are essentially discriminated depending on a way of affecting on fensing surface: opening by a key, knock, blows, drilling locking mechanisms and skeleton key application. Experiment data are resulted in Table 1 [6, 8].

Table 1. Experiment results

Type of affecting		The maximum width of a spectrum signal, Hz	The most expressed frequency components of a signal (in decreasing order of intensity), Hz
Blow		0…2500	300…900 1300…1900
The instrumental affecting		600…6000	1300…2000 5000…5300 600…700
Opening by a key	Free moving of a key	500…3000	1500…1900 500…700
	Movement of lever mechanizm	500…6000	1500…1800 1300 5000…5600
	Movement of collar beam	0…11500	1300…2000 2300…2700 500…700 3000…3500 5500…6000

The analysis of data of Table 1 confirms results of the mathematical modeling testifying to sufficiency of processing of a signal only in the field of low frequencies of a signal for identification of attempt of supernumerary baring of a door [6].

3. Results

At regular opening of the lock fixed on a door leaf, the signal oscillogram is characterised by specific parameters to which it is possible to refer to duration of observation of signals with certain amplitude, sequence of an interleaving of intervals of the frequencies, caused by consecutive change of a position of components of the locking mechanism. In Figure 3 the oscillogram of a signal from the piezoelectic transducer at opening the lever locking mechanism by a key in a regular regime is shown. The time gap 1 matches to a key insert in a locking mechanism larva, time gaps 2 and 6 match to free moving of a key. On a time gap 3 locking mechanisms fixing a collar beam are freed, and on a period 4 the signal matching to moving of a collar beam at change of a position of a key is observed, on a gap 5 installation of lock mechanisms in the fixed position is carried out. All periods, except first, match to one turn of a key in the locking mechanism, and repeat at more quantity of turns.

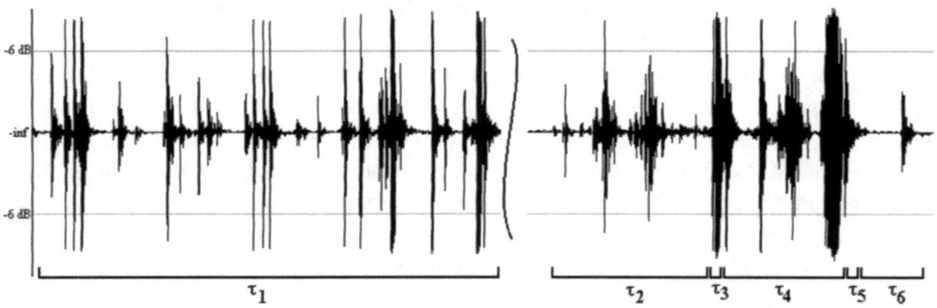

Figure 3. The signal oscillogram at opening the lever locking mechanism in a regular regime.

For locks with the cylinder mechanism on the oscillogram absence of sections of a free moving is observed at a locking mechanism rendering the conducting in a regular regime as collar beam moving do not occur. At conducting of a series of experiments for locking mechanisms of various designs of different producers delimitation of the time gaps matching to the above-stated stages of a rendering the conducting of lock mechanisms is executed.

In Figure 4 the order of occurrence of the characteristic frequency emissions appearing in a spectrum of a signal at opening of the locking mechanism in a regular regime is shown.

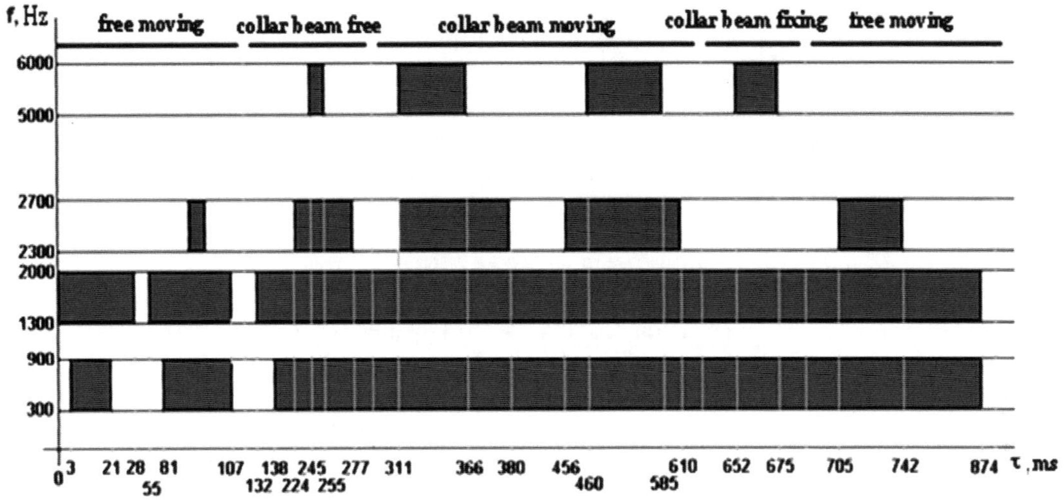

Figure 4. The signal oscillogram at locking lever mechanism opening in a regular regime.

When locking mechanism drilling the signal from the piezoelectic transducer is influenced by a rotational speed installed in the drilling apparatus. The amplitude of a signal registered by experimental arrangement belongs to an interval from 150 to 250 mV. By the spent researches it is established, that such affectings on barring surface as knock and blows, lead to fixing same on behaviour of a signal of oscillograms, difference consists only in crest values. At knock of value of amplitude make 20…100 mV, and at value blow lay over the range 250…300 mV. Duration of a gap 1 makes in both cases from 20 ms to 200 ms, finishes a gap a section of attenuation of a signal [2]. At supernumerary opening the locking mechanism on the oscillogram it is possible to observe sections, characteristic for a gap 4 (Figure 5).

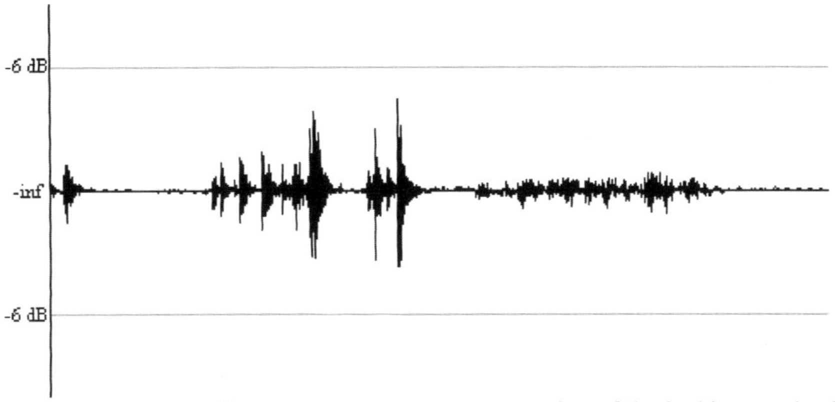

Figure 5. The signal oscillogram at supernumerary opening of the locking mechanism.

IOP Conf. Series: Materials Science and Engineering **289** (2017) 012028 doi:10.1088/1757-899X/289/1/012028

By the spent researches it is installed, that for classification of type of affecting on fencing surface the complex analysis of value of amplitude of a signal from the primary measuring converter and time intervals on which characteristic emissions of amplitude of a signal are observed is required [9,10]. So, at a regular rendering the conducting of the locking mechanism on the oscillogram some series of peak emissions with the value exceeding 100 mV are fixed, and duration of each series does not exceed 3 seconds. At trying of supernumerary opening by a skeleton key duration of excess by a signal of level 100 mV makes 3 seconds and more.

For identification of concrete type of affecting on fencing surface was possibly application of various methods of the intellectual data analysis, for example, a method of basic vectors, method of main components, neural networks added with methods spectral or cepstral analysis, including with application wavelet transformations [11–17]. In the given work for these purposes it is offered to use the hybrid approach merging fuzzy-temporal and linguistic methods of the analysis [18–21]. The analysis of the time rows which readout are values of amplitudes of a signal from the piezoelectric primary measuring converter, was carried out on the basis of a matrix predicting pattern of behaviour which is presented by a set [21]:

$$\boldsymbol{P =<Before, FBefore, DBefore, After, FAfter, DAfter, DPattern, R>.}$$

Let's observe appointment of each component of a matrix pattern of behaviour. The pattern basis is made by a matrix – template ***Before***, consisting of numbers and-or terms of linguistic variables, a lines of a matrix are compared with time processes observed during some interval. Matrix-template columns match to time countdowns, last column is formed for current moment of a time. Matrix ***Before*** is necessary for comparison with really observed on the controllable object values of time rows which can be preliminary processed by means of functions from assemblage ***FBefore,*** forming auxiliary matrix ***B'.*** Matrix ***After*** is called as predicting, it contains numbers and-or the terms of the linguistic variables matching to processes on controllable object during the moments of a time, following a present situation. As elements of matrix ***After*** can be formed by the additional analysis of values of observable time rows depending on various conditions the matrix pattern includes functions ***FAfter*** used for additional transformations of sample predicted values.

DBefore and ***DAfter*** represent descriptors of matrixes ***Before*** and ***After,*** presenting ways of transformation of elements of these matrixes by means of a set of functions of a pattern. ***DPattern*** represents a vector containing the general properties of a matrix pattern. Each pattern has label ***R***, comparing to a pattern some condition of controllable object [6, 20].

So, with reference to guarding systems discriminate, first, a regular condition when on barring construction there are no affectings, or these affectings represent noise, handicapes, procedural affectings for the purpose of regular penetration on guarded object, by, for example, a rendering the conducting of the locking mechanism and door leaf opening; and, secondly, the supernumerary condition called by disturbing mechanical affecting.

It is established, that for identification of the audible signal simple and fixed in a time the method of identification of supernumerary situations on the basis of behaviour patterns causes smaller quantity of errors in comparison, for example, with a differential method. If the signal structure is difficult, and contains the elementary fragments which sequence for the same type of behaviour of system can vary, is expedient apply to identification of an aspect of affecting the gibrid linguistic approach combining flexibility with simplicity and a high speed of identification [21].

It provides processing of values of linguistic variables. For system of identification of affectings on fensing surfaces the linguistic variables characterising duration, amplitude, a rate of speed of change of fast-head and back leading edges with reference to frequency channels have been chosen. For linguistic variables forms matching to features and other parametres of fragments of a signal terms were set. So, for the description of amplitude bending around a signal from the piezoelectic transducer the variable with five terms – “to 20”, “from 20 to 100”, “from 100 to 150”, “from 150 to 250”, “more 250” was used, fitting functions had the trapezoid form. The general number of the variables analysed by algorithms of identification with realisation of operations over fuzzy data, made 20, for four variables the information from five channels (3 frequency and 2 peak) undertook.

As a result of experiments it is established, that probabilities of delivery by security system of false alarms, the dropping of the purpose and an identification error at use of gibrid linguistic patterns of the analysis temporal regularity have made accordingly 0.7 %, 0.7 %, 1.0 %, that essentially exceeds analogous parameters of other approaches to implementation of algorithmic maintenance of devices of protection. Such devices were possibly to use as a part of adaptive subsystems of protection as they have a small number of a false wears.

4. Conclusions

The advantage of using patterns to analyze the behavior of a time series is the independence of algorithms for testing the applicability of the left part of the pattern from the pattern itself. It is possible to add new patterns without adjusting the software of the computing device that performs identification of regular and abnormal situations.

In the integrated control system for the life support of a building, in the event of an abnormal situation, it is possible to generate control actions for other subsystems. In the integrated control system for the life support of a building, in the event of an abnormal situation, it is possible to generate control actions for other subsystems. For example, we can send a signal to the lighting control subsystem to turn on the illumination of the entrance door.

Acknowledgement

Authors express profound gratitude to a management of Altay State Technical University named by I.I. Polzunov (AltSTU) for financial support of the publication of article, firm «Promautomatica» for the technical help in preparation of article and graduate AltSTU Chumakov I.A. for the help at experiment carrying out.

References

[1] Fundamentals of Information System Security. Access Control Systems [Electronic resource] URL: https://en.wikibooks.org/wiki/Fundamentals_of_Information_Systems_Security/Access_Control_Systems
[2] Doors Hinges and Security [Electronic resource] URL: https://www.statefarm.com/simple-insights/residence/3-safer-door-hinges-to-make-your-home-more-secure
[3] Scripnik D 2016 *The general questions of technical protection of the information* (Moscow: INTUIT)
[4] Avtsinov V 2000 *Working out and research of devices of the control of the security alarm system on the basis of the capacitor converter with a branching of currents in a measuring chain* [Electronic resource] URL: http://www.dissercat.com/content/razrabotka-i-issledovanie-priborov-kontrolya-okhrannoi-signalizatsii-na-osnove-emkostnogo-pr#ixzz4uzvR3rAm (in Russian)
[5] Yakunin A, Avtsinov V, Chernov K 2000 *Measurement, control, automation-2000: Proceedings of the International Scientific Conference* 172–173 (in Russian)
[6] Suchkova L, Chumakov I, Yakunin A 2013 *Identification of affectings in devices of protection of anticipatory type* (Deutschland, Saarbrücken, Palmarium Academic Publishing)
[7] Gibson S 2001 *The Illustrated Dictionary of Electronics* (USA,McGraw-Hill Publications)
[8] Yakunin A, Chumakov I 2003 *61th AltSTU scientific conference: Proceedings of the Conference. Part II.* 26–27 (in Russian)
[9] Chen J, Benesty J, Huang Y 2006 *EURASIP J. Appl. Signal Processing* **2006** 26503 doi: 10.1155/ASP/2006/26503
[10] Knapp C, Carter G C 1976 *IEEE Transactions on Acoustic, Speech, and Signal Processing* **24** 320–327 doi: 10.1109/TASSP.1976.1162830
[11] Nwankwo P N, Nsionu I I, Ezeilo C J 2013 *Journal of Automation and Control Engeneering* **1** 65–69

[12] Ogri U J, Bassey D E 2013 *International Journal of Engineering and Computer Science* **2** 2235–2257
[13] Chung E, Saltzman A Smart locks could make your home less secure [Electronic resource] URL:http://www.cbc.ca/news/technology/smart-locks-could-make-your-home-less-secure-1.3057872
[14] Bamisaye A J, Samuel A 2016 *J Electr Electron Syst* **5** 167 doi:10.4172/2332-0796.1000167
[15] Sarp B, Karalar T, Kusetogullari H 2015 *International Journal of Scientific Research in Information Systems and Engineering* (IJSRISE) **1**
[16] Song K-T, Chen J-L 2003 *IEEE International Symposium. Computational Intelligence in Robotics and Automation* Istanbul TURKEY **3** 1445–1450 doi: 10.1109/CIRA.2003.1222210
[17] Freund Y, Schapire R E 1997 *Journal of Computer and System Sciences* **55** 119–139
[18] Kovalev S M 2007 *Proceedings of the* IV *International Scientific Conference* Moscow 354 (in Russian)
[19] Batyrshin I, Kacprzyk J, Sheremetov L et al 2007 *Series: Studies in Computational Intelligence* **36**
[20] Suchkova L, Abdenov A 2013 *Scientific Herald of Novosibirsk State Technical University* **3** 78–83 (in Russian)
[21] Suchkova L 2014 *Polzunov Herald* 45–51 (in Russian)

The selection criteria elements of X-ray optics system

I V Plotnikova[1], N V Chicherina[2], S S Bays[2], R G Bildanov[3,4] and O Stary[5]
[1]Associate Professor, National Research Tomsk Polytechnic University, Tomsk, Russia
[2]Graduate student, Kazan National Research Technical University named after A.N. Tupolev, Kazan, Russia
[3]Senior lecture, Ural Federal University, Ekaterinburg, Russia
[4]Leading researcher, Technological Research Institute of Ulyanovsk State University, Ulyanovsk, Russia
[5]Professor, Czech Technical University in Prague, Prague, Czech Republic
E-mail: inna@tpu.ru

Abstract. At the design of new modifications of x-ray tomography there are difficulties in the right choice of elements of X-ray optical system. Now this problem is solved by practical consideration, selection of values of the corresponding parameters – tension on an x-ray tube taking into account the thickness and type of the studied material. For reduction of time and labor input of design it is necessary to create the criteria of the choice, to determine key parameters and characteristics of elements. In the article two main elements of X-ray optical system – an x-ray tube and the detector of x-ray radiation - are considered. Criteria of the choice of elements, their key characteristics, the main dependences of parameters, quality indicators and also recommendations according to the choice of elements of x-ray systems are received.

1. Introduction

In the modern world, the most widespread systems for the industrial nondestructive control are x-ray tomography, which are divided on several types: the sizes of the studied quantity, technical characteristics, a scope, etc. As a popular direction of systems' development can be considered microtomography [1]. These systems allow investigating internal structure of objects with micron sizes. These systems are characterized by tension on an x-ray tube from 10 to 160 kV, the small size of a focus spot - from 1 to 10 microns [2, 3]. As betatrons belong to another class of x-ray systems due to other principle of realization and technical characteristics they aren't considered in this article.

At the moment four companies Matrix (Great Britain), SkyScan (Belgium), Feinfocus (Germany), Fraunhofer (Germany) which offer the micro tomography meeting the modern advanced requirements in the field of energy saving, functional characteristics, the accuracy of measurement and a bunch of x-ray radiation are presented at the market [4].

By the creation of new systems of the x-ray tomography the design stage is one of fundamental. At this stage key parameters of technical system according to the specification pay off, essentially new technical solutions are developed [5, 6], configuration of system is made [7].

The activities for development and design of x-ray systems have a skilled character, the choice of a complete set and parameters of system is made by means of theoretical formulas and also intuitive selection of accessories which correspond to these parameters [8]. Before the research scientists and engineers there is always a sensitive issue about the choice of elements of an X-ray optical system [9].

IOP Conf. Series: Materials Science and Engineering **289** (2017) 012029 doi:10.1088/1757-899X/289/1/012029

At design of new systems first of all are guided by the following components: technical parameters of system's elements, a exterior and operating modes of the detector, the size and a type of the studied material [10], requirements to accuracy [11], prime cost, admissible working conditions [12], energy consumption [13–15].

Therefore, it is extremely important to define and mark out the main choice criteria of elements of X-ray optical system.

2. Methodology

The X-ray optical system intended for nondestructive control consists of four elements: an X-ray tube, the detector of X-ray radiation, the system of positioning of a desktop, the studied material [16].

In this article bigger attention is paid to two key elements of X-ray optical system – to an X-ray tube, the detector of X-ray radiation and also the geometrical and spatial sizes of the most X-ray optical system. The final quality of the shadow image, in particular, its sharpness, clearness, optimum scale, an opportunity to recreate the three-dimensional image depends on the choice of the correct distance.

As it has been marked above, a basic element of X-ray optical system is the x-ray tube. We will consider her in more detail. An X-ray tube – the electric vacuum device with a source of radiation of electrons (cathode) and a target in which they are braked (anode) [16]. The main technical characteristics of sources of X-ray radiation are: type of an X-ray tube; range of tension; step of the choice of tension; range of currents on the anode; the maximum power on the anode; power of a x-ray tube; size of a focal spot, mm; radiation angle of emergence, degree.

The key characteristics of X-ray tubes defining them opportunities:

•The focusing elements are electromagnetic or other elements in a tube which "squeeze" the accelerated electrons into so small point on a target as far as it is possible. This point on a target is called a focus spot. The less focus spot, the better permission of the final image is.

•By the size of a focus spot the tubes a is divided into two types: microfocus and nanofocus. The size of a focus spot in microfocus tubes is 3 mkm.

•Exterior and type of a target. The type of a target exerts the direct impact on the size of the shadow image [17]. It also should be noted that the minimum distance between a focal spot and the studied material for various types of targets it is various.

•The target material and material thickness. for increase in service life the good density of a stream of x-ray radiation and at the same time its weak absorption when passing through a target is necessary. Most often the tungsten is used for a target.

Wavelength of X-ray radiation defines his energy, i.e. ability to get into the object. The X-ray radiation with shorter wavelength formed at higher value of kW has bigger penetration in comparison with X-ray radiation with the bigger wavelength (less energy radiation). It should be taken into account as an important nuance that in X-ray tubes with tension to 60 kV only 0.1% of energy will be transformed to energy of X-ray radiation. At a voltage of 100 kV the efficiency of a tube increases to 1%.

We will consider the behavior of X-ray radiation. An object of a research is the most important data source for determination of such parameters as energy of X-ray radiation (tension on a tube), type of a source of X-ray radiation, thickness of protective lead screens (from disseminated) and the scheme of raying. All these parameters are chosen depending on the geometrical sizes of a controlled product that the sensitivity of control didn't exceed a half of the size on depth minimum of unacceptable defects. In the Figure 1 the dependence of thickness of the illuminated material on tension on an X-ray tube is represented. The studied material in this case is iron, an X-ray tube of Hammatsu. These values are received in the empirical way. It is extremely important.

Activities for determination of system's optimum parameters have a skilled character, the choice of a complete set and the modes of system is made by means of theoretical formulas and also intuitive settings. These calculations characterize only a situational (static) condition of X-ray optical system and don't allow estimating the fully behavior of system (change of output parameter) at change of factors' values. The problem is solved with the help of mathematical dependence. As it is shown in the Figure 1 this dependence has the form of a curve and is described by $y = 0.0005x^2 + 0.01x + 0.278$ equation. The share of this factor is 99.68%. It says that the equation has very high rates of accuracy.

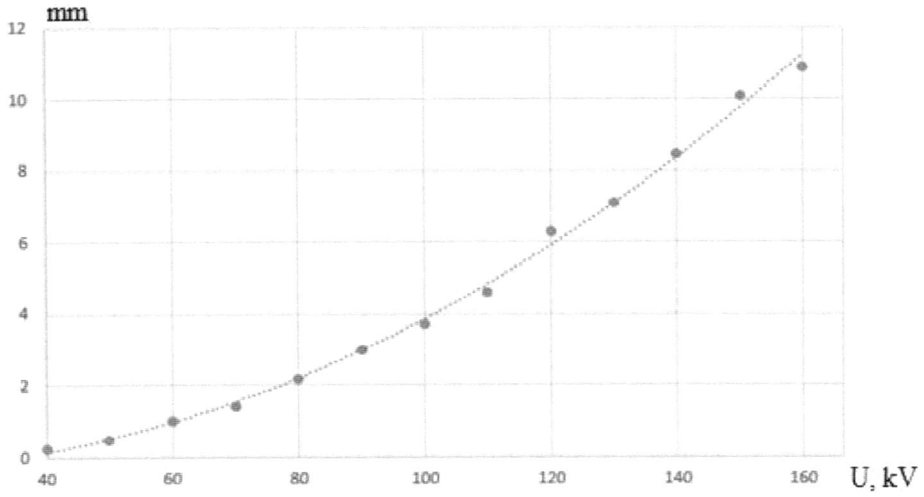

Figure 1. The dependence of thickness of the illuminated material on the tension.

An important role in X-ray optical system it is allocated for the detector of the X-ray radiation and also the distance between an X-ray tube investigated by material and the detector. In the majority of computer X-ray tomography the digital detectors are used. The X-ray digital detector is a device which is intended for obtaining digital X-ray images.

Digital detectors for X-ray have the following main technical characteristics:

• Size of active area of the detector.
• Factor of optical susceptibility of the detector's surface.
• Size/step of a point (pixel) of digital images.
• Extreme spatial resolution.
• Matrix size of the image.
• Graded resolution.

Key parameters of quality of the image, received by means of digital detectors are:

• Unsharpness of the received pictures. This property is characterized by the relation of geometrical unsharpness on projective increase and unsharpness of the detector which are described by the main spatial resolution.

• A contrast ratio to noise. The given characteristic is based on the relation of a signal and noise and also depends on the coefficient of effective absorption of a research object.

• Ratio of a signal to noise. The given characteristic is based on the rated relation of a signal to noise as a dependence of radiation quality and an exposition dose).

• Basic spatial resolution. It exerts impact on the size of effective pixel of a matrix.

• Contrast sensitivity. The characteristic is described by the size equal to the contrast relation to image noise level. Calculates taking into account coefficient of dispersion and coefficient of absorption μ. The characteristic influences the sizes of the studied material and size of an error of measurements.

• Spatial resolution. The increase in the physical sizes of detectors leads to increase in the size of a semiconductor photosensitive matrix, and it always influences the price. Producers differently try to resolve this contradiction. For example, the detector sometimes moves along a sample [16]. The received parts of the image "are sewed" in a common 3D model by means of specialized software. Despite seeming simplicity, it is connected with some technical difficulties. The successful stitching of the image requires the precision system of the detector's movement. And again it should be noted that the equipping the tomography with the similar system of the detector's movement doesn't lead to reduction of its cost at all.

For the analysis nine digital X-ray tomography, which are presented, at the market have been chosen: Hammatsu, GEDXR250, GE DXR250RT, Perkin Elmer XRD0840, Varian PaxScan 1313, Toshiba E5877J, Toshiba E5876J, SkyScan 210, RadEye200.

The possible to define that the dependence of width of an object on distances on a tube to an axis of rotation of an object changes under the linear law. As a result, the more the value of length of an object at the fixed distance, the broader the range of the studied sizes is that does detectors with these parameters better than competitors [18].

Mathematically projective increase in M can be calculated on the following relations:

$$M = \frac{D_F}{D_O} = \frac{d_f}{d_o},$$

where d_0 is a distance from a focus spot of an X-ray tube to an object, d_f is distance from a focus spot of an X-ray tube to the detector, D_o– the research object size; D_f– projective increase in an object of a research.

From the Figure 2 it is seen that geometrical increase decreases under the hyperbolic law with increase of values of distance from a tube to an object axis.

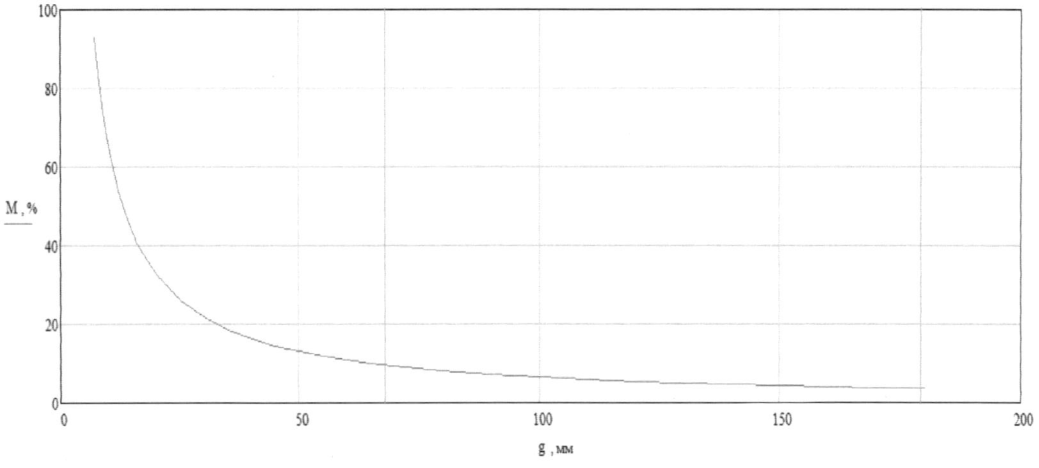

Figure 2. The schedule of geometrical increase from distances from a tube to an object.

The size of voxel calculates as the relation of the pixel size to the geometrical increase of M

$$\left(M = \frac{H}{g}\right): \; v = \frac{p}{M} = \frac{pg}{H}.$$

The size of voxel is an important characteristic of the detector for obtaining the shadow tiff-images and the subsequent three-dimensional reconstruction. She determines the quality of the picture, its specification. Therefore, it is very important to design and choose the sizes of X-ray optical system correctly. The focus length should be chosen so that the geometrical unsharpness was commensurable with value of internal unsharpness. Increase in focus length allows to reduce unsharpness, but at the same time the intensity of radiation decreases and exposition time increases. In X-ray tubes the small focal spot allows to get better permission. The bigger the spot is, the more indistinct contours are and permission of the received image is more limited.

From the Figure 3 it is possible to define, that the size of the detector's voxel changes under the linear law depending on changes of values of distances from an X-ray tube to an axis of an object rotation. The best indicators from the detectors given in this calculation were shown by SkyScan and Hammatsu.

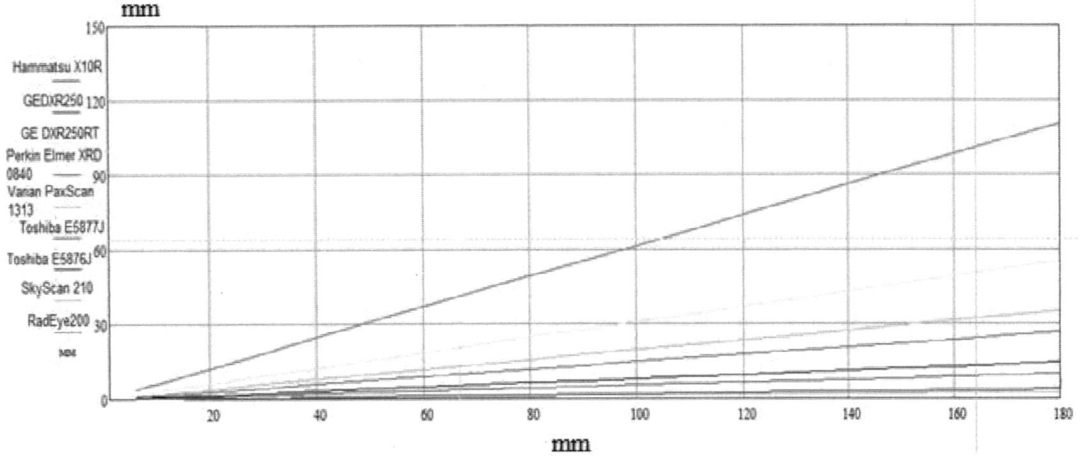

Figure 3. The schedule of dependence of the size of voxel from distances to a rotation axis.

The possibility of reliable reconstruction of the dose field and, therefore, structure of an object of diagnostics significantly depends on the mode of carrying out a tomography. The quality of the received image and the resolution of system depend on the choice correctness of such parameters as distance, tension, current, etc. For example, the image removed at a voltage of 80 kV, noisier, than the removed one at 140 kV. It's extremely important in the choice of elements of X-ray optical system to rely on the criteria offered in the article.

3. Summary

In the conclusion, there is a wish to note that in the article the choice criteria of the elements, their key characteristics, the main dependences of parameters, the quality indicators and also recommendations after the choice of elements of X-ray systems are received. Knowing behavior, the dependences of parameters of X-ray optical system received in the empirical way, it is possible to predict easily and quickly the size of output intensity, tension, current on the anode, the size of a focal spot and voxel. Knowing the dependence obtained from practical data, can accurately and fully describe the system. This fact distinguishes this mathematical dependence on the available theoretical means of calculation. This model has the importance in practice as it allows to receive better shadow images for the creation of three-dimensional images and also facilitates work of engineers and constructs on development of new modifications of X-ray systems. The offered criteria not only will facilitate a set of works on design of X-ray systems, but also it will become useful at the choice of necessary parameters of system for a concrete X-ray research.

References

[1] Sasov A et al 1998 *J Microsc* **191** 151–158 doi: 10.1046/j.1365-2818.1998.00367.x

[2] Galtseva O V *et al* 2016 *IOP Conf. Ser.: Mater. Sci. Eng.* **110** 012094 doi: 10.1088/1757-899X/110/1/012094

[3] Landisa E N, Keane D T 2010 *Mater Charact* **61** 1305–1316 doi: 10.1016/j.matchar.2010.09.012

[4] Goldshtein A E *et al* 2016 *J. of Physics: Conf. Ser.* **671** 012062 doi: 10.1088/1742-6596/671/1/012062

[5] Bemš J, Králík T, Kubančák J, Vašíček J, Starý O 2014 *Radiation Physics and Chemistry* **104** 398–403 doi:10.1016/j.radphyschem.2014.02.008

[6] Fedorov E M, Koba A A 2017 *2016 Dynamics of Systems, Mechanisms and Machines, Dynamics 2016* 7819008 doi: 10.1109/Dynamics.2016.7819008

[7] Nizhegorodov A et al 2016 *J Vibroeng* **18** 3734–3742 doi: 10.21595/jve.2016.16994
[8] Luchnikov P A et al 2017 *IOP Conf. Ser.: Mater. Sci. Eng.* **189** 012014 doi: 10.1088/1757-899X/189/1/012014
[9] Glazyrina E 2004 *8th Korea-Russia International Symposium on Science and Technology - Proceedings: KORUS* 123–124
[10] Goldshtein A E, Fedorov E M 2010 *Russian Journal of Nondestructive Testing* **46** 424–430 doi: 10.1134/S1061830910060069
[11] Goldshtein A E et al 2016 *MATEC Web of Conferences* **79** doi: 10.1051/matecconf/20167901009
[12] Gavrilin A, Moyzes B, Zharkevich O 2015 *Journal of Vibroengineering* **17** 3495–3504
[13] Obukhov S et al 2016 *IOP Conf. Ser.: Mater. Sci. Eng.* **132** 012017 doi: 10.1088/1757-899X/132/1/012017
[14] Obukhov S G et al 2017 *IOP Conf. Ser.: Mater. Sci. Eng.* **189** 012008 doi: 10.1088/1757-899X/189/1/012008
[15] Akhmetbayev D S et al 2017 *Results in Physics* **7** 1644–1649 doi: 10.1016/j.rinp.2017.03.010
[16] Dunsmuir J et al 2006 *Powder Diffraction* **21** 125–131 doi: 10.1154/1.2204956
[17] Kerridge B 2002 *Sharpen X-ray images*. Retrieved from [Electronic resourse] URL:https://www.edn.com/design/test-and-measurement/4385159/Sharpen-x-ray-images
[18] Li Ya 1992 *NDT & E International* **25**183–189 doi: 10.1016/0963-8695(92)90159-E

IOP Conf. Series: Materials Science and Engineering **289** (2017) 012030 doi:10.1088/1757-899X/289/1/012030

On the connection of the generalized cascade-probability function with the Boltzmann equation

N A Voronova[1], V I Kirdyashkin[2], A I Kupchishin[3,4], A A Kupchishin[5]
[1] Senior researcher, Abay Kazakh National Pedagogical University, Almaty, Kazakhstan
[2] Senior researcher, Abay Kazakh National Pedagogical University
[3] Head of laboratory, Abay Kazakh National Pedagogical University
[4] Al-Farabi Kazakh National University, Almaty, Kazakhstan
[5] Senior researcher, Abay Kazakh National Pedagogical University
E-mail: ankupchishin@mail.ru

Abstract. It was revealed that the generalized cascade-probability function (GCPF) can be obtained from the three-dimensional equation of the cascade process (of the Boltzmann type). It was established that this function, depending on the number of interactions, first increases, reaches a maximum, and then decreases. With the increase in H, the maximum of GCPF is shifted to the right, and its value decreases.

1. Introduction
When modeling radiation-physical processes, there is often a problem of choosing a theoretical calculation method, developing physical and mathematical models of research. At present, the most well-known and widely used theoretical methods for calculating the passage of particles through matter in physics are the Monte Carlo method, the Boltzmann kinetic method, and so on [1,2].

The analytical-numerical method of calculation, called cascade-probabilistic (CP), was developed in the process of many years of work in the field of applied mathematics [3], elementary particle physics, cosmic ray physics [4,5], radiation physics of materials [6–9], microelectronics [10, 11] and positron physics [12], and also on the basis of analysis and generalization of these studies. Its validity has been verified on a large number of specific problems from various fields of nuclear physics and solid state physics [6]. In the framework of the cascade-probabilistic method [12], we also succeeded in obtaining an analytical expression and calculating the threshold energy of displacement for various elements. The history of the CP-method is connected with the name of the great mathematician of the last century, S.D. Poisson, in particular with the distribution of random variables, named in his honor [13]. The Poisson distribution for certain values of the parameters is the limiting transition of the CP-functions, which are the basis of the CP-method [6].

2. Physical model
Let us consider the passage of particles of one kind through the material. Let a particle (nucleon, electron, positron, primary-knocked out atom, etc.) be generated at a certain variable depth x at the zenith and azimuth angles γ and φ to the chosen direction (relative to the perpendicular to the sample surface or with respect to the zenith). The most realistic description of the process is the three-dimensional model of particle distribution in matter, since the trajectory of particle motion is twisted strongly in space.

In the balance equation for the particle flow at the depth of x we will have a decrease in the particles due to their absorption, on the other hand, formation due to generation. In this case, the equation of the cascade process (of the Boltzmann type) for the stationary case can be written in the form of:

$$\frac{dN(E,\theta,\varphi,x)}{dx}=-\frac{N(E,\theta,\varphi,x)}{\lambda(E)\cos\theta}+\frac{1}{2\pi}\int_{E}^{E_m}\int_{\theta_1}^{\theta_1'}\int_{\varphi_1}^{\varphi_1'}N(E',\theta',\varphi',x)\frac{n(E',E,\theta',\theta)}{\lambda(E')\cos\theta'}dE'd\theta'd\varphi', \tag{1}$$

with the initial condition $N(E,\theta,\varphi,x)=N(E,0)$.

Here, $N(E,\theta,\varphi,x)$ is the particle flow at the depth of x; $dN(E,\theta,\varphi,x)|dx$ is the change in the radiation flow in an infinitesimally small dx layer, $\lambda(E)$ and γ are the interaction range and the emission angle of the particle relative to the vertical after the interaction, $n(E',E,\theta',\theta)$ is the normalized differential energy spectrum of the secondary particles in an elementary act, E is energy.

3. Results and discussion

Let us find the solution of equation (1) in a general form. After the Laplace transform, we have:

$$N(E,\theta,\varphi,p)=\frac{1}{p+\frac{1}{\lambda\cos\theta}}\frac{1}{2\pi}\int_{E}^{E_m}\int_{\theta_1}^{\theta_1'}\int_{\varphi_1}^{\varphi_1'}N(E',\theta',\varphi',p)\frac{n_p(E',E,\theta',\theta,\varphi',\varphi)}{\lambda(E')\cos\theta'}dE'd\theta'd\varphi'+\frac{N(E,\theta,\varphi,0)}{P+\frac{1}{\lambda\cos\theta}}. \tag{2}$$

where p is the transformation parameter.

In the general case, the $N(E,\theta,\varphi,p)$ value can be represented as follows:

$$N(E,\theta,\varphi,p)=\sum_{n=0}^{\infty}N_n(E,\theta,\varphi,p), \tag{3}$$

where

$$N_n(E,\theta,\varphi,p)=\int_{E}^{E_m}\int_{\theta_1}^{\theta_1'}\int_{\varphi_1}^{\varphi_1'}\ldots\int_{E_{n-1}}^{E_m}\int_{\theta_{1n-1}}^{\theta_{1n-1}'}\int_{\varphi_{1n-1}}^{\varphi_{1n-1}'}N(O,E_n,\theta_n,\varphi_n)\frac{n_p(E_n,E_{n-1},\theta_n,\theta_{n-1},\varphi_n,\varphi_{n-1})}{2\pi}\ldots*$$
$$*\frac{n_p(E_1,E,\theta_1,\theta,\varphi_1,\varphi)}{2\pi}\prod_{i=0}^{n}\frac{1}{p+\frac{1}{\lambda(E_i)\cos\theta_i}}\frac{dE_nd\theta_nd\varphi_n}{\lambda(E_n)\cos\varphi_n}\ldots\frac{dE_1d\theta_1d\varphi_1}{\lambda(E_1)\cos\varphi_1}. \tag{4}$$

Making the inverse Laplace transform, we find the original. After the appropriate calculations, we obtain:

$$N(E,\theta,\varphi,x)=\sum_{n=0}^{n}Y_n(E,\theta,\varphi,x), \tag{5}$$

$$J_n(E,\theta,\varphi,x)=\int_{E}^{E_m}\int_{\theta_1}^{\theta_1'}\int_{\varphi_1}^{\varphi_1'}\ldots\int_{E_{n-1}}^{E_m}\int_{\theta_{n-1}}^{\theta_{n-1}'}\int_{\varphi_{n-1}}^{\varphi_{n-1}'}N(E_1,\theta_1,\varphi_1,0)\frac{n_p(E_n,E_{n-1},\theta_n,\theta_{n-1},\varphi_n,\varphi_{n-1})}{2\pi}\ldots\times$$
$$\times\frac{n_p(E_1,E,\theta_1,\theta,\varphi_1,\varphi)}{2\pi}\psi_n(x,\lambda_0\div\lambda_n)dE_nd\theta_nd\varphi_n\ldots dE_1d\theta_1d\varphi. \tag{6}$$

where

$$\psi_n(x, \alpha_0 \div \alpha_n) = \sum_{j=0}^{n} \frac{\alpha_n \alpha_j^{n-1}}{\prod\limits_{\substack{i=0 \\ i \neq j}}^{n} (\alpha_j - \alpha_i)} exp\left(-\frac{x}{\alpha_j}\right). \tag{7}$$

$\alpha_n = \lambda_n(E)\cos\theta_n$; $\alpha_i = \lambda_i(E)\cos\theta_i$; $\alpha_j = \lambda_j(E)\cos\theta_j$.

As can be seen from the solution of (6), it includes a generalized cascade-probability function that takes into account the distribution of radiation in three-dimensional space. It has the sense of the probability that a particle generated at the depth of x' will reach the depth of x, after n number of collisions. Thus, GCPF can be obtained from the three-dimensional Boltzmann equation. Next, we developed an algorithm, compiled a program and performed calculations of this function on a computer. The results of the calculations are shown in Figure 1 as the dependence of $\psi_n(x, \alpha_0 \div \alpha_n)$ on n for different values of $H = (x - x)/\alpha_n$. It follows that this function, depending on the number of interactions, first increases, reaches a maximum, and then decreases (solid lines). With increasing H, the maximum of the generalized cascade-probability function shifts to the right and decreases, and its maximum value is at $n = (x - x')/\alpha_n$. It also follows from the Figure 1 that a slight change in H (by 1%) leads to a change in the CP function by more than 40% (dashed lines).

We also carried out a mathematical analysis of GCPF, which shows that the equation (7) satisfies all the properties of probability theory and can be obtained not only from the Boltzmann equation, but also from the recurrence relation:

$$\psi_i(h', h, \alpha_0) = \int_{h'}^{h} \psi_{i-1}(h', h'', \alpha_0)\, \psi_0(h'', h, \alpha_0) \frac{dh''}{\alpha_0}$$

$$\text{at } \psi_0(h', h, \alpha_0) = \exp\left(-\frac{h - h'}{\alpha_0}\right)$$

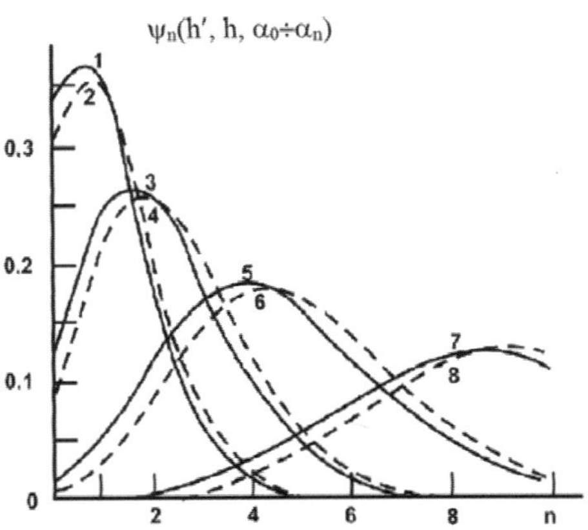

Figure 1. Dependence of the generalized CPF on the number of interactions n for different values of H = (h – h') / α_n = 1, 2, 4 and 7, respectively: 1, 3, 5, 7 – generalized CPF at $\alpha_n = 0.99\alpha_{n-1}$: 2, 4, 6, 8 – the simplest CPF [6].

4. Conclusions

1. The connection between the generalized cascade-probability function (GCPF) and the three-dimensional Boltzmann equation is established. It is shown that GCPF can be obtained from this equation by the Laplace transform method.
2. The calculation of GCPF on a computer was made. It is revealed that this function, depending on

the number of interactions, first increases, reaches a maximum, and then decreases. With increasing H, the maximum of the generalized cascade-probability function shifts to the right, and its value decreases, with the most probable number of interactions $n = (h - h') / \alpha_n$.

Acknowledgments
The study was supported by the Grant of the Ministry of Education and Science of the Republic of Kazakhstan.

References

[1] Ed. AM Prokhorov 1983 *Physical encyclopaedic dictionary* (Moscow: Soviet Encyclopedia)
[2] Ed. IM Vinogradov 1984 *Mathematical Encyclopedia* (Moscow: Soviet Encyclopedia)
[3] Kupchishin A A, Kupchishin A I, Shmygaleva T A at al. 1994 *Modelling, Measurement & Control, A, AMSE* **55** 49 – 55
[4] Efimov A K, Kolomeets E V, Boos E G at al 1973 *Int. conf. of Cosmic Rays Denver* **2** 863 – 868
[5] Baysakalova A B, Bogdanova O A, Iskakov T Z at al. 1975 *Proceedings of AS USSR Physical series* **39** 264 – 271
[6] Kupchishin A I, Kupchishin A A, Shmygaleva T A 2015 *Cascade-probabilistic method. Solution of radiation-physical problems, Boltzmann equations. Connection with Markov chains* (Almaty Pibl. Abay KazNPU, Kama LLP, SRINCTaM Al-Farabi KazNU)
[7] Voronova N A, Kirdyashkin V I, Gyngazov V A at al. 2016 *IOP Conf. Series: Material Science and Engineering* **110** 012039 doi: 10.1088/1757-899X/110/1/012039
[8] Aleutdinov A D, Ghyngazov S A, Mylnikova T S, Luchnikov P A 2015 *IOP Conf. Series: Material Science and Engineering* **81** 012069 doi: 10.1088/1757-899X/81/1/012069
[9] Ghyngazov S A, Vasil'ev I P, Surzhikov A P, Frangulyan T S and Chernyavskii A V 2015 *Technical Physics* **60** 128 – 132 doi: 10.1134/S1063784215010120
[10] Luchnikov P A, Vetrova V V, Zubkov N P, Surzhikova O A, Chernova N I 2017 *IOP Conf. Series: Material Science and Engineering* **189** 012027 doi: 10.1088/1757-899X/189/1/012027
[11] Luchnikov P A, Surzhikova O A 2016 *IOP Conf. Series: Material Science and Engineering* **189** 012014 doi: 10.1088/1757-899X/189/1/012014
[12] Kupchishin A I, Kupchishin A A, Voronova N A at al. 2016 *IOP Conf. Series: Material Science and Engineering* **110** 012040 doi: 10.1088/1757-899X/110/1/012040
[13] Poisson S D Recherches sur la probabilite des jugements en matiere crimenelle et en matiere cioile, precedees des regles generales calcul des probabilites 1837

IOP Conf. Series: Materials Science and Engineering **289** (2017) 012031 doi:10.1088/1757-899X/289/1/012031

Composite material for optical oxygen sensor

A P Antropov[1], A V Ragutkin[1], P V Melnikov[1], P A Luchnikov[2], N K Zaitsev[1]
[1]Associate professor, Moscow Technological University (MIREA), Moscow, Russia
[2]Lead engineer, Moscow Technological University (MIREA), Moscow, Russia
E-mail: xamdex@gmail.com

Abstract. A new composite material for use in optical molecular oxygen sensors is proposed. The absence of pores on the surface of the material avoids microbiological fouling and concomitant deterioration of the characteristics with time, and the presence of the mesoporous phase results in a linear calibration and acceptable response times, even for layers that are significant in thickness.

1. Introduction

A very urgent task for aquaculture and other production processes associated with the saturation of solutions with molecular oxygen, for example microbiological processes in wastewater treatment, is the creation of dissolved oxygen (DO) analyzers and biosensors based on them, which are resistant to poisoning and microbiological fouling. The introduction of such operational monitoring devices will allow the creation of self-regulating systems with significantly increased energy efficiency.

At present the optical method of DO measurement is considered to be the best for these purposes in comparison with the traditional Winkler titration and amperometric Clark electrode [1]. However optical sensor poses a number of requirements for the matrix material in which the sensing dye is distributed. First of all, it must be resistant to singlet oxygen, an unavoidable by-product in the optical measurement. It is also extremely desirable to linearize the calibration curve, as this facilitates the operation of the sensor. Exploitation in a medium with a high content of biomass requires protection of the surface of the sensing element from biological fouling.

Mesoporous structures obtained by the sol-gel process [2, 3] satisfy most of the requirements. They are widely used including biosensors [4], but their exploitation in environments with a high content of biomass is associated with the inevitable contamination of voids. As a result, the performance of the sensor deteriorates.

Perfluorinated polymers are the alternative materials with good oxygen permeability [5] and almost omniphobic surface that prevents the adhesion of the biomaterial. Their drawbacks include the nonlinearity of calibration curve, which sometimes is very significant [6], possible aggregation of dye molecules due to poor solubility in the polymer, as well as poor adhesion to the substrate material.

Another approach is the chemical immobilization of the dye molecule on the substrate. It allows one to obtain linear calibration and makes the sensor resistant to many solvents, but worsens the photostability of the indicator [7].

In this paper we propose the composite material combining good oxygen permeability of fluorinated polymers with a uniform distribution of the photostable dye on a large surface and linear calibration, which are the case for mesoporous structures.

IOP Conf. Series: Materials Science and Engineering **289** (2017) 012031 doi:10.1088/1757-899X/289/1/012031

2. Experimental

2.1. Sensor fabrication

The schematic structure of the composite material is shown in the Figure 1. It consists of two phases with different functionalities. The gas-permeable layer (2) is made of amorphous fluorinated polymer. In present work we used fluoroplastic 42V (similar to Kynar 7200), because it is highly stable, has good adhesion to the substrate and is soluble in a wide range of solvents. Another phase is presented by microparticles of nanostructured SiO_2 (3), which are used as centers for the distribution of the sensing dye. Pt(II) complex of meso-tetra(pentafluorophenyl)porphyrin (Pt-TFPP) was chosen as an indicator due to excellent photostability, good quantum yields [8] and commercial availability. Microparticles of SiO_2 were synthesized according to a known procedure [2, 3]. The average particle size was 30 μm.

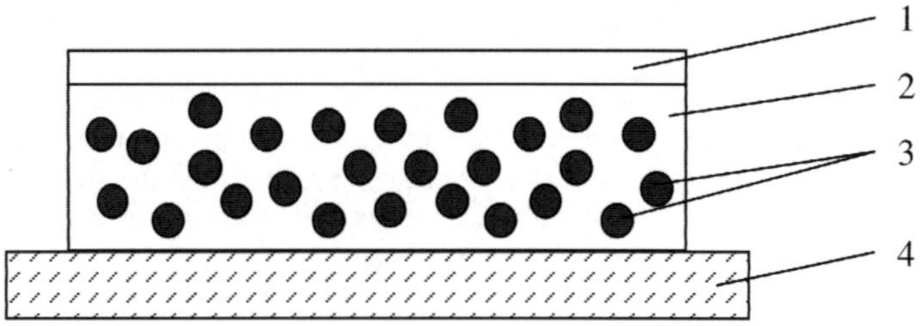

Figure 1. The schematic structure of the composite material.

The particles with adsorbed dye were added to the polymer solution in a suitable solvent, and then the mixture was knife-coated on the circular sand blasted glass substrate (4). The tablet then was dried at 60 °C for 1 h. After that the additional white reflecting layer (1) based on the same polymer as in the case of layer (2) was deposited on the outer surface to increase the signal intensity. Finally the prepared tablet was dried at 60 °C for 24 h. Protective coating (1) from oxygen-transmitting fluoropolymer can be applied with thickness of 0.5…3 μm using vacuum technologies [9–11].

2.2. Instrumentation

The characteristics of the SiO_2 microparticles were determined using Micromeritics Gemini VII 2390t surface area analyzer.

The phosphorescence lifetime was measured by means of analyzer Expert-009 (Econix-Expert, Russia). The design of the analyzer and the principle of operation are described in detail earlier [12]. The measurement is based on determination of the phase shift of the LED sine wave modulation. This value can then be converted into a phosphorescence lifetime τ (assuming mono-exponential behavior of the decay curve) with the Equation:

$$\tau = tan(\phi/2\pi f) \tag{1}$$

in which ϕ is the phase of the modulation signal, f is the modulation frequency.

The experimental setup used to characterize the composite matrix sensor is shown in Figure 2. The optical sensor (1) with sensor layer to be examined is fixed in the sealed chamber (2) by means of a special fixing element (3). The assembled unit is placed in a liquid thermostat (4) with accuracy of temperature maintenance ±0.1 °C, controlled by the reference thermometer (5). The readouts are displayed on the display analyser (6) and transmitted to the PC. The chamber (2) is connected by a gas line with a certified vacuum gauge (7) (accuracy class 0.25) and a three-way valve (8) with a buffer volume (9) and vacuum pump (10). This experimental setup allows to set the internal chamber pressure in the range from atmospheric (P_{atm}) to vacuum 1 at, i.e. up to $P = P_{atm} - 98.0665$ kPa with an error of no more than 0.250 kPa. The partial pressure of oxygen was determined, taking into account its percentage in air (20.9476 % by volume). Measurements at the residual pressure of the vacuum pump (passport value $2 \cdot 10^{-3}$ kPa), were taken as sensor readings in an oxygen-free environment.

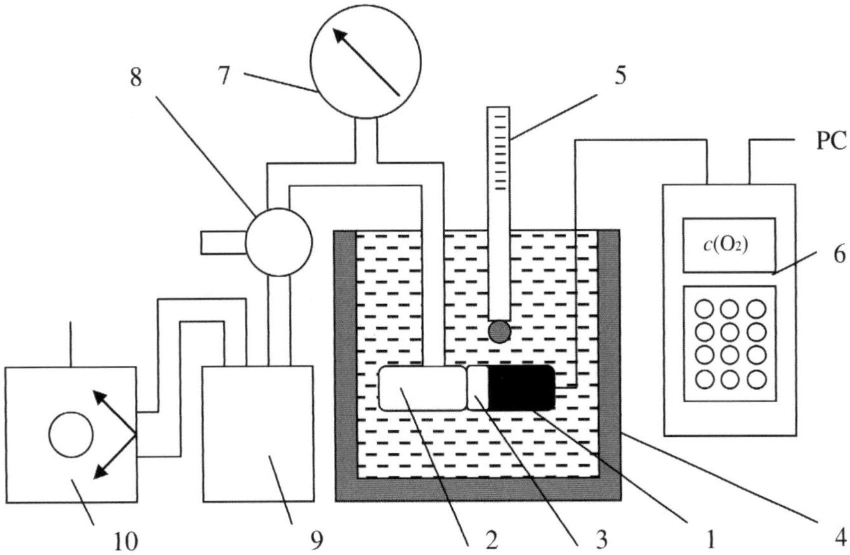

Figure 2. Experimental setup for the sensor calibration.

3. Results and discussions

3.1. SiO_2 microparicles properties

The particles were examined before and after adsorption of the dye. According to IUPAC, the pores are divided into three types: micro (< 2 nm), meso (2…50 nm) and macro (> 50 nm). It was found that the surface area of the micropores is 382 ± 2 m^2/g (micropore volume 0.20 cm^3/g) and does not change with the adsorption of the dye. This behavior is expected, since Pt-TFPP molecule is too large to pass into the micropores.

The surface of mesopores decreased up to 20 % in comparison with the initial particles (from 374 to 299 m^2/g). Further increase in dye content led to self-quenching, that is, the limiting distribution of the dye monolayer was reached for this material.

3.2. Oxygen sensing properties

The Stern-Volmer calibration plots of the sensor are shown in Figure 3. In contrast to earlier created sensor based on poly(hexafluoropropylene) (PHFP) [6], the composite structure has a nearly linear dependence in the whole temperature range studied. This allows us to use for calculation the simple form of the Stern-Volmer Equation:

$$[O_2] = (\tau_0/\tau - 1)/\, k_q \cdot \tau_0, \qquad (2)$$

where τ_0 is the excited state phosphorescence lifetime in the absence of oxygen, k_q is the bimolecular rate constant. Two-point calibration is sufficient to calibrate such a sensor.

The response time of the sensor (establishing stable readings with accuracy not worse than 5 %) was about 40 s in a gaseous medium and not more than 2 minutes in an aqueous medium. The increase in the response time in the latter case can be explained by the lower diffusion rate in solution as compared with the gas phase. These sensor characteristics are slightly worse than those obtained earlier for a sensor based on PHFP [6].

Swelling of the polymer matrix can lead to dye leaching, which causes the change in the calibration dependence or the failure of the sensor. To test the stability of the sensor material against the effects of organic solvents, the sensor was immersed in chloroform, toluene, and cyclohexane. Even after two weeks of continuous contact, migration of the dye to the solution was not observed.

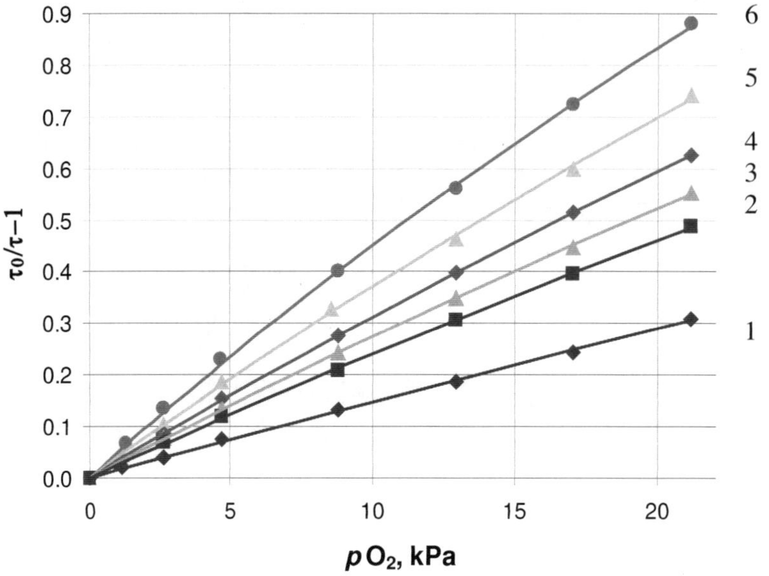

Figure 3. Temperature dependence of the Stern–Volmer calibration for the composite matrix based sensor. The series temperatures in degrees Celsius are as follows: (1) 0.1; (2) 14.0; (3) 19.9; (4) 24.8; (5) 29.6; (6) 34.6 °C.

Photostability of the sensor is similar to that observed for other sensors based on Pt-TFPP [3], because physical adsorption with properly selected dye concentration does not lead to properties degradation, unlike the covalent coupling [7].

4. Conclusion

The created composite material with the use of structured SiO_2 is a suitable matrix for manufacturing the sensing element of an optical oxygen sensor, which greatly simplifies the operation of the analyzer in comparison with polymer matrix sensor [6]. It is resistant to the action of a number of organic solvents and their vapors, which allows us to quickly determine the oxygen content even in non-aqueous media. The absence of pores on the surface of the material avoids microbiological fouling and dye leaching, preventing degradation in time.

Acknowledgments

The research was carried out with financial support of the Russian Science Foundation (project No. 17-79-10439).

References

[1] Wolfbeis O S 2015 *BioEssays* **37** 921–928 doi: 10.1002/bies.201500002
[2] Chu C-S, Lo Y-L 2007 *Sensors and Actuators, B: Chemical* **124** 376–382 doi: 10.1016/j.snb.2006.12.049
[3] Chu C-S, Lin C-A 2014 *Sensors and Actuators B: Chemical* **195** 259–265 doi: 10.1016/j.snb.2014.01.032

[4] Bolivar J M et al 2015 *ACS Catal.* **5** 5984–5993 doi: 10.1021/acscatal.5b01601
[5] Lehner P et al 2014 *Nature Communications* **5** 4460 doi: 10.1038/ncomms5460
[6] Zaitsev N K et al 2016 *Procedia Engineering* **168** 309–312. doi: 10.1016/j.proeng.2016.11.203
[7] Müller B J et al 2015 *Sensors and Actuators* B **216** 527–534 doi: 10.1016/j.snb.2015.04.067
[8] Quaranta M et al 2012 *Bioanal. Rev.* **4** 115–157 doi:10.1007/s12566-012-0032-y
[9] Luchnikov P A et al 2017 *IOP Conf. Ser.: Mater. Sci. Eng.* **168** 012092 doi: 10.1088/1757-899X/168/1/012092
[10] Luchnikov P A et al 2017 *IOP Conf. Ser.: Mater. Sci. Eng.* **189** 012014 doi: 10.1088/1757-899X/189/1/012014
[11] Luchnikov P A et al 2017 *IOP Conf. Ser.: Mater. Sci. Eng.* **189** 012027 doi: 10.1088/1757-899X/189/1/012027
[12] Zaitsev N K et al 2018 *J. Anal Chem* **73** 102–108

Improvement of statistical methods for detecting anomalies in climate and environmental monitoring systems

A G Yakunin[1], H M Hussein[2]
[1]Head of chair, Altai State Technical University, Barnaul, Russia
[2]Assistant professor, Minia University, Minia, Egypt
E-mail: yakunin@agtu.secna.ru

Abstract. The article shows how the known statistical methods, which are widely used in solving financial problems and a number of other fields of science and technology, can be effectively applied after minor modification for solving such problems in climate and environment monitoring systems, as the detection of anomalies in the form of abrupt changes in signal levels, the occurrence of positive and negative outliers and the violation of the cycle form in periodic processes.

1. Introduction

Recently, continuous on-line monitoring with the ability to automatically detect emergency situations is gaining relevance in solving many technical and environmental monitoring applications. However, existing solutions for identifying such situations are quite limited possibilities [1–5].

Furthermore, with low-cost devices for primary data collection, the volume of measurements has increased dramatically that has led to the problem of its timely stream processing, transmission, storing, and its immediate access [6–10].

In this context, it becomes urgent to develop software-hardware and algorithmic solutions to effectively detect anomalies, i.e., violations of laws in controlled processes based on their features and minimize the amount of transmitted and stored information without losing its pragmatic values. As was shown in [11–13], among all types of anomalies in monitoring systems, the most common and unambiguously interpreted are the fluctuations, outliers and cyclical violations of the recorded signals in monitoring systems, and statistical methods are the most suitable for their identification. And in the development of data compression algorithms can be used such a signal property as smooth character of their changes.

This paper deals only with issues devoted to solving the problem of identifying the most characteristic anomalies. It offers various options for improvements of well-known statistical methods of processing the data and their application for solving such problems of climate and environmental monitoring systems, as the detection of violations of the regularities of the flow of controlled processes.

2. Model, methodology and equipment of the experiment

The proposed methodology is a computational experiment conducted on model signal capable to reproduce the previously listed anomalies types. The source data for determining the parameters of the model signal were obtained by using measured data of technical and meteorological monitoring system in campus of Altai State Technical University. Hardware and software used in the system have repeatedly described in [14–19]. The same system was used for full-scale testing and effectiveness evaluation of the developed algorithmic solutions.

Since a large amount of information in the system was the temperature data, special attention was paid to data processing from temperature sensors.

To investigate the algorithms for detecting anomalies and compressing data using a computational experiment, the following model has been proposed to describe the measured signals from sensors:

$$x(t) = x_s(t)+x_p(t) \cdot k_{am}(t)+x_a(t)+x_n(t), \qquad (1)$$

wherein $x_s(t)$ – non-periodic quasi deterministic component of a useful informative signal, which is not generally a random process and provides information about changing the value of the controlled parameter over time; $x_p(t)$ – a periodic component of the information signal; $k_{am}(t)$ and $x_a(t)$ – respectively, multiplicative and additive components of the abnormal and $x_n(t)$ – the noise component.

To describe the informative signal, a model function was used in the form of sine wave with an angular frequency ω_o and amplitude A_o:

$$x_s(t) = A_o \cdot Sin(\omega_o \cdot t). \qquad (2)$$

The frequency of this sine wave $f_o = \omega_o / 2\pi$ was selected about 50 ... 5000 times lower than the sampling frequency of the signal $x(t)$. A higher frequency was used when there was a periodic component in the signal. Signal amplitude A_o determines the maximum possible rate of rise or fall of the signal, i.e., limit for its trend.

The component $x_p(t)$ was used for modeling the periodic processes. It was implemented using two types of modeling functions: sinusoidal and triangular shapes with the same frequency f_p, and amplitude A_p:

$$x_p(t) = A_p \cdot Sin(2\pi f_p \cdot t); \qquad (3)$$

$$x_p(t) = F(t)*A_p \cdot (t/2f_p - 2i), \text{ if } i/f_p < t <= (i+0.5)/f_p; \qquad (4a)$$

$$x_p(t) = F(t)*A_p \cdot (2i - t/2f_p), \text{ if } (i+0.5)/f_p < t <= (i+1)/f_p, \qquad (4b)$$

where i – nonnegative integer that specifies the number of triangular "teeth", the sign "*" means convolution function and smoothing sharp peaks – conversion is described by Gaussian function with a time constant of $\tau = 0.1 \cdot f_p$:

$$F(t) = 1/(\tau \sqrt{2\pi}) \cdot exp(-t^2/2\tau^2) \qquad (5)$$

Cycles frequency f_p was chosen 20–100 times higher than the frequency F_o and also cycle time had to at least 20–100 counts.

The function of a rectangular window with an amplitude lower unity was used to describe the multiplicative abnormal component. The beginning and end of the window coincide with the beginning and end of the ***i-th*** period:

$$k_{am}(t) = k_{mo} \cdot rect(f_p \cdot t - i + 0.5), \qquad (6)$$

where $rect(t) = 0$ for $|t| > 0.5$, 0.5 when $|t| = 0.5$ and 0 otherwise.

The additive type component of abnormal depended on the type of anomaly. In the simplest case for single level difference by the amount A_a occurring at time t_o, it was represented as a Heaviside function:

$$x_a(t) = A_a \cdot h(t - t_o), \qquad (7)$$

where $h(t - t_o) = 1$ for $t > t_o$, 0.5 at $t = t_o$ and to 0 otherwise.

For a more accurate description of the real signals level changes, an approximation of the following form was used

$$x_a(t) = F(t)*\{A_a \cdot [(t - t_o\, sign(A_a)) / \tau_f + 0.5] \cdot rect(t - t_o / \tau_f)\}, \qquad (8)$$

where τ_f – the length of the front level changes, and the function ***F(t)*** described by equation (5), wherein the constant τ was selected to be from 0.1 to 0.3 of the duration of τ_f. The duration of the front level changes themselves varied within 3 ... 20 times the sampling rate of the signal Δt.

In order to accumulate statistics on error detection of level changes, they interfered with the information signal model in the form of a series in which the interval between the level changes varied according to the uniform law between the minimum and maximum values, with constant alternation of positive and negative level changes.

The minimum repetition time for level changes lay in the range of 0.5 to three periods of the information signal x_s, but not less than 30 ... 100 durations level changes fronts. The maximum repetition time exceeded the minimum value by 2– 3 times so that the total duration of a series of observations took a reasonable time.

For outlier modeling, it is either used the Gaussian Function ***F(t)*** described by expression (5) multiplied by the required amplitude of level deference A_a, or the outlier can be formed of two bipolar level changes with same amplitude as described in the expression (7) or (8) and the interval between them was chosen to be within 10 -50 times the signal sampling rate.

To select an adequate model to describe the noise component of the signal $x_n(t)$ describing the fluctuation of the information signal due to the influence of many difficulties of the destabilizing factors, series of field experiments was carried out to determine its statistical properties. Since, according to the model (1) it was assumed that the dynamic impact of such factors is much higher than dynamics of informative component and is comparable with the dynamics of changes for abnormal components, to extract the informative component, the signal was subjected to low-frequency filter by double-smoothing it with a rectangular window of 50–100 samples width, and the result was subtracted from the original signal. A smaller width of the smoothing window was used to analyze periodic signals when the length of the cycle contained about 500 samples. If the number of samples was smaller, a signal averaged over 10–20 periods was subtracted from each period of the original signal (2–10 before and the same after the current period). Thus, regions that contained anomalies and that did not contain were analyzed separately.

As a result of the conducted studies, it was found that in the absence of anomalies, the properties of the stochastic signal component extracted in the manner described above are close to the properties of normal white noise, since the distribution law of the readings obtained is close to normal, and the autocorrelation function has a well-defined peak. As an example, Figure 1 shows the distribution laws of $x_n(t)$ samples for the temperature graphs and graphs of pressure changes.

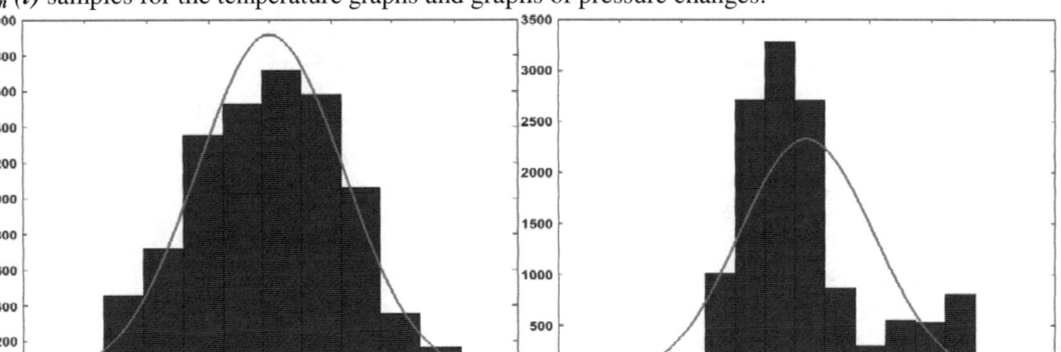

a) b)

Figure 1. Distribution histogram of temperature random fluctuations in the absence (*a*) and existence (*b*) of anomaly. The width of the smoothing window – 50 samples, a series of observations on 07–10 February 2014.

As seen from Figure 1, for an anomaly-free portion, the distribution law is close to normal, whereas, in the presence of anomalies on the side lobes of the distribution histogram, additional ups are appearing which is quite natural, since the anomalies are a non-stationary and nonergodic process.

The described model was used to carry out computational experiments to investigate proposed and modified known processing algorithms, which consisted in analyzing the dependencies of identification errors of various types of anomalies for their parameters, information signal parameters, signal-to-noise ratio and parameters included in the detection algorithm. The ultimate goal of these studies was the parameters selection for algorithms and estimating their possible effective application. Only at the final stage of the study, computational experiment was replaced by field experiment in order to confirm the results obtained in the computational experiment, and their comparison with reference methods of identification. Let us first discuss in detail algorithms for identifying violations of regularity of controlled processes.

3. Algorithm for detection downs and ups of signal level

The essence of the proposed algorithm (Table 1) is based on comparing of average speeds changes in adjacent with equal time intervals and containing $\boldsymbol{n}$ samples into which the results of measurements of *x (t)* informative signal was divided. the average value of controlled variable is computed for each of these intervals. Then, with these values, the averaged derivative of *x (t)* is calculated using the two-point scheme, that is actually average speed of change of controlled parameter on time. The threshold tolerance value $\boldsymbol{\delta}$ of the speed deviation is calculated as a modified standard deviation of speed at individual intervals. Depending on the ratio of threshold tolerance value and the current change rate of the signal, the analyzed time zone can be classified into a normal, critical zone or a zone of anomalies.

Table 1. Algorithm for identifying anomalies in the form of signal level changes.

Calculating the average value of the measured data ***x(t)*** at the k-interval that includes $\boldsymbol{n}$ discrete samples::	$x_k = \frac{1}{n}\sum_{i=(k-1)*n+1}^{k*n} x(t_i)$; (9)
Calculating the difference $\boldsymbol{v}$ of the average values between adjacent intervals (or between the maximum and minimum of the candles):	$v_k = (x_k - x_{k-1})/n$ or $v_k = (high_k - Low_k)$ (10)
Calculating the average $\bar{v}$ for the difference $\boldsymbol{v}$ in $\boldsymbol{n}$ intervals:	$\overline{v_k} = \frac{1}{N}\sum_{i=k-N+1}^{k} \lvert v_i \rvert$; (11)
Calculating the deviation $\boldsymbol{\Delta v}$ of speed changes:	$\Delta v_k = v_k - \bar{v}_k .sign(v_k)$ (12)
Calculating the threshold of tolerance δ:	$\delta = \frac{1}{N}\sqrt{\sum_{i=k-N+1}^{k} \Delta v_k^2}$ (13)
Making a decision on the admissibility of $\boldsymbol{\Delta v}$ values:	

$$\forall \Delta v_k : \begin{cases} \lvert \Delta v_k \rvert \le \delta & : \text{Normal zone,} \\ \delta < \lvert \Delta v_k \rvert < k_\varepsilon \delta : & \text{Critical zone ,} \\ \lvert \Delta v_k \rvert \ge k_\varepsilon \delta : & \text{Zone of anomalies.} \end{cases} \qquad (14)$$

As an example, Figure 2*a* shows the result of applying the proposed algorithm to real data coming from digital thermal sensor of an information system. To assess the effect of noise on the value of the probability of errors (as type I and type II errors) of identifying level changes, additional studies have been conducted, the results of which are shown in Figure 2*b*. And Figure 2*c* shows comparison of noise effect on error of level changes detection for Bayesian method, which is often used for similar purposes. As seen from the Figure 2*c*, the Bayesian method is more sensitive to noise than proposed.

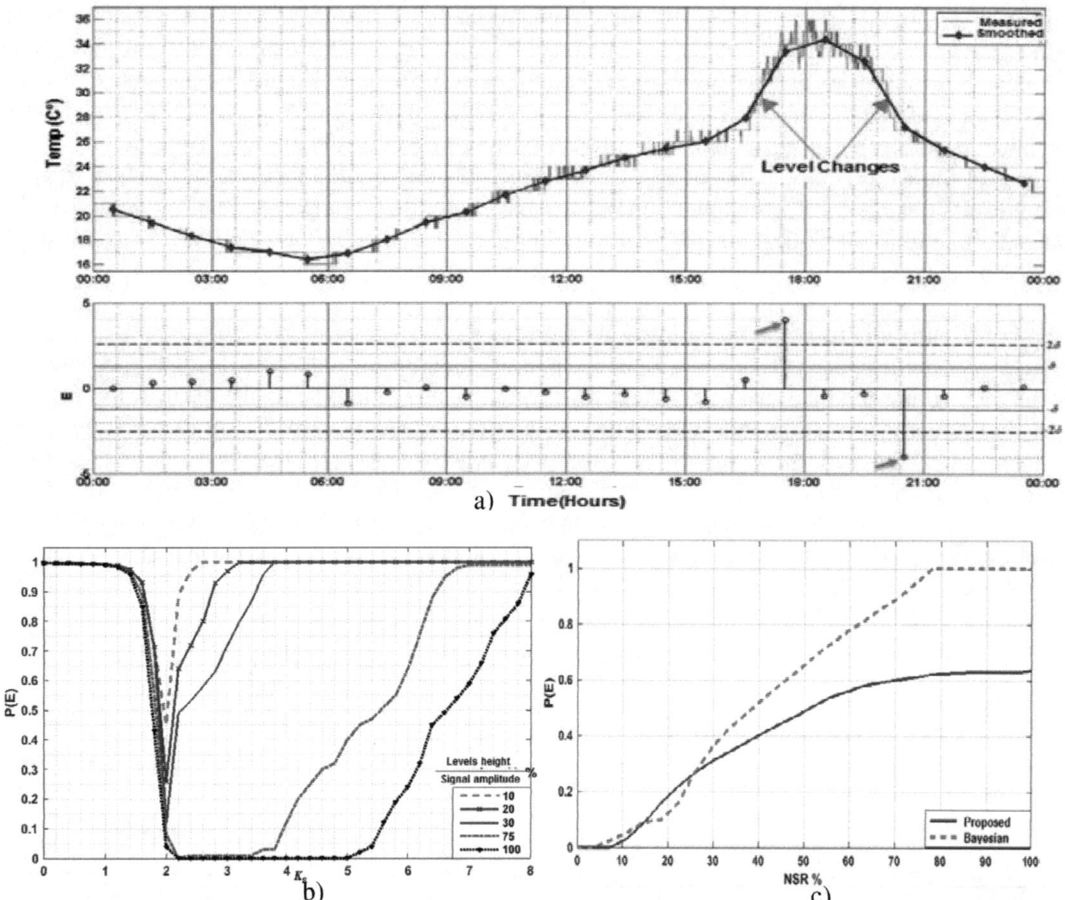

Figure 2. The results of the average speed comparison algorithm for detecting level changes in temperature (*a*), dependence of the error probability for level change identification on kε for different level change amplitudes (*b*) and the dependence of the error probability on the signal to noise ratio for the proposed and Bayesian methods (*c*).

4. Algorithm for detection signal outliers

To detect outliers, a method has been developed based on the analysis of clusters in the trend of the time series. The method is able to identify two types of outliers: type A when observed anomalous trend lies above the expected measured data boundaries and Type B when the emission falls below the expected limits. In contrast to the level changes detection method, in this case, the changes of average speed for adjacent intervals is analyzed by the three-point scheme:

$$\mathrm{a}_k = (x_k + x_{k-2} - x_{k-1}), \tag{15}$$

The physical analogue of a_k is the change in the magnitude of the function's trend, that is, the acceleration. In contrast to (11), in expression (15) division into the number of samples N in the average interval is omitted, which reduces the computational complexity, but does not affect the performance of the method. All the other equations of average speed comparison method (12) ... (14) valid with the only difference that the symbol of speed $\boldsymbol{v}$ should be replaced with the acceleration symbol $\boldsymbol{a}$:

$$\bar{a}_k = \frac{1}{N}\sum_{i=k-N+1}^{k} |a_i|. \tag{16}$$

$$\Delta a_k = a_k - \bar{a}_k sign(a_k), \tag{17}$$

$$\Delta a = \delta = \frac{1}{N}\sqrt{\sum_{i=k-N+1}^{k} \Delta a_k^{\ 2}}. \tag{18}$$

$$\forall \Delta a_k : \begin{cases} |\Delta a_k| \leq \ \ \delta : Normal\ zone, \\ \delta < |\Delta a_k| < K_\varepsilon \delta : Critical\ zone; \\ |\Delta a_k| \geq \ K_\varepsilon \delta : Zone\ of\ outliers. \end{cases} \tag{19}$$

According to the mathematical analysis rules, if the second derivative of a function is positive, and the first equals zero, then there is a minimum, otherwise (a <0 and v = 0) is the maximum of the function. However, the control of the condition v = 0, reducible to the given case, to checking divergence of the signs of average speeds $\boldsymbol{v_k}$ and $\boldsymbol{v_{k\_-1}}$ on the last two intervals is not usually required, since when the last two inequalities in system (19) are satisfied, this condition is practically always satisfied. An example of operation of the proposed algorithm is shown in Figure 3.

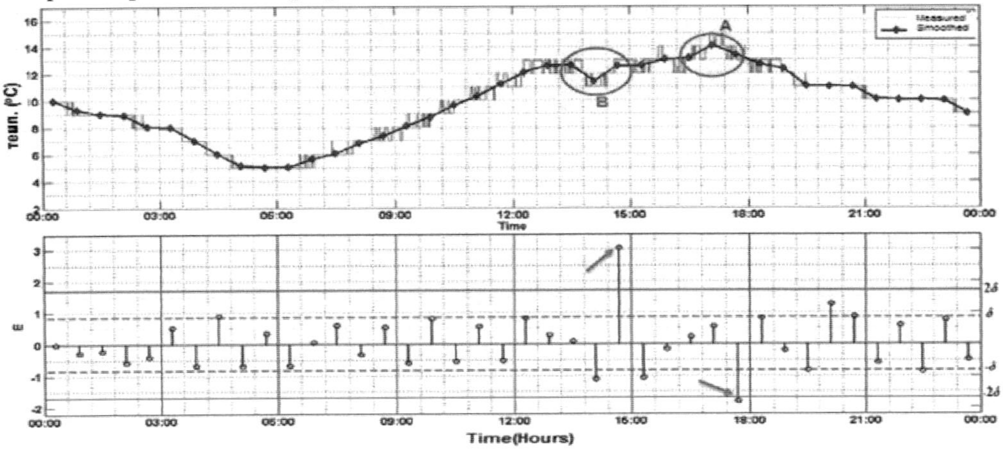

a)

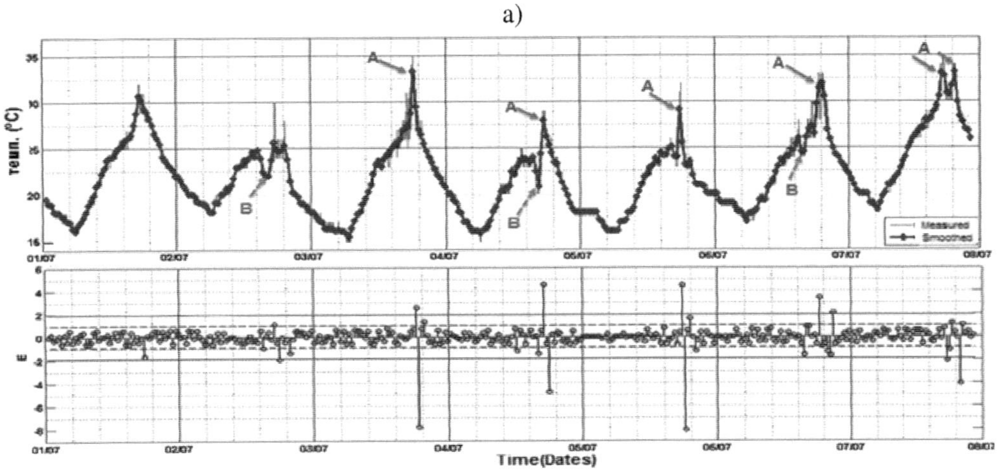

b)

Figure 3. Detection of outliers' type B and A on daily fluctuations of outdoor temperature on 20 June 2014 (*a*) and during a week from 1 to 7 July 2014 (*b*)

In order to objectively evaluate the effectiveness of the proposed method, a computation experiment was conducted using the developed model of the information signal. The dependence of the total error of outliers' detection (type I and II), depending on the signal to noise ratio for various values of the outliers' amplitudes to the signal amplitude ratio, the ratio of the outlier duration to the period of harmonic oscillations, and the various time intervals of the observation was investigated. As an example, Figure 4 shows such dependencies for the ratio of amplitudes is equal to 0.2 and the ratio of periods is equal to 0.1. The period of harmonic oscillations contained 250 samples.

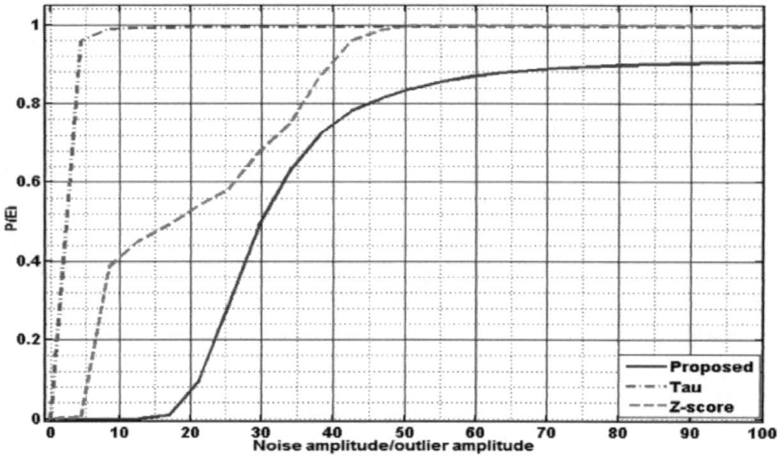

Figure 4. Dependence of outliers' detection error on signal to noise ratio (in %) for the proposed method based on trend comparison, Z-score method and τ – Thompson method. A_O / A_a = 30%.

For comparison, the same Figure 4 shows the dependencies for the methods Z-score [20,21] and τ – Thompson [22], which are also widely used for outlier detection. However, these methods did not work efficiently without certain modifications. For example, in Z-score method after normalization of original sequence according to the formula [21]

$$y_i = \frac{0.6745[x(t_i) - x_m]}{y_m}, \tag{20}$$

where $\boldsymbol{x_m} = \boldsymbol{median[x(t_i)]}$– median value of the analyzed sequence, and $\boldsymbol{y_m} = \boldsymbol{median[|x(t_i) - x_m|]}$ – normalizing coefficient, the obtained values were compared not with a fixed value but with a floating threshold given by the expression

$$\boldsymbol{y_o = median[||x(t_i) - x_m| - y_m|]} \tag{21}$$

As seen in Figure 4, the proposed algorithm has clear advantages, particularly at high noise levels.

5. Algorithm for distortions detection of periodical signals

Many controlled processes in the monitoring systems are characterized cyclic behavior. It is therefore very important to detect in a timely and reliable way the deviations of the process's real behavior from its pattern. To detect these cycling violations, a modification of the method based on the pattern forms was proposed, described by the following algorithm [12].

Step 1. Normalization of the measured data on the cycle period by Z-scores:

$$y(t) = \frac{x(t) - \overline{x(t)}}{\sigma}; \tag{22}$$

Step 2. Finding a trend for an aperiodic signal component $x_s(t)$:

$$x_{s_i}^{k-1} = \begin{cases} \frac{\frac{m}{2}-i}{m}\bar{x}_{k-2} + \frac{\frac{m}{2}+i}{m}\bar{x}_{k-1} \; for \; i = 1 \dots \frac{m}{2} \\ \frac{3m-2i}{2m}\bar{x}_{k-1} - \frac{\frac{m}{2}-i}{m}\bar{x}_k \; for \; i = \frac{m}{2}+1 \dots m \end{cases} \tag{23}$$

Step 3. Recovery values purely periodic signal component:

$$x_{p_i}^{k-1} = x_i^{k-1} - x_{s_i}^{k-1} \tag{24}$$

Step 4. Evaluation of the signal amplitude in the previous cycle: D_{k-1}

IOP Conf. Series: Materials Science and Engineering **289** (2017) 012032 doi:10.1088/1757-899X/289/1/012032

$$D_{k-1} = max_{i=1...m} {x_p}_i^{k-1} - min_{i=1...m} {x_p}_i^{k-1} \quad (25)$$

Step 5: Detection of cyclicity violations if the condition ***k*> *3*** is satisfied:

$$max_{i=1...m}\left(\frac{{x_p}_i^{k-1}}{D_{k-1}} - P_i^{k-2}\right) > \delta\, , P_i^{k-1} = \alpha \frac{{x_p}_i^{k-1}}{D_{k-1}} + (1-\alpha)P_i^{k-2}.$$
$$\overline{D}_{k-1} = \alpha D_{k-1} + (1-\alpha)\overline{D}_{k-2} \quad (26)$$

where $\boldsymbol{\delta}$ is the detection criterion specifying the value of the maximum permissible relative deviation, $\boldsymbol{\alpha}$ is the exponential smoothing coefficient, approximately equals to $\frac{1}{n}$.

The algorithm includes data smoothing, normalizing them, forming a data period template with non-zero coefficients in the trend line, calculating the actual deviation of the actual data from the template, and comparing the deviation with the acceptable threshold. If the cycle period is unknown, before applying the algorithm it must first be found, for example, using the discrete Fourier transform. Unlike other similar solutions, during the formation of the template, a compensation for the trend of the monitored signal was applied, which allowed increasing the authenticity of the identification.

The results of practical application of the method for detecting violations in cyclic processes are shown in Figure 5. For calculations, the temperature values measured every 30 seconds were taken. The measured values were smoothed using the moving average method with an hour window. The normalized data were divided into one-day periods that were used to form the primary template and calculate the correction factor.

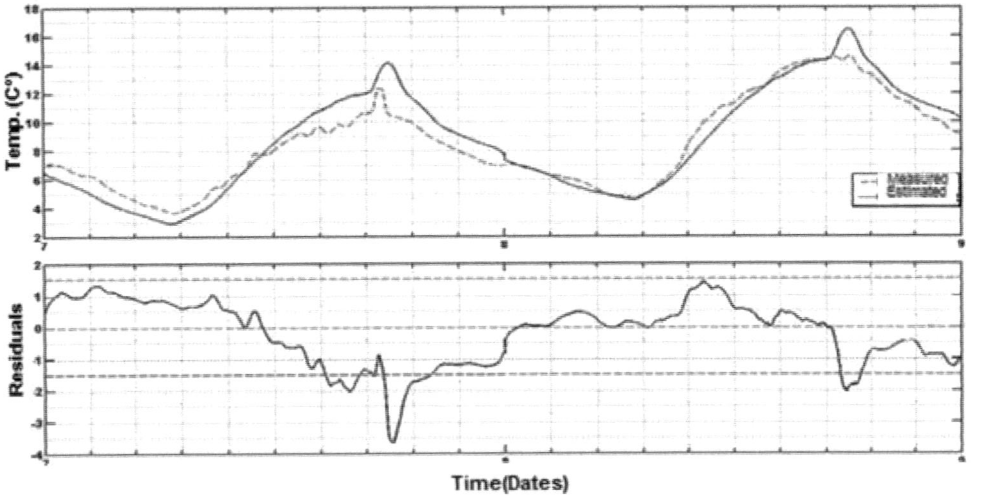

Figure 5- The original data, the corresponding patterns and the relative normalized differences between the original signal and the patterns.

6. Conclusions

In the course of the studies it was found that the use of the proposed modifications of known statistical methods for the detection of anomalies makes it possible, because of their low computational complexity, it is easy to implement methods using the simplest computing equipment in the vicinity of the sensor. As a result, in some cases it is no longer necessary to transfer the entire stream of recorded information, replacing it with the transmission of information only at intervals of the time on which the anomaly is occurring. At the same time, it is important to note that the reliability of the detection of anomalies in the proposed methods is as good as, and often better than the reference methods. All this makes it very promising to apply the proposed solutions in different systems not only to environmental monitoring or data collection systems but also to technical monitoring systems as well as to SCADA systems of various types.

Acknowledgement
The authors express deep gratitude to the management of Altai State Technical University. II Polzunov for financial support of this publication and OOO "Promavtomatika" for technical assistance in the preparation of article, as well as former students of the department computer science, computer engineering and information security department of the aforementioned University for their active participation in the development of hardware and software, which has allowed to realize the described researches.

References

[1] Ye M, Li X and Orlowska M E 2009 *Expert Syst. Appl.* **36** 7104–7113 doi: 10.1016/j.eswa.2008.08.030
[2] Huang Y, Hu L, Yang D and Liu H 2017 *IOP Conf. Ser. Earth Environ. Sci.* **81** 012208 doi: 10.1088/1755-1315/81/1/012208
[3] Branch J, Szymanski B, Giannella C et al 2006 *Proceedings - International Conference on Distributed Computing Systems* **2006** 1648838 doi: 10.1109/ICDCS.2006.49
[4] Ruggieri E 2013 *Int. J. Climatol.* **33** 520–528 doi: 10.1002/joc.3447
[5] Subramaniam S, Palpanas T, Papadopoulos D, Kalogeraki V and Gunopulos D 2006 *VLDB '06 Proc. 32nd Int. Conf. Very large data bases* 187–198
[6] Graefe G and Shapiro L D *Computer Science Symp. on Applied Compurting, Kansas City,Apr, 1991 - Proceedings of ACM/IEEE, 199.* 22–27
[7] Zhen C and Ren B 2009 *CiSE 2009* 5366670 doi: 10.1109/CISE.2009.5366670
[8] Ester M, Kriegel H P, Sander J and Xu X 1996 *Second International Conference on Knowledge Discovery and Data Mining* pp 226–31 URL: http://www.cse.ust.hk/~raywong/comp5331/References/ADensityBasedAlgorithmForDiscoveringClustersInLargeSpatialDatabasesWithNoise.pdf
[9] Hussein H M and Yakunin A G 2013 *Polzunovsky Vestn.* 65–68 (in Russian)
[10] Hussein H M, Suchkova L I and Yakunin M A 2012 *Polzunovsky Alm. AltSTU* 48–50 (in Russian)
[11] Hussein H and Yakunin A 2015 *Mater. XI Int. Res. Pract. Conf. «Modern Sci. potential-2015». Tech. Sci. - Sheff.* **38** 33–38 (in Russian)
[12] Hussein H S and Yakunin A G 2015 *J. Sib. Fed. Univ. Math. Phys.* **8** 157–164
[13] Hunt B 2007 *Int. J. Climatology* **27** 139–156 doi: 10.1002/joc.1379
[14] Hussein H M, Kuntz R V, Suchkova L I, Yakunin A G and Yurchenko A V 2013 *IV Scientific-practical conference "Information and measuring equipment and technology"* 160–169 (in Russian)
[15] Hussein H M, Kuntz R V, Suchkova L I and Yakunin A G 2013 *Izv. Altai State Univ.* **1** 210–4
[16] Suchkova L, Hussein H M, Yakunin M A, Yakunin A G and Yurchenko A V 2013 *Kontrol'. Diagnostika* **13** 95–101(in Russian)
[17] Yakunin A G 2017 *J. Phys. Conf. Ser.* **881** 012030 doi: 10.1088/1742-6596/881/1/012030
[18] Suchkova L I, Hussein H M, Yakunin M A and Yakunin A G 2013 *Polzunovsky Alm.* 75–82
[19] Yakunin A G and Hussein H M 2017 *J. Phys. Conf. Ser.* **881** 012028 doi: 10.1088/1742-6596/881/1/012028
[20] Shiffler R E 1988 *Am. Stat.* **42** 79–80 doi: 10.1080/00031305.1988.10475530
[21] Iglewicz B and Hoaglin D C 1993 *How to Detect and Handle Outliers* (United States:ASQC Quality Press)
[22] Thompson W R 1935 *Ann. Math. Stat.* **6** 214–219

IOP Conf. Series: Materials Science and Engineering **289** (2017) 012033 doi:10.1088/1757-899X/289/1/012033

Phenomenological model of mechanoelectric transformations in rocks

P I Fedotov[1], E A Kargina[2]
[1]Reseacher, National Research Tomsk Polytechnic University, Tomsk, Russia
[2]Graduate student, National Research Tomsk Polytechnic University, Tomsk, Russia
E-mail: pif@tpu.ru

Abstract. A phenomenological model is proposed on the example of the rock destruction development in underground mines. The characteristics of the electromagnetic signal generated due to the appearance and change of the dipole moment of cracks, whose beads are charged when the discontinuity is disturbed, are analytically investigated. The model is constructed using the theory of reliability and percolation theory, which allows to take into account the non-synchronism of the mechanical converters.

1. Introduction

The complexity of the qualitative explanation of the electric emission signals responses, which are an indicator of destruction, to mechanical or acoustic effects is due to the fact that the mechanisms of the phenomenon of mechanoelectric transformations are caused by a wide range of effects occurring at various levels from micro to macro-level. This may be manifestations of the piezoelectric effect, electrical polarization, electromagnetic-elasticity in the propagation of waves, etc [1, 2]. In rocks having a complex texture, porosity and fracturing, there are double electrical layers. Mechanoelectric transformations take place in them under mechanical or acoustic action on such rocks.

The main task of all the studies is practicing of the developed monitoring method of the change in the rock massive stress-strain state under conditions of an operating mine and a forecast of geodynamic manifestations.

During the mining of solid minerals by an underground method, the rock massive stress-strain state changes, that results to geodynamic manifestations activation, such as rock burst, rock- tectonic burst, shakes, and rock falls. This changes the physical fields. The influence of physical fields on the defects accumulation rate in a loaded rock is proved by numerous geophysical observations and in laboratory experiments. In addition, as a result of the mining of ore bodies, seismic effects are manifested by explosions. All these geodynamic manifestations are accompanied by mechanoelectric transformations, the consequence of which is electromagnetic emission [3–7]. The interrelation of these phenomena was proved in experiments on rock samples, which revealed a high sensitivity of electromagnetic emission to the action of acoustic waves [8–10]. The effect of electromagnetic emission activation as a response to the effect of physical fields is established for samples of rocks with different petrophysical properties [11, 12]. The regularities of the electromagnetic response of rocks of different sizes are revealed and this is due to the fact that in all cases electromagnetic emission arises from destruction [13].

A phenomenological model is proposed on the example of the rock destruction development in underground mines. The characteristics of the electromagnetic signal generated due to the appearance and change of the dipole moment of cracks, whose beads are charged when the discontinuity is disturbed, are analytically investigated. The model is constructed using the theory of reliability and percolation theory [14], which allows to take into account the non-synchronism of the mechanical converters.

2. Phenomenological model

The equation of electrodynamics, provided that the wavelength λ is large in comparison with the size of the radiator system L ($L < \lambda$ or $v < c$), allows to obtain the electric field strength [15, 16]:

$$E = (E_1 + E_2 + E_3) \exp(-r/\delta), \tag{1}$$

where $E_1 = \dfrac{\left[(3 + r/\delta)(rD)r - Dr^2\right]}{4\pi\varepsilon r^5}$ characterizes the field strength in the near zone ($r < \lambda$), E_2 - field strength in the far field ($r > \lambda$), E_3 - field strength in the intermediate zone ($r \approx \lambda$). If the "correction for the delay" [17] is partially taken into account, neglecting it in calculating the potential $\phi = -div\dfrac{D}{r} \cong \dfrac{(rD)}{r^3}$, but taking it into account in calculating field strength $E = -grad\,\phi - \dfrac{1}{c}\dfrac{\partial}{\partial t}A$, *than*

$$E_2 = \frac{\left[n\times\left[n\times\ddot{D}\right]\right]}{4\pi\varepsilon c^2 r} - \frac{n\left(n\ddot{D}\right)}{4\pi\varepsilon c^2 r}; \; E_3 = n(n\dot{D})/4\pi\varepsilon c r^2. \tag{2}$$

In the general case, in accordance with $\phi = \dfrac{(rD)}{r^3} + \dfrac{(r\dot{D})}{cr^2}$.

The factor exp ($-r$ / δ) takes into account the weakening of the field in the medium and $\delta = c(2\pi\sigma f)^{-1/2}$, σ - is the conductivity, f – is the frequency of the signal, c – is the light speed of (at f = 10 kHz, $\sigma = 10^{-4}$ (Ohm·m)$^{-1}$, $\delta \approx 1$ km). If we use, for the prediction of rock disruption, already existing equipment and the operating frequency range is 1–10^2 kHz, then for distances smaller than the thickness of the skin layer r <δ, the ratio r / δ <1 for $\sigma = 10^{-2}$-10^{-5} (Ohm· m)$^{-1}$. Therefore, electromagnetic effects in the mines can be described by the first term E_1. In this case, after averaging over the angles for the isotropic distribution of dipoles, we obtain that for a set of cracks

$$\left\langle E^2\right\rangle = \frac{1}{4\pi}\int E_1^2 d\Omega = 2N\left\langle D^2\right\rangle / r^6 \tag{3}$$

where N = nV is the number of cracks in the volume V, n - is the volume concentration of cracks, $\langle D\rangle = D\sqrt{N}$, D - is the dipole moment of a single crack. We get the integral when integrating expression (3) over the volume occupied by elementary sources of the signal,

$$I = \int\frac{dU'}{r^6} = 2\pi\int_0^{\pi}\int_0^{R_{\max}}\frac{r'^2 dr' \sin v dv}{|r - r'|^6} = \frac{2\pi}{r^3}\iint\frac{x^2 dx d\cos v}{(1 - x\cos v)^6},$$

where $x = r/r'$, r - is the radius vector of the observation point, r' - is the radius vector of the elementary source of the signal. As

$$\int_0^{\pi}\frac{\sin v dv}{(1 - x\cos v)^6} = \frac{1}{5x}\left\{\frac{1}{(1+x)^5} - \frac{1}{(1-x)^5}\right\},$$

then finally we get that

$$I = 2\pi\frac{40x^3 + 8x^5}{60(1 - x^2)^4 r^3} \cong \frac{4\pi}{3}R_{\max}^3\frac{1}{r^6}.$$

Thus,

$$|E| = \sqrt{\left\langle E^2\right\rangle} = \sqrt{2n_* P_0 P_1 P_2 U}\, D/|\varepsilon| h^3, \tag{4}$$

Where, h – is the distance from the center of the investigated region to the antenna; n_* - is the critical crack concentration and $k = 2 \div 5$ [18]; $n_*^{-1/3} = k \cdot l$; l - is the linear dimension of the crack; P_0 - is the scalar function characterizing the damage to the medium, $U = \frac{4\pi R_{max}^3}{3}$; R_{max} – is the radius of the investigated region; $D = \sigma_0 U_{src}$ – is the crack dipole moment, U_{src} – volume of a single crack, σ_0 – surface charge density on the cracks sides; P_1 – is the probability that at a time t the system of elementary sources will be able to radiate; P_2 – is the probability that the system of signal sources will work reliably on the time interval $(t, t + \tau)$ - the time interval during which the measurement is made.

From (4) we obtain for the magnitude of the electric field strength change rate

$$dE / dt = E_* \dot{P}_0 / \sqrt{P_0} \tag{5}$$

where $E_* = \sqrt{2n_* P_1 P_2 U}\rangle D\langle / 2|\varepsilon|h^3$.

Investigation of the destruction process of a geomaterial in time within the framework of reliability theory makes [16] it possible to obtain the probability of the formation of a system of elementary sources at time t

$$P_0 = \left\{1 - \exp\left(-\int_0^t \lambda(t)dt\right)\right\}, \tag{6}$$

where $\lambda(t)$ is the rate of formation of elementary sources (ES).

The probability $P_2 = \exp(-\tau/T_1)$, and

$$P_1 = 1 - F(t) + \int_0^t [1 - F(1-x)h(x)dx],$$

$$h(x) = \sum_{n=1}^{\infty} \phi(n), \phi_n(x) = \Phi_n'(x), \Phi_n = \int_0^t F_n(t-x)dG_n(x),$$

$$F_n(t) = \int_0^t F_{n-1}(t-x)dF(x), G_n(t) = \int_0^t G_{n-1}(t-x)dG(x);$$

$$G_1(t) \equiv G(t); F_1(t) \equiv F(t);$$

$F(t)$ – is the distribution function of the working time duration (the system radiation time);
$G(t)$ – is the distribution function of the system pause duration.

The process of crack formation in geomaterials is random [14]. Such processes are well described by an exponential law [16]: $F(t)=1-\exp(-\lambda t)$; $G(t) = 1-\exp(-\mu t)$.

Wherein

$$P_1(t) = \frac{\mu + \lambda \exp-(\lambda+\mu)t}{\mu+\lambda} = \frac{T_1 + T_2 \exp(-t/(T_1+T_2)}{T_1+T_2} \tag{7}$$

and $\lim_{t\to\infty} P_1(t) = T_1/(T_1+T_2)$; here $T_1 = \left[\sum_{m=1}^{n} T_m^{-1}\right]^{-1}$ – average duration of the system operating period; T_m – average duration of the working period of the m-th element (ES); T_2 - average system pause time.

For the destruction process in mines [19]: $\lambda(\tau) = \omega/(1-\omega)$, $P_0 = \omega$,
where the damage

$$\omega = (2-m)\, D(t)/[r/a)^{m+1} - 3n/(m+1)], \tag{8}$$

here $D(t) = D_1' \exp(-vt) = 3nx^{-(m+1)} / (m+1)(2-m)$,

$$m = \left(2 + \sqrt{\frac{2\,a}{3\,\beta}}\right) / \left(1 - \sqrt{\frac{2\,a}{3\,\beta}}\right);$$

$$n=\gamma/\left(\beta-a\sqrt{\frac{2}{3}}\right);$$

α, β, μ_0, μ_∞ - medium constants; $v=\tau^{-1}$, τ - shear relaxation time; $x(t)=a/b(t)$; a – passage linear dimension; $b(t)$ - thickness of the fractured layer near the working surface; breakdown time t_* is found from condition $\omega(a,t_*)=0$ and

$$t_*=-\frac{\tau\mu_\infty}{\mu_0}\ln\left[1-\frac{\mu_\infty}{\mu_0}\left(\sqrt{\frac{2}{3}}\frac{P_*}{\Delta P}-1\right)\right]\cong\tau\left(\sqrt{\frac{2}{3}}\frac{P_*}{\Delta P}-1\right)\cong 0.63\tau\,; \tag{9}$$

$$P_*=2\mu_0\gamma/a;$$

$$\Delta P=P_\infty-P_0$$

ΔP – pressure drop, P_∞– hydrostatic pressure; $\alpha^2/\beta-2\mu_0=M$ – modulus of decline.
The value of $x(t)$ is found from Equation

$$\dot{x}=\frac{n(v-\overline{v})x}{(2-m)}+\frac{9n\zeta(x-x^{-m})}{(2-m)(m+1)^2}+x\left(\frac{\Delta\dot{P}+v\Delta P}{4\mu_0}+\frac{3\mu_\infty}{2\mu_0}v\ln x\right)/\left(\frac{9n\zeta x^{-(m+1)}}{(2-m)(m+1)}-\frac{3}{2}\varepsilon_*\right), \tag{10}$$

where,

$$\overline{v}=v\left(1-\frac{\mu_\infty}{\mu_0}\right),\ \varepsilon_*=\frac{\alpha\gamma}{2\mu_0\beta}\sqrt{\frac{2}{3}},$$

$$\zeta=\left(1-\frac{\alpha^2}{2\mu_0\beta}\right)/\left(1-\frac{\alpha}{\beta}\sqrt{\frac{2}{3}}\right).$$

Since $\mu_0/\mu_\infty<<1$, the main role in the Equation (10) is played by the last term

$$(\dot{x/x})\cong\frac{\Delta P v}{4\mu_0}/\left(\frac{9n\zeta x^{-(m+1)}}{(2-m)(m+1)}-\frac{3}{2}\varepsilon_*\right). \tag{11}$$

Equation (11) leads to

$$b(t)=b_c^{(0)}\left[1-\sqrt{\frac{(m+1)\Delta P v}{3\mu_0\varepsilon_*}\Delta t}\right]^{\frac{1}{m+1}}, \tag{12}$$

$\Delta t=t_{\max}-t$, $t_{\max}\cong\tau$.
The value $b_c^{(0)}$ can be found from the condition $\omega=\omega_c$, where ω_c is the percolation limit [20], so

$$b_c^{(0)}=a\left[1+\omega_c(m+1)/3n\right]^{1/(m+1)} \tag{13}$$

Thus, from the relations (8), (11), (12) we finally obtain that

$$\frac{\dot{E}}{E_*}=\frac{\dot{\omega}}{\sqrt{\omega}}=\frac{\Delta P v}{6\mu_0\varepsilon_*}\left(\frac{b_c^{(1)}}{b}\right)^{m+1}\left[\frac{b_c^{(1)}}{b}-1\right]^{-1}\sqrt{\frac{3n(m+1)}{\left[\left(\frac{b}{a}\right)^{m+1}-1\right]}}=, \tag{14}$$

$$=\sqrt{\frac{n\Delta p v}{4\mu_0\varepsilon_*}}/\sqrt{\Delta t}\left(-1\left(\frac{a}{b}\right)^{m+1}\right)^{1/2}$$

here $b_c^{(1)}=\left[\frac{(2-m)(m+1)\varepsilon_*}{6\zeta n}\right]^{1/(m+1)}$, a - is the critical value of $b(t)$ at which a sharp increase in the damage rate $\dot{\omega}$ occurs. The value $b_c^{(1)}$ does not depend on ΔP and the occurrence of failure can be determined only by the material parameters α, β, μ_o, μ_∞, γ. The pressure differential ΔP determines only the beginning of the collapse process in accordance with the expression (9).

Since $\gamma/\alpha = \varepsilon_c << 1$,

$$\frac{\alpha^2}{2\mu_0\beta} > \text{и } \frac{a}{\beta}\sqrt{\frac{2}{3}} < 1 \text{, than } \frac{b_c^{(1)}}{b_c^{(0)}} < 1 .$$

Therefore, the moment of speed increase $\dot{\omega}$ precedes the moment of damage increase ω, so that fracture prediction can be based on the measurement of $\dot{\Gamma}$ magnitude. In accordance with (14), we estimate the possibility of recording signals during the development of collapse:

$\dot{\Gamma} = E_* \dot{\omega}/\sqrt{\omega}$, where $E_* = \frac{D\sqrt{2n_*U}}{2|\varepsilon|h^3} = E_0 \frac{l_U^3}{h^3}\sqrt{\frac{8\pi n_*}{3}R_*^3}$, ε - the dielectric constant, l_U - linear dimension (ES), $R_* = l_U^3\sqrt{\frac{E_0}{E_n}}$ - radius of signal collection, E_Π - instrument sensitivity, $\Gamma_0 = \sigma_0 / 2|\varepsilon|$; $\dot{\omega}/\sqrt{\omega}$ - is defined by (14).

Taking into account the concentration criterion $n_*^{-1/3} = kl_U$, we obtain

$$\frac{E_*}{E_n} = \frac{2\sqrt{\frac{2\pi}{3}}}{h^3 n_* k^{3/2}}\left(\frac{E_0}{E_n}\right)^{3/2} . \tag{15}$$

The right-hand side of (15) must be greater than one, for which it is necessary that $h \leq 10^{-1}$ m, $\sigma_o \geq 10^{-8}$ C/m^2, $n_* \leq 10^9$ 1/m^3. Since $\sigma_o = 10^{-2} \div 10^{-6}$ C/m^2 [21], the electromagnetic signal generated during the preparation of the destruction is available for registration by the currently used equipment.

3. Result

Thus, the developed phenomenological model of rock failure allows qualitative and quantitative prediction of it by the precursor of the electromagnetic nature. Within the framework of the reliability theory and the percolation theory, the problem of the non-synchronization of the elementary radiation sources is solved, which is important for the practice of predicting geodynamic phenomena based on precursors of electromagnetic nature. It is proved that fracture prediction can be based on a characteristic change in the electric field strength derivative.

Acknowledgments

The work was supported by the Ministry of Education and Science of the Russian Federation within the framework of the State task in the field of scientific activity, project No. 11.980.2017 / 4.6.

References

[1] Akhmetbayev D S *et al* 2017 *Results in Physics* **7** 1644–1649 doi: 10.1016/j.rinp.2017.03.010

[2] Mustafina R M, Plotnikov I A, Plotnikova I V,Tchaikovskaya O N 2016 *Journal of Physics: Conference Series* **671** 012046 doi: 10.1088/1742-6596/671/1/012046

[3] Frid V, Vozoff K 2005 *International Journal of Coal Geology* **64** 57–65 doi: 10.1016/j.coal.2005.03.005

[4] Li X *et al* 2016 *Rock Mech Rock Eng* **49** 4393–4406 doi: 10.1007/s00603-016-1037-6.

[5] Hagag W, Obermeyer H 2016 *Journal of African Earth Sciences* **121** 136–153 doi: 10.1016/j.jafrearsci.2016.06.005

[6] Bespal'ko A A , Yavorovich L V, Fedotov P I, Shtirts V A 2010 *Journal of Mining Science* **46** 136–142 doi: 10.1007/s10913-010-0018-5

[7] Lebedev N I, Sigov A S 1994 *Integrated Ferroelectrics* **4** 21–24 doi: 10.1080/10584589408018656
[8] Bespal'ko A A, Lyukshin B A, Utsyn G E, Yavorovich L V 2015 *Russian Physics Journal* **58** 567–573 doi: 10.1007/s11182-015-0535-4
[9] Bespal'ko A A, Isaev Y N, Yavorovich L V 2016 *Journal of Mining Science* **52** 279–285 doi: 10.1134/S1062739116020418
[10] Biao Kong *et al* 2016, *Rock Mech Rock Eng* **49** 4911–4918 doi: 10.1007/s00603-016-1011-3
[11] Yavorovich L V, Bespalko A A, Fedotov P I, Baksht R B 2016 *Acta Geophysica* **64** 1446–1461 doi: 10.1515/acgeo-2016-0081
[12] Vorotilov K A, Sigov A S 2012 Physics of the Solid State **54** 894–899 doi: 10.1134/S1063783412050460
[13] Eremenko A A, Bespalko A A, Eremenko V A, Yavorovich L V 2016 *Diagnostics of geophysical precursors of geodynamic phenomena and development of geotechnology for the development of iron ore deposits* (Novosibirsk: Nayka) (in Russia)
[14] Chelidze T L 1987 *Methods of percolation theory in the mechanics of geomaterials* (Moscow: Nauka) (in Russia)
[15] Landau L D, Lifshitz E M 1990 *Field theory* (Moscow: Nauka) (in Russia)
[16] Sokolov A A, Ternov I M 1974 *The relativistic electron* (Moscow: Nauka) (in Russia)
[17] Feynman R, Layton R, Sande M 1977 *Feynman lectures in physics* (Moscow: Mir) (in Russia)
[18] Sobolev G A, Zavialov A D 1980 *Dokl. AS USSR* **254** 567–569 (in Russia)
[19] Gnedenko B V, Belyaev Yu K, Soloviev A D 1987 *Mathematical methods in the theory of reliability*, (Moscow: Nauka) (in Russia)
[20] Borchik E Yu, Kondaurov V I 1990 *FTPRPI* **1** 45–52 (in Russia)
[21] *Catalog of pulsed electromagnetic precursors of earthquakes* 1991 (Moscow: Academy of Sciences SSSR)(in Russia)

IOP Conf. Series: Materials Science and Engineering **289** (2017) 012034 doi:10.1088/1757-899X/289/1/012034

Study of the effect of electron irradiation on the density of the activated sludge in aqueous solution

A I Kupchishin[1,2], M N Niyazov[3],B G Taipova[3], N A Voronova[1], N N Khodarina[3]
[1]Head of laboratory, Abay Kazakh National Pedagogical University, Almaty, Kazakhstan
[2]Al-Farabi Kazakh National University, 71 al-Farabi ave., Almaty, Kazakhstan
[3]Researcher, Abay Kazakh National Pedagogical University
E-mail: ankupchishin@mail.ru

Abstract. Complex experimental studies on the effect of electron irradiation on the deposition rate of active sludge in aqueous systems by the optical method have been carried out. The obtained dependences of density (ρ) on time (t) are of the same nature for different radiation sources. The experimental curves of the dependence of the active sludge density on time are satisfactorily described by an exponential model.

1. Introduction

As it is known, industrial, agricultural production and the accompanying urbanization, which are developing at a very rapid pace, have led to an unprecedented degree of environmental pollution in the history of mankind. The increase in the number of consumers of natural resources caused a jump in the level of sewage and the degree of their ecological danger, which led to an increase in their direct discharge into natural watercourses, but their biological decomposition and self-purification is very slow [1]. Sewage treatment is now increasingly replaced by artificially created and industrial facilities, as they are a protective barrier to natural hydrobionts [2]. The present state of the quality of water in natural reservoirs is largely determined by the degree of purity of wastewater that has undergone biological treatment, since they constitute the bulk of the total volume of sewage. The efficiency of purification is tightly related to the processes of self-purification in natural reservoirs. However, the various methods of intensifying biological purification that are being developed open up the possibility of a significant improvement in the latter. Biological purification should be considered as an obligatory final link before discharge of sewage into the water body, therefore, as a powerful protective barrier against pollution of the natural environment [3].

The increased requirements to the quality of treated wastewater discharged into water bodies, in connection with the approval of the maximum permissible discharges (MPD) of pollutants and the solution of issues of re-use of biologically treated waters in production, necessitates the intensification of the operation of existing treatment facilities, the search for new opportunities to improve the quality of treated sewage waters [4]. Intensification of biological purification through reconstruction of existing facilities is associated with capital costs and technical difficulties, so now the development of hydrobiological, microbiological and biochemical methods for improving the operation of structures is more promising [5]. In recent years, the work related to the study of the effect of various types of irradiation on the structure and physico-chemical properties of various materials has been of great practical interest [6–8]. We have begun study on the effect of electron irradiation on the structure and properties of various biomaterials, which under such impacts vary greatly [9, 10]. At the same time, the

IOP Conf. Series: Materials Science and Engineering **289** (2017) 012034 doi:10.1088/1757-899X/289/1/012034

quality of systems also changes significantly. The quality of the treatment facilities can be determined by various methods, including by means of a sludge index and sedimentation rate. A well-precipitated sludge has a sludge index of about 60 mL/g, a less dense – 80 – 90 mL/g, and the sludge index higher than 300 mL/g indicates a failure of the treatment plant. This paper is devoted to determining the density of unirradiated and irradiated sludge by the optical method in aqueous solutions.

2. Experimental procedure

To determine the parameters of activated sludge (biomaterial), an experimental setup was created, the scheme of which is shown in Figure 1. The installation consists of a laser source, radiation detector, electronic unit, racks and various holders. The sludge, placed in a flask with water, was carefully shaken and installed between the laser and the detector. The dependence of the illumination (intensity) of light (green and red lines of the laser) on time was recorded.

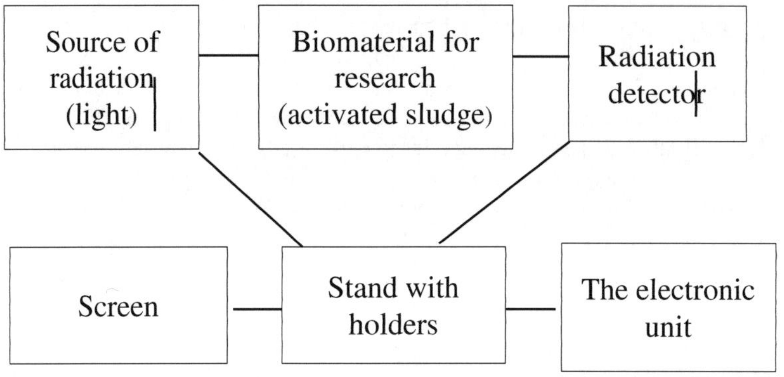

Figure 1. Scheme of an experiment on the passage of radiation through the biomaterial.

3. Discussion of results

As a result of the studies (green and red ray), it was found that the total sedimentation of the sludge occurs for about 100 seconds. Figures 2 and 3 show the time dependences of the illumination of laser radiation. As follows from these Figures, the magnitude of the illumination, first increases sharply, and then gradually goes to saturation.

To determine the sludge index and sedimentation rate of activated sludge, the dependence of the active sludge density in water on time was used. The relationship between the illumination and the density of activated sludge in water was determined by the equation: (1)

$$\rho = \rho_{\max} * \frac{E_{\max} - E}{E_{\max}}, \tag{1}$$

where $\rho_{\max}$ is the maximum density of activated sludge in water; $E_{\max}$ is illumination with complete sedimentation of the activated sludge; E is illumination at a given time.

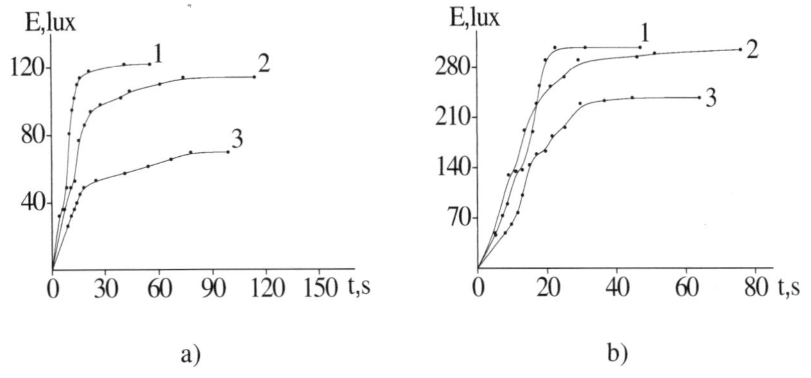

Figure 2. Time dependences of the illumination of green (*a*) and red (*b*) laser radiations as they pass through the activated sludge.

Figure 2 shows the experimental data on the dependence of the activated sludge density on time, obtained using the dependence of E on t under irradiation with red and green lasers (dots). It is seen that in time the density decreases gradually in both cases. Similar experiments were performed on electrons and photons (gamma rays and X-rays) irradiated activated sludge (Figures 3–5). In this case, irradiation affects significantly the experimental dependences. As can be seen from the Figures, the nature of the ρ dependence on t after irradiation does not change. The experimental curves are explained within the framework of the exponential model proposed by the authors. The essence of it is as follows. The active sludge, which is in solution, is affected constantly by gravity. In the first approximation, the density change $d\rho$ at some (almost any) depth is proportional to the density ρ itself (with the minus sign, since the density decreases with time) and the time variation dt. Proceeding from this assumption, we will write down the differential equation for density variation $d\rho$. Thus:

$$d\rho = -\rho dt / t_0 . \tag{2}$$

Integrating t from 0 to t_0 and from ρ to 0, we obtain (3).

$$\rho = \rho_{\max} * \exp(-t/t_0). \tag{3}$$

In equation (3), the value of t_0 is the time during which the density of the activated sludge decreases by *e* –fold.

Figures 3 – 5 show the dependence of the activated sludge density on time in comparison with the experimental data. From the curves of the dependence of the density (ρ) on time (t) it follows that they are of the same nature for different radiation sources. At high radiation doses (Figure 4), the density of activated sludge decreases (the deposition rate increases), as it becomes less active. When irradiated with low doses of 3 R (Figure 5), the sedimentation rate decreases, indicating an increase in the activity of sludge in the aqueous solution. According to the calculations, the activated sludge index from Lake Sorbulak was 90 mL/g, and the average sedimentation rate was 2/3 mL/min.

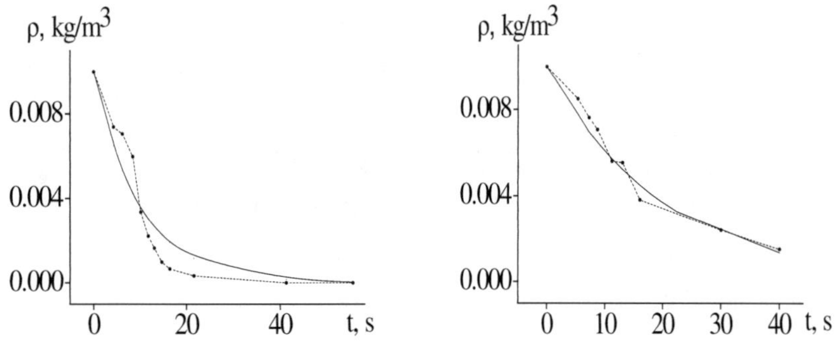

a) b)

Figure 3. Dependence of the density of non-irradiated sludge on time during the passage of green (*a*) and red (*b*) laser radiations through activated sludge.

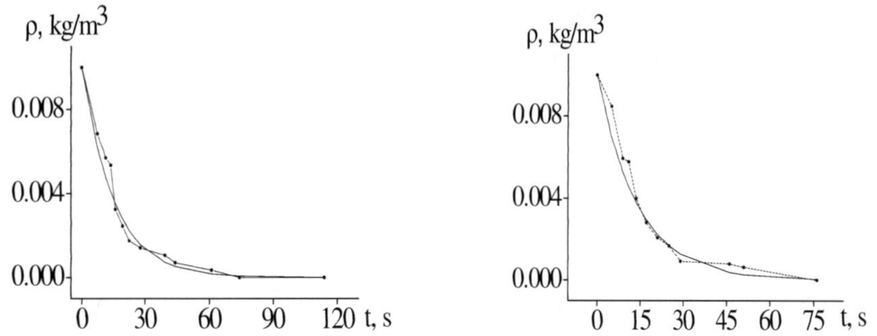

a) b)

Figure 4. Dependence of the density of sludge irradiated with electrons at the dose of 500 kGy on time as green (*a*) and red (*b*) laser radiations pass through the activated sludge.

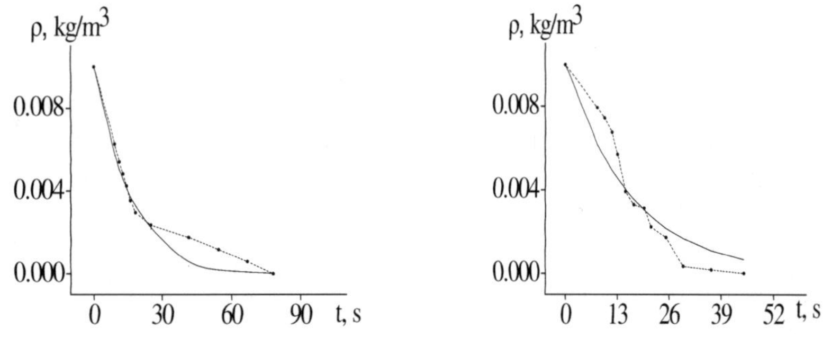

a) b)

Figure 5. Dependence of the density of sludge irradiated with photons with the dose of 3 R on time as green (*a*) and red (*b*) laser radiations pass through the activated sludge.

4. Conclusions

1. From the curves of the dependence of the density (ρ) on time (t), patterns having the same nature for different radiation sources were obtained. According to the calculations, the activated sludge index from Lake Sorbulak was 90 mL/g, and the average sedimentation rate was 2/3 mL/min.
2. Different types of irradiation of biomaterial samples lead to an increase in the sludge index, which is associated with a decrease in natural purification. The reason for this is the reduction in the number of living microorganisms.
3. The curves of the dependence of the density (ρ) on time (t) are described satisfactorily in the framework of the exponential model.

Acknowledgments

The study was supported by the Grant of the Ministry of Education and Science of the Republic of Kazakhstan.

References

[1] Zakhtaeva N V, Shelomkov A S 2013 *Active sludge as a managed ecological system* (Moscow: Expo-media-press)
[2] Benina N N 1990 *Ecology of marine and freshwater free-living protozoa; Research works collection* (L.: Nauka) 143 – 153
[3] Orosov N S 1987 *Biocenosis in nature and industrial conditions* 20 – 32
[4] Alimov A F 1986 *Materials of the Fifth Congress of the All-Union Hydrobiological Society.* (Kuibyshev: Publishing House of the USSR Academy of Sciences) 132 – 133
[5] Andreyuk Ye I, Yanover S V, Kopteva Zh P at al 1981 *Microbiol. journal* **43** 683 – 686
[6] Krasikov E, Nikolaenko V 2016 *IOP Conf. Series: Materials Science and Engineering **110** 012001* doi: 10.1088/1757-899X/110/1/012001
[7] Surzhikov A P, Galtseva O V, Vasendina E A, Vlasov V A, Nikolaev E V, 2016 *IOP Conf. Series: Materials Science and Engineering **110** 012002* doi: 10.1088/1757-899X/110/1/012002
[8] Ivchenko V A 2016 *IOP Conf. Series: Materials Science and Engineering* **110** 012003 doi: 10.1088/1757-899X/110/1/012003
[9] Yesyrev O V, Kupchishin A I, Abdukhairova A T, Nauryzbayev M K, Khodarina N N Cherednichenko V C 2017 *IOP Conf. Series: Materials Science and Engineering* **168** 012017 doi: 10.1088/1757-899X/168/1/012017
[10] Esyrev O V, Kupchishin A A, Kupchishin A I, Voronova N A 2016 *IOP Conf. Series: Material Science and Engineering* **110** 012036 doi: 10.1088/1757-899X/110/1/012036

Study of the effective filtration method applicability for the control of composite materials stress-strain state

V P Surzhikov[1], P I Fedotov[2], P N Khorsov[3]
[1]Engineer, Tomsk Polytechnic University, Tomsk, Russia
[2]Researcher, Tomsk Polytechnic University, Tomsk, Russia
[3]Junior researcher, Tomsk Polytechnic University, Tomsk, Russia
E-mail: iva1954@mail.ru

Abstract. The paper presents the results of experimental studies on the possibility of applying the effective filtration principle in the developing non-destructive testing methods. The presented results make it possible to conclude that the optimal filtration principle can be applied in the further development of methods for determining the absolute values of the stress-strain state of similar objects. It is necessary for its implementation to have impulse responses of similar objects for given stress-strain state values.

1. Introduction

The development of a method based on the mechanoelectric transformations (MET) phenomenon for controlling structural disturbances in dielectric composites first began at the Tomsk Polytechnic Institute [1]. The essence of the method is as follows [2]. A pulsed mechanical method excites in a dielectric sample an acoustic wave, propagating in the sample and experiencing multiple reflections. The acoustic wave, interacting with the centers of MET, induces an alternating electromagnetic field whose time realizations are recorded as a response signal by means of capacitive or inductive receivers.

Studies were conducted aimed at elucidating the possible use of the MET phenomenon to create methods for nondestructive testing and evaluation of the stress-strain state (SSS) of composite materials and their strength.

The method sensitivity to the size of the filler in concrete was analyzed in the paper [2]. The dependence of the time response amplitude-frequency characteristics on the geometry of the location of inclusions in concrete and their dimensions was established in the paper [3], in concretes under shock excitation. In the paper [4], under uniaxial compression, the effect of contact between rock samples of different mineralogical composition on the response parameters was investigated. The presence of such a contact has a significant effect on the spectral characteristics of the response. In work [5] acoustic sounding of crystalline quartz and rocks samples, represented by quartzites, was carried out. The study showed a significant change in the amplitude-frequency response characteristics from the distance between the capacitive sensor and the sample surface. The connection between the frequency characteristics of electromagnetic signals and the characteristics of the acting pulse is also established. Works [6–10] show the efficiency of using the MET phenomenon and the high sensitivity in the responses variations on changes in the SSS and on the degree of samples defectiveness for pulsed acoustic probing.

The revealed dependencies and high sensitivity of the response parameters to the change in the stress-strain state are prerequisites for using the effective filtering method in determining the absolute values of the control object's SSS by the set of time realizations obtained at the given SSS values.

The optimal filtering method [11] is as follows.

If a short pulse (in the limit of infinitesimal duration) is applied to the input of a linear system (a system in which the superposition principle is satisfied), then it will have a response, which is called the impulse response $h(t)$.

If now the signal of the given form $s(t)$ is fed to the input of the same system, then the output $u(t)$ is received. There is a mathematical expression relating $h(t)$, $s(t)$ and $u(t)$, which is the convolution of the signal $s(t)$ with $h(t)$:

$$u(t) = \int s(\tau) \cdot h(t - \tau) \cdot d\tau$$

An input signal matched with a system and having this impulse response is a signal of the form

$$s^*(t) = h(t - T) \tag{1}$$

on condition $0 < t < T$, where T – is the observation interval.

As follows from (1), the matched input signal is a mirror image of the impulse response of the system. In this case, the response will have the maximum signal-to-noise ratio at time T and, in addition, in the time interval from 0 to T will be the autocorrelation function $h(t)$.

In our case, the system to be monitored is the sample itself, whose input is fed by an acoustic excitation pulse, and an electrical response arises at the output. The impulse response of this system should depend on the degree of the sample SSS, so if you have a set of impulse characteristics at different levels of the sample's SSS, it becomes possible to estimate the absolute value of the sample's SSS under an unknown load. It is possible to do this by submitting mirror images of these impulse responses to the input and to evaluate according to the responses at which of the impulse characteristics the optimality conditions will be best satisfied. The problem is that a piezoelectric transducer is used as the excitation device, by means of which it is impossible to form a short, and therefore broadband, pulse, because of its selectivity in frequency.

Therefore, it is necessary to consider the impulse response of a system including the characteristics of the excitation device. In such a "composite" system, under the influence of external influences, only the characteristics associated with the sample will change.

If a short electrical impulse is applied to the input of the piezoelectric transducer, then the response can be written in the form

$$u(t) = \int h_0(\tau) \cdot h(t - \tau) \cdot d\tau$$

where $h_0(t)$ – is the impulse response of the piezoelectric transducer.

It is necessary to evaluate the effect of the spectral characteristics of the excitation device on the response parameters when using the optimal filtering.

If an inverse impulse response of duration T is chosen as the input signal, the output signal will have a maximum value at time T. The time dependence of the response is the autocorrelation function of the input signal when it is fully matched to the impulse response.

2. Materials and experimental procedure

A description of the measurement procedure is given in [8, 9]. A block diagram of the measurements is shown in Figure 1.

We used in the experiments samples from epoxy resin with a size (6×7.8×9.2) cm^3. Quartz sand was used as a filler. The propagation velocity of longitudinal waves in these samples was 2440 m/s. The piezoelectric source was located on the edge of the sample with a size (6×9.2) cm^2. Parallel to the face of the sample measuring 9.2×7.8 cm^2 at a distance of 0.1 cm from the surface, a capacitive sensor was fixed.

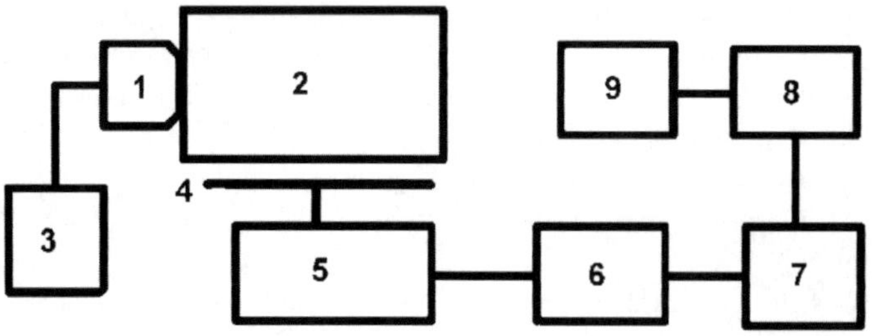

Figure 1. Block diagram of the experiments: 1 – source of acoustic oscillations on the piezoelectric ceramics of ZTS19; 2 – test sample; 3 – master generator of sinus-like voltages; 4 – capacitive sensor; 5–input amplifier; 6 – bandpass filter; 7 – low-frequency amplifier; 8 – emitter repeater; 9 – data input/output board.

The measuring cell with the samples was located on a fixed punch of the MIS-500K test machine. The samples were subjected to uniaxial compression from the side of the sample face with the size (6×8) cm^2. The load stepwise in increments of 20 kN increased from 0 kN to the samples destruction. In the experiments, time realizations of the responses were recorded for each fixed load value using the NI PCI – 6133 data acquisition board. Time realizations, averaged over 140 responses, were analyzed. Sounding acoustic pulses followed with a period of 7 ms. The frequencies of the acoustic pulse spectrum range from 40kHz to 125kHz. Dispersion in the experiments did not exceed 5%.

3. The experimental results

The measurements results are shown in Figures 2 – 4. Figure 2 shows the pulse characteristics for the samples under uniaxial compression at a pressure of 0 MPa, 3.2 MPa and 8 MPa.

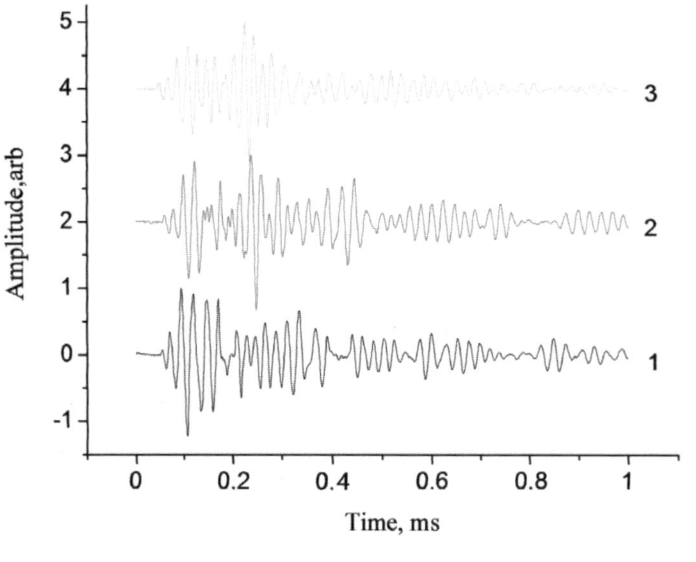

Figure 2. Time realizations of responses at a pressure of 1–0MPa; 2–3.2 MPa, 3–8MPa. Excitation pulse duration – 45 μs, frequency – 44 kHz.

The inverted values of the impulse responses were converted into electrical signals, which were fed to the piezoelectric transducer to form an acoustic probe pulse. The corresponding responses are shown in Figure 3. For this, the objects of investigation were at the same pressure at which impulse responses were recorded. It is evident that the form of the responses has a symmetrical form with respect to the maximum value.

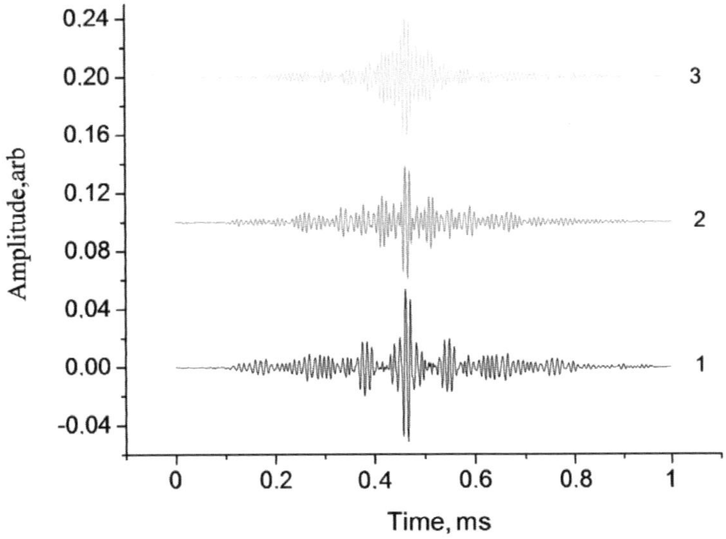

Figure 3. Probe results from inverse impulse responses at a pressure of 1–0 MPa; 2–3.2 MPa; 3–8MPa.

Figure 4 shows the response from the sample, which is under uniaxial compression with a pressure of 8 MPa. The probe acoustic pulses were formed by inverted impulse characteristics, measured at 0MPa, 3.2 MPa and 8 MPa. It can be seen that the non-coincidence of the pressure, at which electrical response is measured, with the pressure, at which the impulse response is recorded, leads to a decrease in the amplitude of the response and to the breaking and disappearance of the symmetry in the form of a response.

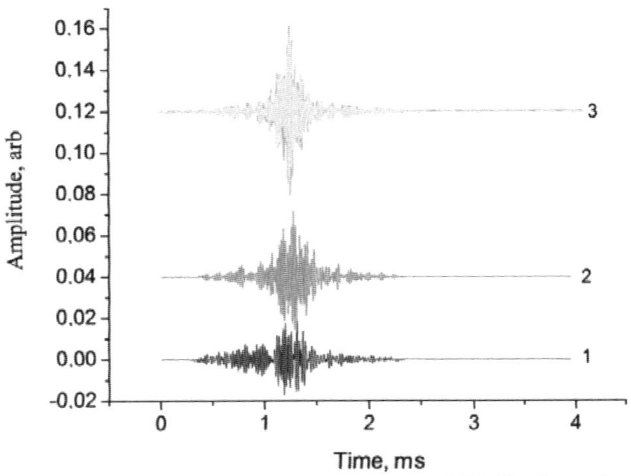

Figure 4. Responses from the samples at a pressure of 8 MPa from the inverse impulse characteristics, measured at loads of 1–0MPa; 2–3.2 MPa; 3–8MPa.

4. Conclusions

The data obtained in the experiments make it possible to conclude that the principle of optimal filtration can be applied to the further development of methods for determining the absolute values of the similar objects SSS. For its implementation, it is necessary to have impulse responses of similar objects for the given SSS values.

Acknowledgments

This work was financially supported by The Ministry of Education and Science of the Russian Federation in part of the science activity program.

References

[1] Vorob'yev A A, Zavadovskaya Ye K, Sal'nikov V N 1973 Proceeding of 4 Vsesoyuznogo simpoz. po mekhanoemissii i mekhanokhimii tverdykh tel. 72–73 (in Russian)

[2] Fursa T V, Gordeyev V F 2000 *Technical Physics Letters* **26** 105–106

[3] Fursa T V, Osipov K Yu 2005 *Russian Physics Journal* **48** 307–311 doi:10.1007/s11182-005-0124-z

[4] Bespal'ko A A, Surzhikov A P, Yavorovich LV 2006 *Tsvetnyye metally* **4** 32–34

[5] Bespal"ko A A, Yavorovich L V, Fedotov P I 2007 *Journal of Mining Science* **43** 472–476 doi: 10.1007/s10913-007-0049-8

[6] Surzhikov V P, Khorsov N N 2011 *Kontrol'. Diagnostika* 132–134 (in Russian)

[7] Surzhikov V P , Khorsov N N 2012 *Kontrol'. Diagnostika* 69–71 (in Russian)

[8] Surzhikov V P , Khorsov N N 2011 *Russian Journal of Nondestructive Testing* 47 687–690 doi: 10.1134/S1061830911100159

[9] Surzhikov V P , Khorsov N N , Khorsov P N 2012 *Russian Journal of Nondestructive Testing* **48** 85–89 doi: 10.1134/S106183091202009X

[10] Surzhikov V P, Khorsov N N 2015 *Russian Physics Journal* **58** 138–140 doi: 10.1007/s11182-015-0473-1

[11] Surzhikov A P, Surzhikov V P, Khorsov N N, Khorsov P N 2014 *Mul'tisensornaya apparatura kontrolya defektnosti i napryazhenno-deformirovannogo sostoyaniya geterogennykh dielektricheskikh struktur* (Tomsk: TPU) (in Russian)

IOP Conf. Series: Materials Science and Engineering **289** (2017) 012036 doi:10.1088/1757-899X/289/1/012036

The influence of the ion plasma flow on the phase composition and microhardness of carbon coatings doped with metals

A S Rudenkov[1], D G Piliptsou[2], P A Luchnikov[3], A V Rogachev[4], Xiaohong Jiang[4], N N Fedosenko[2]

[1]Research scientist, International Chinese-Belarusian scientific laboratory on vacuum-plasma technology: Francisk Skorina Gomel State University, Belarus, Gomel, Nanjing University of Science and Technology, China

[2]Associate professor, International Chinese-Belarusian scientific laboratory on vacuum-plasma technology: Francisk Skorina Gomel State University, Belarus, Gomel, Nanjing University of Science and Technology, China

[3]Lead engineer, Moscow Technological Universtity, Russia, Moscow

[4]Professor, International Chinese-Belarusian scientific laboratory on vacuum-plasma technology: Francisk Skorina Gomel State University, Belarus, Gomel, Nanjing University of Science and Technology, China

E-mail: xamdex@gmail.com

Abstract. The Raman spectroscopy method established that subsequent treatment with ion ions leads to a change in the microstructure of coatings based on carbon and metals. It is established that the structure change is determined by the changes occurring at the interface and depends on the sp3 → sp2 phase transition due to the appearance of local thermal peaks at the end of tracks of nitrogen ions implanted in the coating volume. It is shown that the microhardness of metal-carbon coatings (α-C: Cu, α-C: Ti, α-C:Al) increases after ion-plasma treatment, which is determined by the formation of solid interstitial phases based on CNx and metal carbonitrides.

1. Introduction

Amorphous carbon coatings (α-C), produced by various methods, have a unique combination of electrical, optical, thermal and mechanical properties. At the same time, they are characterized by a number of defects, including low temperature stability, high internal stresses, and low adhesion to dielectric substrates [1]. The most effective method of enhancing the properties of α-C is their doping with nitrogen and various metals, which have different chemical activity with respect to carbon [2–4]. For the production and usage of such coatings it is very important to control the properties of the produced nanocomposite systems, in which the chemical interaction processes can lead to the formation of carbide and nitride phases in the carbon matrix. The choice of individual components and their concentration in the coating layer, as well as the effect of the ion-plasma treatment method and regimes on the carbon structure, largely determine its basic performance characteristics. At ionic processing of composite carbon coatings, selective etching of a certain coating component occurs, the concentration and character of distribution in the volume of alloying elements change. With such treatment, the processes of chemical interaction during heating, radiation modification of components become more complex [1, 2]. An increase in the operational properties of metal-carbon coatings is possible with a directed change in the structure of the layer due to a change in the modes and parameters of doping or ion treatment during their formation [1–4].

The matter of topical interest is to select the composition of the individual components of the coating and develop technological recommendations for directed changes in the structure and mechanical properties of metal-carbon coatings formed under the conditions of ions assisting of different nature and energies with subsequent technological ion treatment.

The effect of the subsequent technological ion-plasma treatment on the structure and microhardness of composite metal-carbon coatings formed by simultaneous deposition of carbon from a pulsed cathode plasma and ions / metal atoms (Cu, Ti, Al) generated by magnetron sputtering is under study.

2. Models and research methods of production and properties

Carbon coatings were formed on a silicon substrate by simultaneously depositing carbon from the plasma source of a pulsed cathode-arc discharge with a graphite cathode and metal from the flow of the generated magnetron sputtering system. As alloying elements, metals possessing different chemical activity with respect to carbon and nitrogen (Cu, Ti, Al) were used. Subsequent processing of the produced coatings with nitrogen ions (ion energy up to 1.5 keV) promotes phase transitions $sp^3 \rightarrow sp^2$ and the formation of metal nitrides and chemical compounds of nitrogen and carbon CN_x in the near-surface layers. The thickness of the coatings was 270–300 nm, the metal concentration was of 2.5 at. %, the microhardness varied from 10 to 12 GPa, depending on the nature of the alloying metal.

The phase composition of the carbon coatings was determined by analyzing the Raman spectra obtained with a Senterra spectrometer with an excitation wavelength of 532 nm and power of 10 mW. The obtained spectra were decomposed into two Gaussians (D- and G-peaks).

Raman spectroscopy is one of the most widely used methods of non-destructive quality control of diamond [1], graphite, carbon nanotubes and carbon-based coatings [2, 4–6].

There are three main reasons for the change in the Raman spectra of α-C coatings [7]:

– conjugated π-bonds increase the polarizability range;
– G and D modes are conjugated;
– the phenomenon of double resonance for the D mode [8].

A large range of polarizability substantially increases the intensity of the G and D modes. Particularly, the intensity of the D mode increases, which is caused by the respiratory mode. In the graphite layer there is an amplifying interference of the eigenvectors of each six-atom ring and weakening interference for rings of other orders.

All three factors increase the intensity of respiratory modes and suppress the modes of other symmetries. In [9] it is noted that the ratio I_D/I_G is inversely proportional to the size of graphite L_a grains:

$$I_D/I_G=c(\lambda)/L_a, \quad (1)$$

where I_D and I_G are the intensities of the corresponding peaks, $c(\lambda)$ is the proportionality coefficient that depends on the wavelength of exciting radiation, and L_a is the grain size of graphite (nm). This means that the ratio I_D/I_G is proportional to the number of rings at the grain boundary. Obviously, this L_a can not tend to zero. For ta-C coatings, it was found that for L_a is less than 2 nm, the ratio decreases according to the expression (2) [10]:

$$I_D/I_G=c(\lambda)\cdot L_a^2 \quad (2)$$

Since the G-peak of the mode exists due to sp^2-hybridized atoms, and the D-peak is only due to the rings, I_D/I_G decreases with a decrease in the number of carbon rings and an increase in the proportion of carbon chains.

Figure 1 shows the causes of the displacement of the G and D peaks, as well as the changes in their relative intensities.

In [6, 11], a three-stage model for increasing the degree of disorder in carbon is presented:

– ideal graphite passes into a nanocrystalline;

– nanocrystalline graphite passes into a-C form;
– sp^2 a-C goes to sp^3 ta-C.

Stage 1 corresponds to a decrease in the grain size of an ordered graphite layer, but with the preservation of aromatic rings. The G-peak shifts to 1600 cm^{-1}. Mode D is banned for an ideal layer of graphite, but the disorder of bonds and the reduction in grain size leads to an increase in its intensity. It is worth noting that the width of the D- and G-peaks is a characteristic of the disorder degree.

Stage 2 corresponds to the topological disorder of graphite, the decrease of aromatic bonds, but with the preservation of the perfect sp^2 matrix. The G-peak shifts to 1520 cm^{-1}.

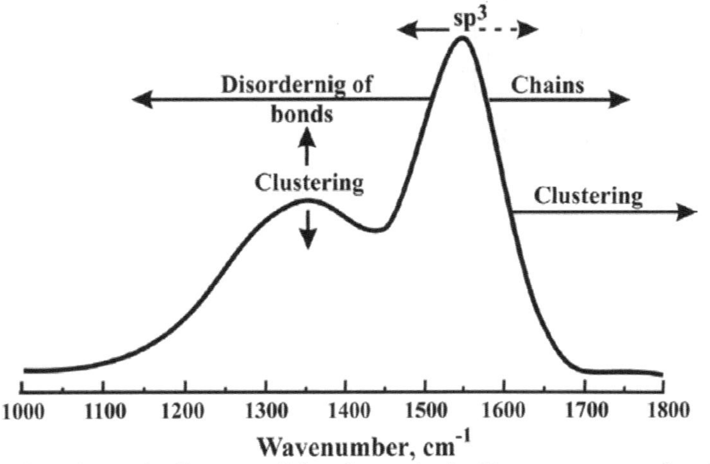

Figure 1. The schematic diagram of the change in the Raman spectra of carbon coatings [7].

At stage 3, the content of sp^3 increases to 100 %. This fact changes the sp^2-configuration from rings to short chains [6]. The length of bonds of chains (olefins) is shorter than that of carbon rings, since their oscillation frequencies are higher. The G-peak shifts to 1580–1600 cm^{-1}, its intensity increases, and the D-peak intensity tends to zero.

An increase in the I_D/I_G ratio indicates a decrease in the size of sp^2 clusters, narrowing of the width of the G peak indicates an increase in the order degree, its shift to the region of large wavenumbers can be explained by an increase in the level of internal stresses, and a shift to the region of smaller wave numbers by a decrease in the content and disordering of sp^3-bonds.

To measure the microhardness according to Knupp, the microfirms *AFFRI DM*-8 were used during experiments. The evaluation of microhardness is based on measuring the length of the diagonal of the reprint produced by indenting a diamond pyramid with a certain load into the material under study. When measuring the microhardness according to Knupp, a pyramid with a square base is used, in which the angle between the opposite lateral faces was 172° 30´.

The sizes of prints were determined with the help of an optical microscope with interchangeable lenses providing an increase in 400^x and 100^x. The load value F is 245 mN; the duration of indentation of the pyramid during measurements by both methods was the same – 10 s. Since the thickness of the coatings analyzed in the research does not exceed 200 nm, the measured microhardness of the samples is a superposition of the hardness of the "coating-substrate" system and was determined according to the expression (1):

$$H_k = 14{,}23 \cdot F/L^2, \quad (3)$$

where L is the diagonal length and F is the load.

3. The results and their analysis

Figure 2*a*–*e* shows the results of statistical treatment of the data of Raman spectroscopy of metal-carbon coatings (ɑ-C: Cu, ɑ-C: Ti, ɑ-C: Al) after ion-plasma treatment.

The analysis of Raman spectroscopy shows that the subsequent ion-plasma treatment of metal-carbon coatings leads to an increase in the I_D/I_G ratio, which indicates a decrease in the size of sp^2 clusters. In addition, a decrease in the width of the G peak is observed, which is explained by an increase in the degree of ordering of the sp^2 clusters.

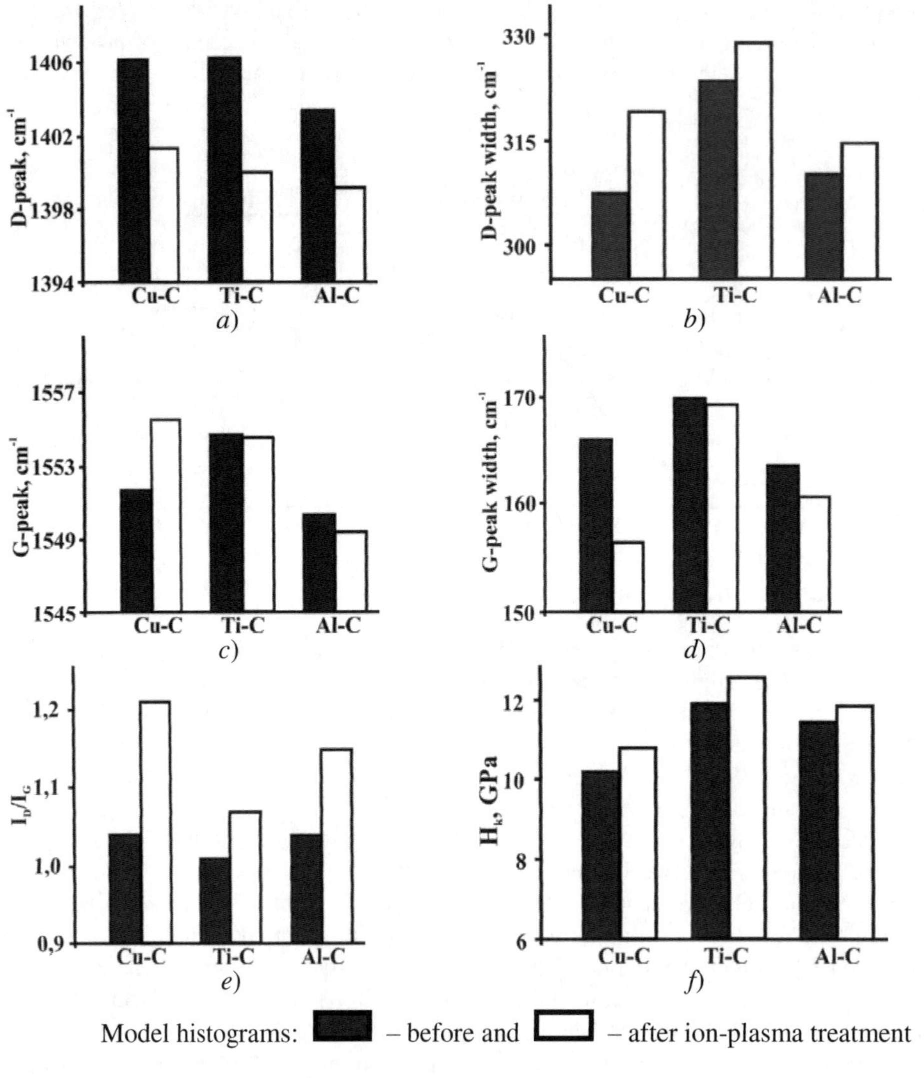

Figure 2. Histograms of values of position (*a, c*), width (*b, d*), intensity of absorption peaks (*e*) of Raman spectra and microhardness (*f*) of poduced metal-carbon coatings (α-C: Cu, α-C: Ti , α-C: Al) after their subsequent technological ion-plasma treatment.

The position of the G peak, in the case of α-C: Al and α-C: Ti coatings, practically does not change. Its shift to the region of large wave peaks after ion-plasma treatment of copper-doped coatings can be caused by an increase in the level of internal stresses due to the formation of a larger number of crystalline inclusions of N-Csp^2, in contrast to titanium and aluminum. Since Co-C coatings initially, even before ion-plasma treatment, are characterized by a high content of sp^2 phase, while sp^2-hybridized carbon atoms in α-C: Al and α-C: Ti coatings are bound in the form of metal carbides. In addition, the level of internal stresses after ion-plasma treatment in α-C: Cu coatings can increase as a result of the formation of Cu3N, however, the formation of such compounds by vacuum methods is unlikely.

The increase in the width of the D-peak and its shift to the region of lower wave numbers may indicate disordering and a decrease in the number of sp^3-clusters. Such changes can be caused by the destruction of the grain boundaries and $sp^3 \rightarrow sp^2$ phase transition, due to the formation of local thermal peaks at the end of tracks of nitrogen ions implanted in the coating volume during ion-plasma treatment.

The results of measuring the microhardness of samples of metal-carbon coatings before and after ion-plasma treatment are introduced in Figure 2*f*, where it is shown that the microhardness of metal-carbon coatings (α-C: Cu, α-C: Ti, α-C: Al) increases after ion-plasma treatment. It can be assumed that this increase in microhardness is mainly due to the formation of solid interstitial phases based on CN_x and metal nitrides. It should be noted that the formation of Cu_3N as a result of ion-plasma treatment is less likely than AlN and TiN.

The previously established fact of decreasing the size of sp^2 clusters due because of ion-plasma treatment also makes it possible to explain the change in microhardness from the point of view of the Hall-Petch law. The relatively small (0.5–0.7 GPa) increase in microhardness can be explained by the formation of CN_x phases in near-surface layers, the thickness of which is much smaller than the depth of penetration of the indenter.

4. Conclusion

Raman spectroscopy showed that the subsequent ion-plasma treatment of metal-carbon coatings leads to the increase in the ratio of the I_D/I_G absorption peaks of their Raman spectra, which indicates the decrease in the size of sp^2 clusters. At the same time, the increase in the width of the D-peak and its shift to the region of lower wave numbers may indicate disordering and the decrease in the number of sp^3-clusters. Such changes can be caused by the destruction of the grain boundaries and the sp3→sp2 phase transition due to the formation of local thermal peaks at the end of tracks of nitrogen ions implanted in the coating volume during ion-plasma treatment. It is shown that the microhardness of metal-carbon coatings (α-C:Cu, α-C:Ti, α-C:Al) increases after ion-plasma treatment, which is possible due to the formation of solid interstitial phases based on CN_x and metal nitrides.

Acknowledgments

The research was carried out with the financial support of the Belarusian Foundation for Basic Research, project T16M-015.
This work was supported by the Ministry of Education and Science of the Russian Federation (№3.8074.2017/8.9 basic part of the State task).

References

[1] Robertson J 2002 *Mater. Sci. Eng. R* **37** 129 – 281 doi: 10.1016/S0927-796X(02)00005-0
[2] Ming M Y *et al* 2016 *Appl. Surf. Sci.* **379** 424–432 doi: 10.1016/j.apsusc.2016.04.112
[3] Piliptsou D G *et al* 2017 *IOP Conf. Ser.: Mater. Sci. Eng.* **168** 012103 doi:10.1088/1757-899X/168/1/012103
[4] Piliptsou D G *et al* 2016 *Nanomaterials and Nanostructures – XXI Century* **7**(3) 35–42 (in Russian)
[5] Robertson J *et al* 1987 *Physical Review B* **35** 2946–2957 doi:10.1103/PhysRevB.35.2946
[6] Shroder R E 1990 *Physical Review B* **41** 3738–3745 doi: 10.1103/PhysRevB.41.3738
[7] Ferrari A C *et al* 2000 *Physical Review B* **61** 14095–14107 doi: 10.1103/PhysRevB.61.14095
[8] Thomsen C, Reich S 2000 *Phys. Rev. Lett.* **85** 5214–5217 doi:10.1103/PhysRevLett.85.5214
[9] Tuinstra F, Koenig G L 1970 *J. Chem. Phys.* **53** 1126–1130 doi: 10.1063/1.1674108
[10] Chowalla M *et al* 2000 *Appl. Phys. Lett.* **76** 1419–1421 doi: 10.1063/1.126050
[11] Luchnikov P A *et al* 2017 *IOP Conf. Ser.: Mater. Sci. Eng.* **168** 012092 doi:10.1088/1757-899X/168/1/012092

IOP Conf. Series: Materials Science and Engineering **289** (2017) 012037 doi:10.1088/1757-899X/289/1/012037

Defect formation in fluoropolymer films at their condensation from a gas phase

P A Luchnikov
Lead engineer, Moscow Technological University, Russia, Moscow
E-mail: xamdex@gmail.com

Abstract. The questions of radiation defects, factors of influence of electronic high-frequency discharge plasma components on the molecular structure and properties of the fluoropolymer vacuum films synthesized on a substrate from a gas phase are considered. It is established that at sedimentation of fluoropolymer coverings from a gas phase in high-frequency discharge plasma in films there are radiation defects in molecular and supramolecular structure because of the influence of active plasma components which significantly influence their main properties.

1. Introduction

Thin polymeric films thanks to the chemical resistance to hostile environment and high dielectric properties of wide are used in microelectronic, micromechanical devices [1] and touch sensors [2–4]. For receiving thin-film coverings on the basis of fluoropolymer sedimentation methods from a gas phase and polymerization of molecules of C_2F_4 monomer on a firm surface under the influence of the F_2 [5] catalyst are used. Here the stimulating radiations such as X-ray and gamma beams can be applied to polymerization of monomer. There is a row of the elionic technological and ecologically safe technologies where the principle of dispersion of an initial fluoropolymer in a vacuum under the influence of the accelerated electrons and ions at the subsequent secondary polymerization on a substrate of products of dispersion is used [6–8]. Dispersion of a block fluoropolymer on gas environment can also be carried out in plasma of high-frequency (HF) gas discharge [9, 10]. Sedimentation of a fluoropolymer covering at the secondary polymerization on a substrate can proceed in the same conditions as well as at polymerization from C_2F_4 monomer. Use of the elionic technologies for such purposes is well combined with vacuum technology of microelectronic devices and creates a possibility of steady production control of drawing films of a fluoropolymer. The received vacuum fluoropolymer films on the molecular structure and a supramolecular structure are close to industrial films of the fluoroplast, however by the structure they differ from industrial films because of the existence of defects in the received films. In the process of sedimentation from the gas environment in the high-frequency discharge on a surface of a solid body as processes synthesis of polymer of the growing layer and its destruction under the influence of active components (electrons, UF-radiation, ions) plasmas at the same time proceed [7, 11]. For the purpose of receiving polymeric coverings with the set properties identification of physical and chemical mechanisms of growth of fluoropolymer films is necessary to define ways of production control of their receiving [12].

In the article factors of influence of electronic high-frequency discharge plasma components on molecular structure and properties of the fluoropolymer vacuum films synthesized on a substrate from a gas phase are investigated.

2. Samples and technique of an experiment

The studied samples of fluoropolymer coverings were formed in a vacuum by sedimentation on a substrate in the high-frequency discharge at a frequency of 13.56 MHz of an active molecular stream of products of dispersion of a target from an industrial polytetrafluoroethylene (PTFE) by the technique stated in [10]. In the process of condensation the growing fluoropolymer covering was exposed to electronic processing by the active component of the high-frequency discharge plasma. As a substrate served the silicon polished plates and monocrystals of KCl. Growth rate of coverings was 0.05…1.0 μm/min. Temperature of a surface of a substrate changed within 300…500 K. Molecular structure of films of fluoropolymer on a substrate from KCl estimated on IK-spectrums and data of electronic paramagnetic resonance (EPR). Molecular, electric and adhesive properties of fluoropolymer with coverings 0.5…7.0 μm were investigated.

Structural properties of films of fluoropolymer were investigated by method of the differential and thermal analysis (DTA). The structure of films was investigated by methods of IK-spectroscopy and X-ray photoelectronic spectroscopy (PES) on substrates of silicon and monocrystals of KCl. The degree of crystallinity was determined by method of X-ray spectroscopy, and concentration of radiation defects and communications by keeping of radicals in structure was determined by method of the EPR.

3. Results of researches and their discussion

In Figure 1 the results of a research of films structure are presented by method of the differential and thermal analysis (DTA). Curves 1, 2 and 3 correspond to samples of the vacuum films received at impact on the growing layer of a film electronic radio-frequency-generated plasma stream components with a current density ~ 0.07, 0.6 and 1.3 A/m^2.

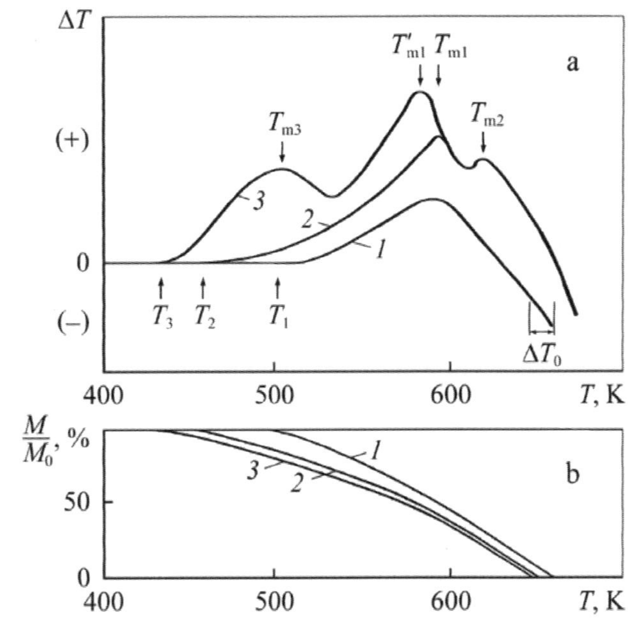

Figure 1. The results of a research of films structure.

According to the ranges of DTA (Figure 1*a*) characteristic maxima of temperatures change at loss of mass (Figure 1*b*) of the substance of a fluoropolymer synthesized on a substrate which was exposed in the course of condensation of a layer to continuous electronic processing by high-frequency discharge particles are observed. So the film of a fluoropolymer besieged on a substrate without the influence of electronic components of the high-frequency discharge plasma has T_{m1} maximum (curve 1, Figure 1*a*) corresponding to decomposition of polymer at 315 °C. At the influence of electronic components of high intensity plasma (curve 3, Figure 1*a* at the density of current of 1.3 A/m^2) are shown two additional maxima of T_{m3} and T_{m2} in the field of temperatures of 225 °C and 339 °C respectively. Influences of the high-frequency discharge plasma of average intensity (at current ~ 0.6 A/м^2) only one high-temperature maximum of T_{m2} in the field of 339 °C is shown. Curve losses of weight show that for samples of vacuum films the 1, 2 and 3 (Figure 1*b*) beginning of process of losses correspond to temperatures: 156 °C, 177 °C and 224 °C.

Researches by DTA method have shown that weak impact of the high-frequency discharge on the besieged layer of a fluoropolymer film leads to formation of cross stitchings of molecular links, which lead to hardening of structure of a fluoropolymer. However, the high intensity of influence leads to an additional radiation inoculation to a polymeric layer of fragments of a molecular chain with radical groups of a gas phase and also to formation of defects at partial decomposition of earlier settled molecules on a surface of the polymerized film [11].

4. Structure of films

In Figure 2 ranges of PES of the fluoropolymer films besieged from a gas phase without influence electronic components (a) of the high-frequency discharge plasma and at her intensive influence (*b*) are presented. Here peaks 1, 2, 3, 4, 5 and 7 which make ranges (*a*) (would) also correspond to communications of the molecular (– CF_3, – CF_3–$\underline{C}F_2$–), (–($\underline{C}F_2$–$\underline{C}F_2$)–), (–$\underline{C}$–F), (–C=O), (–$\underline{C}$–CF–), (–C–O–C–) and (–$\underline{C}$–C–, –$\underline{C}$–) groups respectively. From PES of films (Fig. 2) it is visible that electronic processing of the growing layer of a fluoropolymer by the active component of the high-frequency discharge plasma leads to essential change of maintenance of molecular complexes in structure towards their increase (Figure 2*b*). It also leads to emergence of high deficiency in molecular and supramolecular structures of the received fluoropolymer.

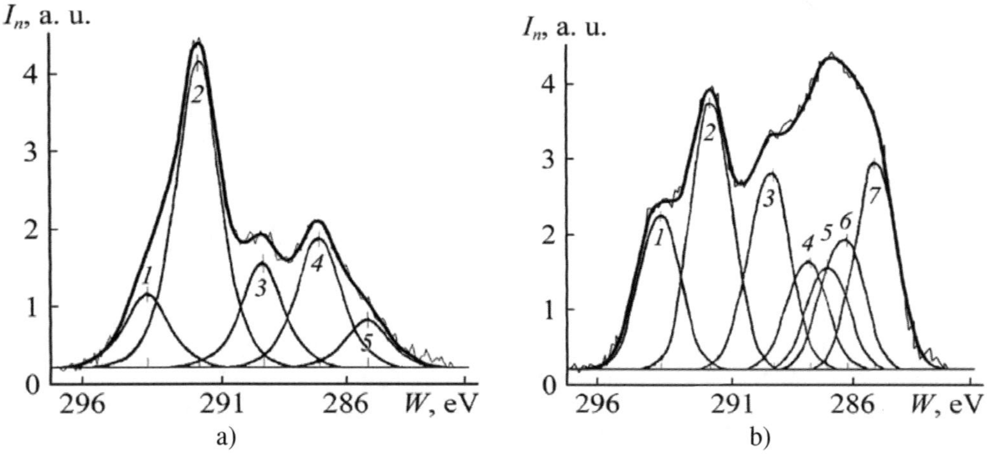

Figure 2. PES-spectrums of the fluoropolymer films besieged from a gas phase without influence electronic components (*a*) of the high-frequency discharge plasma and at her intensive influence (*b*) are presented.

In Figure 3 IK-spectrums of the fluoropolymer films received in the high-frequency field are presented: 1 – without plasma influence; 2 – at weak influence electronic radio-frequency-generated plasma components; 3 – when processing by a strong stream of charged particles of plasma.

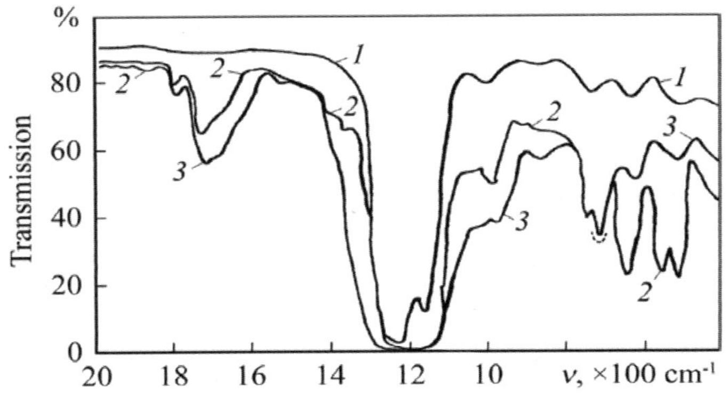

Figure 3. IK-spectrums of the fluoropolymer films received in the high-frequency field.

In Figure 4 the dependence of intensity of strips of absorption of 735 and 1050 cm^{-1} in IK-spectrum, the depreciation of structure of a film of a fluoropolymer corresponding to peaks, from degree of crystallinity of χ is presented.

In Figure 5 the dependence of intensity of an absorption strip of 1710 cm^{-1} in IK-spectrum from concentration of C_p of radical backs in the besieged film of a fluoropolymer is presented.

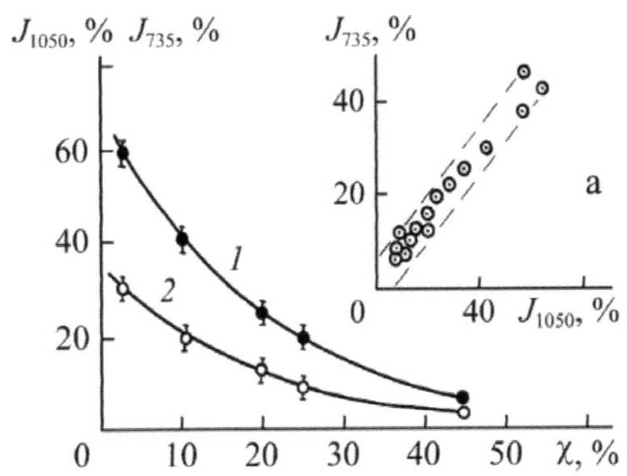

Figure 4. The graph of the dependence between J1050 (1) and J735 peaks (2) of the IR spectrum absorption band intensity of the fluoropolymer film on the degree of the supramolecular structure crystallinity degree χ (1): a) the proportional dependence of these peaks that correspond to 1050 cm^{-1} absorption peaks and 735 cm^{-1} (2) of the amorphization of the fluoropolymer structure

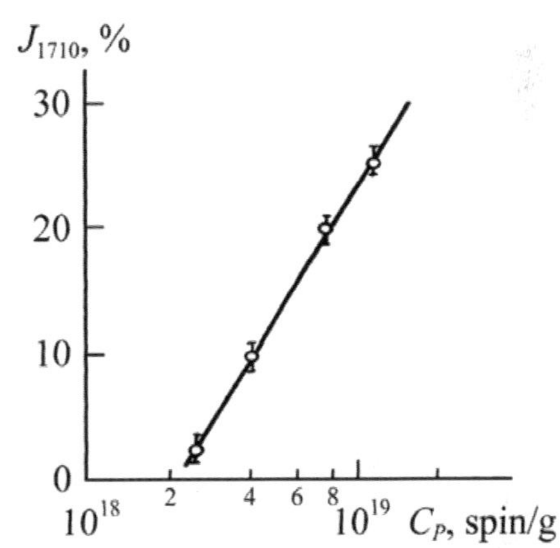

Figure 5. The graph of the dependence of peak J_{1710} of the IR spectrum absorption band intensity of the fluoropolymer film on concentration Cp of spin radicals (defects) in the molecular structure.

The analysis of the obtained experimental data of IK spectrums and a PES ranges (Figure 2–5) shows that with increase in intensity of influence by the active components of a plasma flow of the high-frequency discharge depreciation of its structure increases by the growing fluoropolymer film. On changes of IK-spectrums it is possible to judge imperfections of the received films which are defined by conditions of their formation. Crystallinity χ of the films received by polymerization in the high-frequency field without plasma stream (curve 1 Figure 3*a*) can reach ~ 25…28 %. In IK-spectrums cross stitching's in polymer correspond absorption peaks in the range of 1680…1780 cm^{-1}. The full IK-spectrum of a fluoropolymer has a number of additional strips of absorption in the area 525, 978, 1660…1710 and 1772 cm^{-1} and also considerable expansion of the main peak of absorption of a fluoropolymer chain –C–C–C– in the area 1100…1300 cm^{-1}.

With increase in intensity of electronic processing of the growing polymer layer (curve 3, Figure 2*a*) increase in content of the C_C defects in a film layer is observed at simultaneous decrease in molecular weight of the main chain –C–C–C– of polymeric links. Also increase in maintenance of cross stitching's in molecular structure of a fluoropolymer is observed. Electronic processing of the growing layer of a fluoropolymer leads to partial destruction of its natural structure with formation of radiation defects. Spin-radicals are formed due to radiation modification of the growing film layer in the course of her synthesis [11], similarly as in a graphite covering [13]. In the absence of plasma processing the received polymeric film in structure has concentration of C_c radicals no more $(1\ldots2)\cdot10^{18}$ spin/g. Increase in maintenance of radical backs in a layer of a fluoropolymer leads to increase in its electric conductivity and deterioration in dielectric properties.

By thickness the fluoropolymer covering on a substrate in the process of its growth finds layered structure and in strong degree depends on the electric and thermal modes of her formation. The layered structure can conditionally be divided into three main types: – the adsorptive with thickness (0.2…1.5 μm); – the main, up to 5…8 μm; – superficial, depending on growth modes.

The polymeric layer grows in the mode of the adsorptive growth of a covering without influence of a plasma stream on the mechanism of kinetics of the first order. At the same time the film with the thickness of 1…1.5 μm repeats a substrate relief at volume uniformity has high uniformity. Then with increase in thickness of a film crystal areas in the form кof "balls" with the primary growth of the developed surface are formed.

The main layer of such covering has pores of 0.05…0.3 μm directed on a normal to a substrate. The surface of polymer has the regional angle of wetting by water ~ 105° at a thickness of 2…5 μm. With the described epitaxial growth the covering has extreme thickness ~ 5.8…6.3 μm.

5. Conclusion

Thus, when receiving vacuum fluoropolymer thin-film coverings from a gas phase in radio-frequency-generated plasma in a polymeric layer there are radiation defects in molecular and supramolecular structure due to influence of active components of particles of the high-frequency discharge plasma which considerably influence their main properties. By the directed impact of a plasma stream on the growing polymer layer in technological process, it is possible to operate properties of the received polymeric covering.

Acknowledgments

This work was supported by the Ministry of Education and Science of the Russian Federation (№3.8074.2017/8.9 basic part of the State task).

References

[1] Skorykh V J et al 2001 *5th Korea-Russia International Symposium on Science and Technology - Proceedings: KORUS 2001* **1** 42–45 doi: 10.1109/KORUS.2001.975049
[2] Sessler G M 1987 *Electrets* (Verlag Berlin Heidelberg: Springer) doi: 10.1007/3-540-17335-8
[3] Pevtsov E Ph et al 2017 *IOP Conf. Ser.: Mater. Sci. Eng.* **168** 012095 doi: 10.1088/1757-899X/168/1/012095
[4] Luchnikov P A et al 2017 *IOP Conf. Ser.: Mater. Sci. Eng.* **189** 012027 doi: 10.1088/1757-899X/189/1/012027
[5] Enslin S E et al 1986 *South African J. of Chemistry* **39** 23–26
[6] Krasovski A M et al 1989 *Preparation of thin films by spraying polymers in vacuum* (Minsk: Science and Technology) (in Belarus)

[7] Rogachev A V 2016 Micro- and Nanocomposite Polymer Coatings Deposited from Active Gas Phase (Moscow) (in Russia)

[8] Sokol R et al 2011 *Proceedings of the 6th International Scientific Symposium on Electrical Power Engineering (ELEKTROENERGETIKA)* 380–383

[9] Plotnikova I V et al 2017 Journal of Physics: Conference Series **881** 012006 doi: 10.1088/1742-6596/881/1/012006

[10] Luchnikov P A et al 2017 *IOP Conf. Ser.: Mater. Sci. Eng.* **189** 012014 doi:10.1088/1757-899X/189/1/012014

[11] Luchnikov P A et al 2017 *IOP Conf. Ser.: Mater. Sci. Eng.* **168** 012092 doi:10.1088/1757-899X/168/1/012092

[12] Petrova A.Y et al 2015 *Technical Physics* **60** 592–594 doi: 10.1134/S1063784215040222

[13] Piliptsou D G et al 2017 *IOP Conf. Ser.: Mater. Sci. Eng.* **168** 012103 doi:10.1088/1757-899X/168/1/012103

IOP Conf. Series: Materials Science and Engineering **289** (2017) 012038 doi:10.1088/1757-899X/289/1/012038

A method and instruments to identify the torque, the power and the efficiency of an internal combustion engine of a wheeled vehicle

A V Egorov[1], K E Kozlov[2] and V N Belogusev[3]
[1]Pro-Rector for Development, Professor, Mari State University, Yoshkar-Ola, Republic of Mari El, Russian Federation
[2]Head of Innovation Center, Mari State University, Yoshkar-Ola, Republic of Mari El, Russian Federation
[3]Associate Professor, Volga State University of Technology, Yoshkar-Ola, Republic of Mari El, Russian Federation
E-Mail: konstantin.k-e@yandex.ru

Abstract. In this paper, we propose a new method and instruments to identify the torque, the power, and the efficiency of internal combustion engines in transient conditions. This method, in contrast to the commonly used non-demounting methods based on inertia and strain gauge dynamometers, allows controlling the main performance parameters of internal combustion engines in transient conditions without inaccuracy connected with the torque loss due to its transfer to the driving wheels, on which the torque is measured with existing methods. In addition, the proposed method is easy to create, and it does not use strain measurement instruments, the application of which does not allow identifying the variable values of the measured parameters with high measurement rate; and therefore the use of them leads to the impossibility of taking into account the actual parameters when engineering the wheeled vehicles. Thus the use of this method can greatly improve the measurement accuracy and reduce costs and laboriousness during testing of internal combustion engines. The results of experiments showed the applicability of the proposed method for identification of the internal combustion engines performance parameters. In this paper, it was determined the most preferred transmission ratio when using the proposed method.

1. Introduction

Identification of the main performance parameters of an internal combustion engine (IC engine), among which are the power, the torque, and the efficiency, is an important task in the way of improving the IC engine characteristics. Moreover, since the actual operating conditions of the IC engine are rapidly varying, i.e. preferably transients take place; therefore it is preferred to identify the above mentioned parameters in these conditions. All this allows us to state the importance and topicality of the research to develop the method to identify the performance parameters of IC engines of wheeled vehicles in transient conditions.

Currently, it is preferably to use non-demounting methods to control the performance parameters, which are less labor and energy intensive. Unfortunately at the moment there is a small amount of works dedicated to the development of new non-demounting methods and instruments to identify the IC engine performance parameters, which are measured nowadays with roller dynamometers [1, 2]. In this case the measurements are carried out on the driving wheels of the vehicle, and this leads to inaccuracy connected with the fact that the obtained values of output parameters differ from the actual values of an internal combustion engine parameters due to the losses in transmission and friction units. To determine the actual values of the unknown parameters, it is necessary to dismantle the engine from the vehicle [1–5].

The purpose of this paper is to develop the method to identify the performance parameters of an internal combustion engine in transient conditions without its demounting from a vehicle and without the torque loss due to transfer of the power to the driving wheels, on which the torque is measured with the existing methods. The developed method is supposed to be implemented without the use of expensive equipment and strain measurement instruments.

2. Materials and Methods

2.1. Method to Identify the Performance Parameters of an IC Engine of a Wheeled Vehicle

Prerequisites for the implementation of the method: a vehicle, located on the solid asphalt road surface; an internal combustion engine, which is equipped with a sensor of the crankshaft position and with a sensor of the fuel consumption; a hardware-software complex for measuring the angular acceleration of an engine crankshaft (based on the products of the company National Instruments, analog-to-digital converter (ADC) of the type USB-6009, software environment LabView Signal Express 2.5). Acceleration of the vehicle should exclude slippage of drive wheels.

Let us consider the kinematic scheme of the 4-wheeled rear-steer vehicle (see Figure 1).

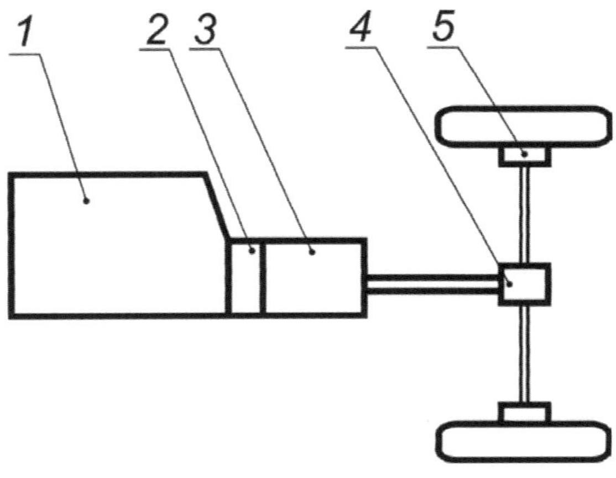

Figure 1. Kinematic scheme of the 4-wheeled rear-steer vehicle: 1 is an IC engine; 2 is a clutch; 3 is a gearbox; 4 is a main gear; 5 is a final drive.

The determination of the moment of inertia normalized with respect to the crankshaft rotation axis is made on the assumption that the IC engine should transfer the mechanical energy, which is necessary to drive the wheels, through a kinematic transmission. Each of the wheels has the weight and the radius taking into account the deformation of a tire, rt. At a first approximation each wheel sustains a quarter of the whole weight both of the vehicle with a driver, mv, and of the weight of a reference body (a passenger), mref.

According to [6] the equivalent moment of inertia of the rotating masses is the moment of inertia of a system consisting of elements, which have kinetic energy equal to the amount of the kinetic energy of the whole system.

It follows from the condition of conservation of the kinetic energy that for a system consisting of four wheels interconnected through gears and rotating at the angular velocity, ωw, which have the total moment of inertia, JΣw, taking into account the vehicle linear speed, and neglecting the effect of air resistance, we get the next Equation:

$$J_{equiv}(\omega)\frac{\omega_e^2}{2} = J_e k_{e.loss}(\omega)\frac{\omega_e^2}{2} + J_{tr} k_{tr.loss}(\omega)\frac{\omega_e^2}{2} + \\ + J_{\Sigma w} k_{w.loss}(\omega)\frac{\omega_w^2}{2} + (m_v + m_{ref})\frac{V_v^2}{2} \tag{1}$$

where, ω_e is the angular velocity of the IC engine, rad/s; m_v is the weight of the vehicle, kg; m_{ref} is the weight of the reference body (a passenger), kg; J_e is the IC engine rotating parts moment of inertia normalized with respect to the crankshaft rotation axis, kg•m^2; $k_{e.loss}$ is the coefficient characterizing the losses in the IC engine ($k_{e.loss} > 1$); J_{tr} is the equivalent moment of inertia of the transmission units, kg•m^2; $k_{tr.loss}$ is the coefficient characterizing the losses in the transmission units ($k_{tr.loss} > 1$); $k_{w.loss}$ is the coefficient characterizing the losses in the wheels ($k_{w.loss} > 1$).

From Equation 1 we can determine the equivalent moment of inertia of the whole system:

$$J_{equiv}(\omega) = J_e k_{e.loss}(\omega) + J_{tr} k_{tr.loss}(\omega) + \\ + J_{\Sigma w} k_{w.loss}(\omega)\cdot\left(\frac{\omega_w}{\omega_e}\right)^2 + \frac{(m_v + m_{ref})V_v^2}{\omega_e^2} \tag{2}$$

The transmission ratio between the engine and the drive wheel is equal to the product of the transmission ratio of the gearbox (k_{gb}), the transmission ratio of the main gear (k_{mg}), and the transmission ratio of the final drive (k_{fd}).

Hence, the Equation 2 can be written as:

$$J_{equiv}(\omega) = J_e k_{e.loss}(\omega) + J_{tr} k_{tr.loss}(\omega) + \\ + J_{\Sigma w} k_{w.loss}(\omega)\cdot\left(\frac{1}{k_{gb} k_{mg} k_{fd}}\right)^2 + \\ + \frac{(m_v + m_{ref})V_v^2}{\omega_e^2} = J_e k_{e.loss}(\omega) + J_{tr} k_{tr.loss}(\omega) + \\ + \left(J_{\Sigma w} k_{w.loss}(\omega) + (m_v + m_{ref})\cdot r_t^2\right)\cdot\left(\frac{1}{k_{gb} k_{mg} k_{fd}}\right)^2. \tag{3}$$

The engine torque at the angular velocity of the crankshaft, ωe, can be determined as:

$$M(\omega) = J_{equiv}(\omega)\cdot\varepsilon(\omega) = \left[\begin{matrix} J_e k_{e.loss}(\omega) + J_{tr} k_{tr.loss}(\omega) + \\ \left(\dfrac{J_{\Sigma w} k_{w.loss}(\omega) + (m_v + m_{ref}) r_t^2}{k_{gb}^2 k_{mg}^2 k_{fd}^2}\right)\end{matrix}\right]\cdot\varepsilon(\omega). \tag{4}$$

The torque during acceleration of the vehicle with the driver (when the accelerator is all the way down) can be defined as:

$$M(\omega)=\begin{bmatrix} J_e k_{e.loss}(\omega)+J_{tr}k_{tr.loss}(\omega)+ \\ \left(\dfrac{J_{\Sigma w}k_{w.loss}(\omega)+m_v r_t^2}{k_{gb}^2 k_{mg}^2 k_{fd}^2}\right) \end{bmatrix}\cdot\varepsilon_1(\omega), \quad (5)$$

where, $\varepsilon_1(\omega)$ is the angular acceleration of the IC engine during speeding up of the vehicle with a driver, rad/s^2.

Estimated Equation of the torque when accelerating the vehicle with a driver and the reference body (a passenger) (when the accelerator is all the way down) can be defined as:

$$M(\omega)=\begin{bmatrix} J_e k_{e.loss}(\omega)+J_{tr}k_{tr.loss}(\omega)+ \\ \left(\dfrac{J_{\Sigma w}k_{w.loss}(\omega)+(m_v+m_{ref})r_t^2}{k_{gb}^2 k_{mg}^2 k_{fd}^2}\right) \end{bmatrix}\cdot\varepsilon_2(\omega), \quad (6)$$

where, $\varepsilon_2(\omega)$ is the angular acceleration of the IC engine during speeding up of the vehicle with a driver and the reference body (a passenger), which has a weight of m_{ref}, rad/s^2.

Equating Equation 5 with Equation 6, one can determine the sum:

$$J_e k_{e.loss}(\omega)+J_{tr}k_{tr.loss}(\omega)= \frac{m_{ref}r_t^2}{k_{gb}^2 k_{mg}^2 k_{fd}^2}\cdot\frac{\varepsilon_2(\omega)}{\varepsilon_1(\omega)-\varepsilon_2(\omega)}-\frac{J_{\Sigma w}k_{w.loss}(\omega)+m_v r_t^2}{k_{gb}^2 k_{mg}^2 k_{fd}^2}. \quad (7)$$

Substituting Equation 7 into Equation 5, one can determine the value of the torque developed by the engine itself as:

$$M(\omega)=\frac{m_{ref}r_t^2}{k_{gb}^2 k_{mg}^2 k_{fd}^2}\cdot\varepsilon_2(\omega)\cdot\left(\frac{\varepsilon_2(\omega)}{\varepsilon_1(\omega)-\varepsilon_2(\omega)}+1\right). \quad (8)$$

The power developed by the engine one can determine as:

$$N(\omega)=M(\omega)\cdot\omega=\frac{m_{ref}r_t^2}{k_{gb}^2 k_{mg}^2 k_{fd}^2}\cdot\frac{\varepsilon_1(\omega)\varepsilon_2(\omega)}{\varepsilon_1(\omega)-\varepsilon_2(\omega)}\omega. \quad (9)$$

After measuring the mass, the fuel consumption, $Q(\omega)$, at the angular accelerations ε_1 and ε_2 (ideally, values of the fuel consumptions must be equal) with the known fuel calorific value, q_{fc}, the effective coefficient of efficiency of the wheeled vehicle is determined as:

$$\eta(\omega)=\frac{N(\omega)}{Q(\omega)\cdot q_{fc}} \quad (10)$$

2.2. *The Instruments to Identify the Mechanical Parameters of an IC Engine.*

For identification of the performance parameters of an internal combustion engine of wheeled vehicles during the full life cycle, it is proposed a hardware-software complex (Figure 2).

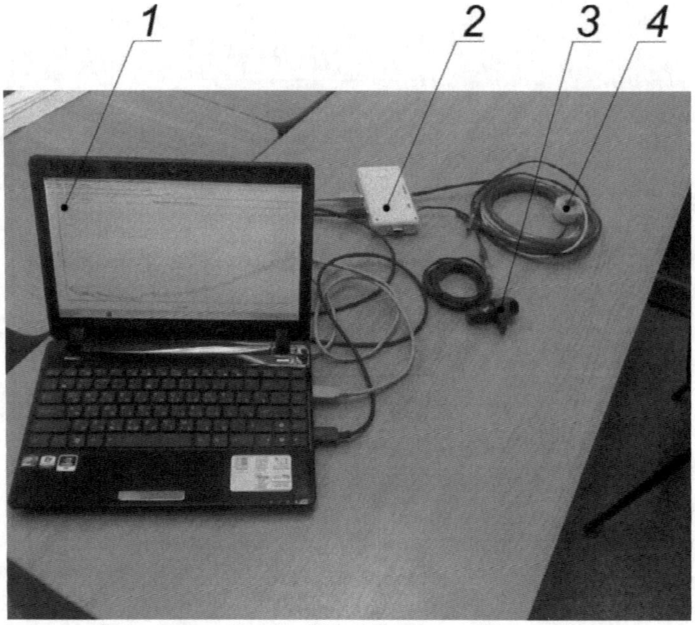

Figure 2. Hardware-software complex for identification of the performance parameters of IC engines of wheeled vehicles: 1 – laptop; 2 – high-speed data acquisition and data processing system; 3 – modified crankshaft position sensor; 4 – high-precision digital fuel flow meter.

The configuration of the hardware-software complex: digital information module (USB-6009) manufactured by National Instuments; laptop Acer Extenca 5210; licensed development environment for virtual instrumentation LabView 7.2 with integrated module Signal Express 2.5; high-precision digital fuel flow meter (11,500 pulses per liter) manufactured by Biotech GmBH; modified crankshaft position sensor. Modifying of the crankshaft position sensor lay in the fact that the two main output leads were connected to the two conductors which were, in turn, connected to the digital information module.

3. Results and Discussion

Figures 3–5 show the results of measurement of the efficiency of the experimental vehicle VAZ-2112 (the engine capacity is 1,500 cm^3, the rated power is 71 HP at 5,600 rev/min, the torque is 104 N×m at 3,400 rev/min, the transmission ratio of the main gear is 3.7, the distance from the axis of rotation of the drive wheel to the center point of the contact is 0.275 m) of the total weight of 1,030 kg. As a reference body was a passenger, the weight of whom together with the clothes was 71 kg. The weight of a car driver together with the clothes was 95 kg. Figures 3–5 show two curves defined by the Equation 11 when the effective values of the torque were determined by the Equation 8 or Equation 9. The dependencies were obtained with a help of the specialized hardware-software complex, as discussed above.

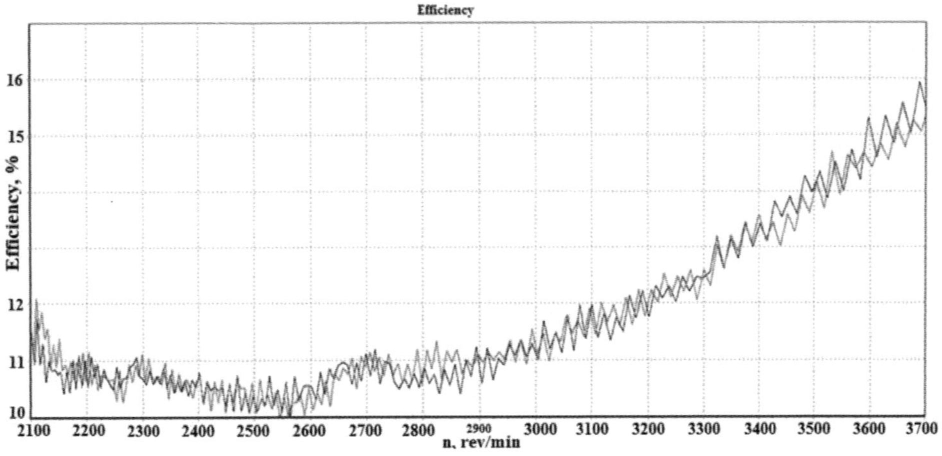

Figure 3. The efficiency of the engine of the wheeled vehicle VAZ-2112 when speeding up in the second gear (transmission ratio is 1.95).

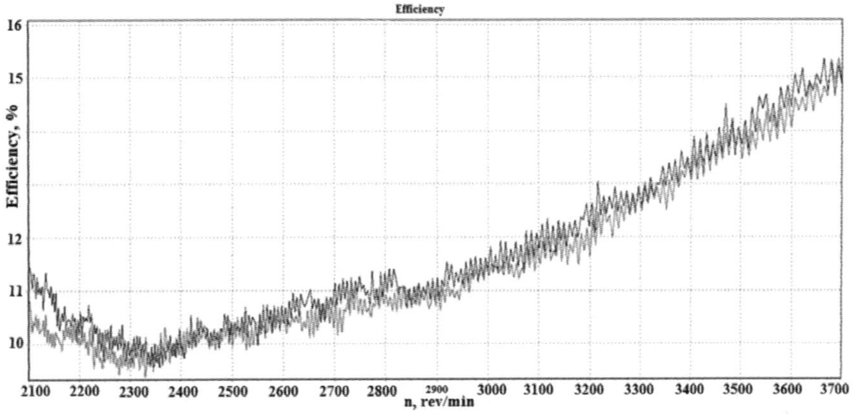

Figure 4. The efficiency of the engine of the wheeled vehicle VAZ-2112 when speeding up in the third gear (transmission ratio is 1.357).

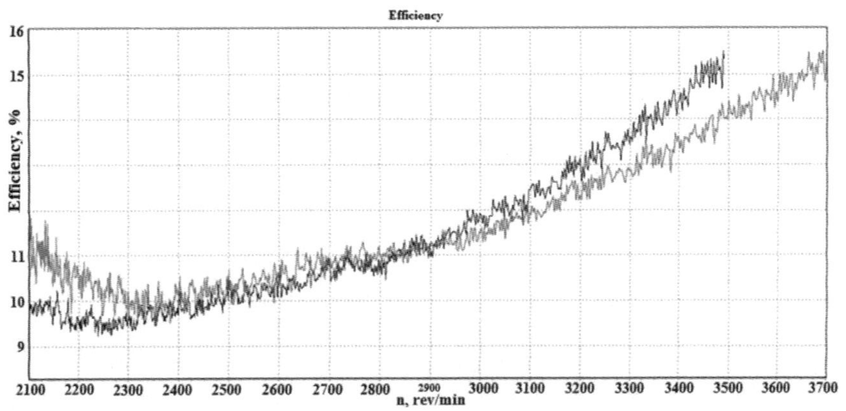

Figure 5. The efficiency of the engine of the wheeled vehicle VAZ-2112 when speeding up in the fourth gear (transmission ratio is 0.941).

IOP Conf. Series: Materials Science and Engineering **289** (2017) 012038 doi:10.1088/1757-899X/289/1/012038

In Figures 3– 5 we can see that the values of the efficiency at the different angular velocities of the IC engine are very similar when measured in the second gear that is when the total transmission ratio is 1.95x3.7 = 7.215. With a lower transmission ratio, there are significant differences, so we recommend using the developed method by the speeding up of the vehicle in the second gear.

Figure 6 and Figure 7 show the speed-torque and the speed-power curves of the IC engine of the wheeled vehicle VAZ-2112 when speeding up in the second gear.

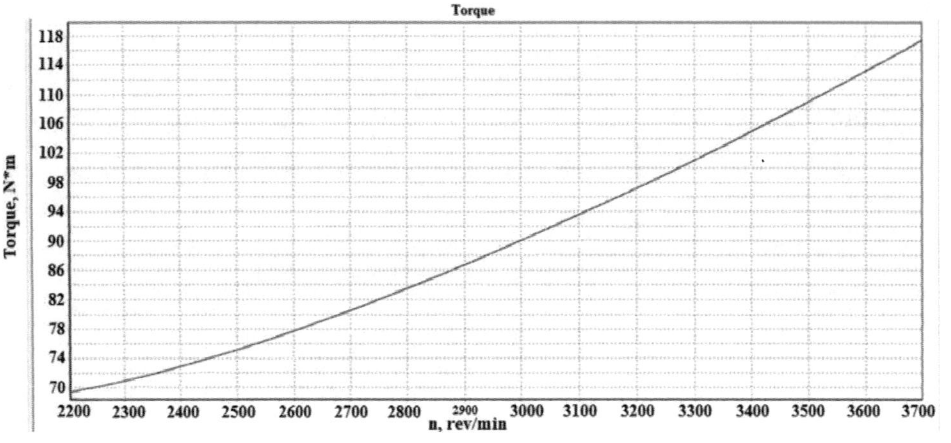

Figure 6. Speed-torque curve of the IC engine of the wheeled vehicle VAZ-2112 when speeding up in the second gear.

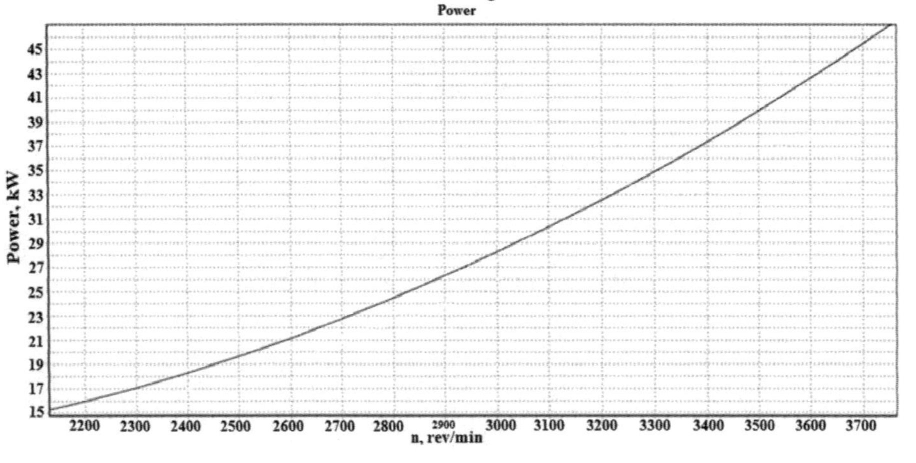

Figure 7. Speed-power curve of the IC engine of the wheeled vehicle VAZ-2112 when speeding up in the second gear.

The speed-torque and the speed-power curves are calculated based on the approximate values of ε_1 and ε_2.

The experiments showed the level of reliability of the results obtained by the proposed method by comparing them with the certificate data of the test IC engine. It should be borne in mind that the torque value specified in the certificate was obtained with the strain gauges by measuring the required value directly on the engine dismounted from the vehicle. According to the certificate of the VAZ-2112 engine with the capacity of 1,499 cm^3, it develops the torque of 104 N×m at the engine speed of 3,400 rev/min (the power is 37.0 kW) using the regular software. According to the experimental data the value of the torque at 3,400 rev/min is about 105 N · m. The discrepancy between the experimentally obtained value and the certificate value is approximately 1%.

These results support the idea that the external characteristic of an IC engine remains practically unchanged while increasing its inertial mass (additional body with the reference moment of inertia). Thus it is confirmed the validity of the Equation 9 and Equation 10.

The experiments have shown the adequacy and the effectiveness of the developed method and have provided new data on the performance parameters of the internal combustion engine of the wheeled vehicle VAZ-2112. The experimental results made it possible to determine the most preferred transmission ratio of the wheeled vehicle VAZ-2112 (the second gear) when speeding up for performance parameters identification with a help of the developed method.

The proposed method solves from our point of view the most important problem of accurate identifying the performance parameters of an IC engine itself without its demounting from a vehicle. In modern mechanical engineering there are two methods to measure the power and the torque of the IC engine: with and without demounting the engine [1, 2, 4, 7]. The first method allows measuring the torque directly on the crankshaft, which gives a more accurate picture of the engine external mechanical characteristics as compared with the second one. If measurements are carried out without demounting the engine, it is currently used roller dynamometers (see Figure 8). Thus the mechanical power delivered to the drive wheels of the vehicle is determined.

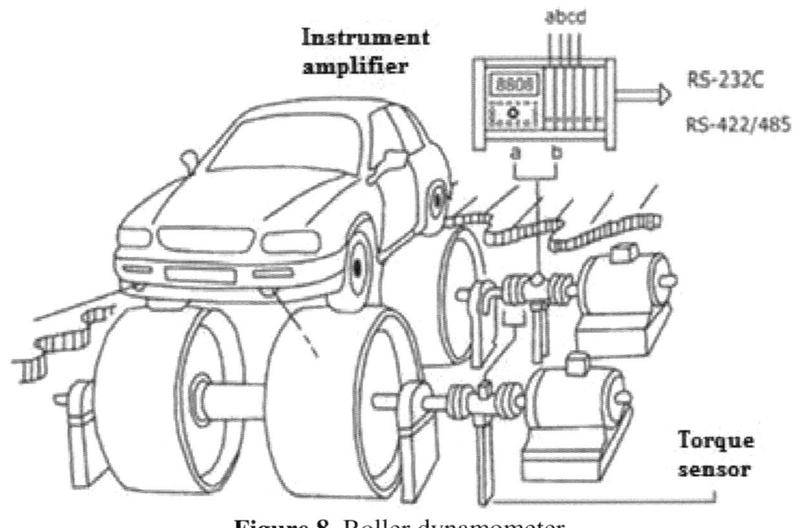

Figure 8. Roller dynamometer.

Roller dynamometers in turn are divided into two much used types: with inertia dynamometers (inertial method) and with strain gauge dynamometers (brake method).

When measured with strain gauge dynamometers [2, 3, 8] for building external mechanical characteristics, it is used the force sensors that should be previously calibrated precisely, in order to determine the torque at each point at a particular rpm value. Application of DC dynamometers on each drum and electrically controlled clutch allows configuring the treadmill depending on the required measurement. Using the integrated inertial mass, one can perform both steady state and dynamic tests [8]. However, the torque control is carried out with relatively large discresity, which is due to the time needed to restore strained state of the piezoresistor. Besides this method requires expensive cooling systems, brake equipment and force sensors. Also, there are such drawbacks as the considerable period of full load, low accuracy as compared with the inertial method; temperature errors and limited brake control leads to deterioration in the accuracy of measurement under load. Much of the work devoted to the study of the performance parameters of IC engines are made on the basis of the strain measuring instruments.

The principle of the inertial method is to determine the mechanical characteristics of the engine when accelerating the whole system of rotating masses with additional inertial mass (flywheel), the moment of inertia of which is known [5]. The drive wheel power is defined as the ability to perform a certain work at a time unit. That is, the work is calculated as the energy transferred from the drive wheel to the flywheel at a time unit:

$$W = \Delta E_c = E_{c2} - E_{c1} = \frac{1}{2} \cdot J \cdot \omega_2^2 - \frac{1}{2} \cdot J \cdot \omega_1^2 \qquad (11)$$

$$P = \frac{W}{\Delta t}, \qquad (12)$$

where, ω_2 is the angular velocity of the flywheel at the moment of time t_1, rad/s; ω_2 is the angular velocity of the flywheel at the moment of time t_2, rad/s; J is the moment of inertia of the flywheel, $kg \cdot m^2$; P is the power of the drive wheel, W.

The inertial method in comparison with the brake one has a number of advantages associated with the simplicity of the method that is easy to reproduce, and with the applicability to internal combustion engines of any power. Also it does not have other drawbacks, which are inherent in the brake method (does not require expensive cooling systems, the measuring time is minimal, etc.).

However, when using both inertial and brake methods without demounting of an IC engine to identify its performance parameters, a problem to determine the power developed by the IC engine itself remains unsolved.

In contrast to the described existing methods the proposed in this paper new method to control the performance parameters of internal combustion engines solves the problem of identifying the required parameters without loss in transmissions and friction units of a vehicle through the use of a body (a passenger) with the reference weight and the assumption that when the moment of inertia of rotating masses (an internal combustion engine, transmission units and drive wheels) is changed, external characteristic (the speed-torque curve) of an internal combustion engine remains unchanged. In addition there is no need for the use of a strain gauge and other expensive equipment that provides an additional advantage of the developed method. Thus the use of this method can greatly improve the measurement accuracy and reduce costs and laboriousness during testing of internal combustion engines.

4. Conclusion

Based on the results obtained we can conclude that the developed method to identify the performance parameters of internal combustion engines allows controlling their performance parameters by measuring the angular acceleration of the rotating masses with a reference body (a passenger) and without it. This method greatly increases the accuracy of the control compared with other existing non-demounting-engine methods due to the lack of the need for strain gauge elements and to the possibility to identify the performance parameters of internal combustion engines without the torque loss in transmission and friction units.

For the first time designed hardware-software complex allows controlling the efficiency of internal combustion engines and identifying the studied dependences in a wide speed range with a high frequency rate.

The results of conducted experiments showed the most preferred transmission ratio of the wheeled vehicle VAZ-2112 (the second gear) to identify the performance parameters with a help of the developed method.

The developed method is easy to create; it differs less labor- and energy-intensity in comparison with existing methods; it does not require a brake tester, and therefore their application significantly reduces material costs for the organization of the test workplace.

The developed method can be used in the companies manufacturing, operating and maintaining wheeled vehicles with IC engines. Defining tolerable levels of the values of the acceleration of rotating masses including internal combustion engines, it is possible to control their power and efficiency.

References

[1] Loskutov A S, Grigoryev A N and Kozhin D V 2007 *Ispytaniya dvigateley vnutrennego sgoraniya* (Yoshkar-Ola: MarGTU)

[2] Martyr A J and Plint M A 2012 Chapter 10 – Dynamometers: The Measurement of Torque, Speed, and Power. *Engine Testing (Fourth Edition)* 227-258 doi: 10.1016/B978-075068439-2/50011-6

[3] Franco J, Franchek M A and Grigoriadis K 2008 *Mechanical Systems and Signal Processing* 22 338–361 doi: 10.1016/j.ymssp.2007.08.002

[4] Kee R. and Blair G 1994 *SAE Motor Sports Engineering Conference Proceedings Dearborn. SAE Technical Paper* doi: 10.4271/942478

[5] Maitree N and Kunanoppadol J 2014 *SISWB VI, Apsara Angkor Resort & Conference*

[6] Yablonskiy A A 1977 *Kurs teoretitsheskoy mekhaniki. Tchast' II. Dinamika* (Moscow: Izdatelskiy centr Akademiya)

[7] Martyr A J and Plint M A 2012 *Engine Testing (Fourth Edition)* (Elsevier) 227–258 doi: 10.1016/B978-0-08-096949-7.00017-0

[8] Thompson J K, Marks A and Rhode D 2002 *SAE International. SAE Technical Paper* doi: 10.4271/2002-01-2601

IOP Conf. Series: Materials Science and Engineering **289** (2017) 012039 doi:10.1088/1757-899X/289/1/012039

Software and hardware complex for research and management of the separation process

A P Borisov
Associate professor, Altai State Technical University, Barnaul, Russia
E-mail:boralp@mail.ru

Abstract. The article is devoted to the development of a program for studying the operation of an asynchronous electric drive using vector-algorithmic switching of windings, as well as the development of a hardware-software complex for controlling parameters and controlling the speed of rotation of an asynchronous electric drive for investigating the operation of a cyclone. To study the operation of an asynchronous electric drive, a method was used in which the average value of flux linkage is found and a method for vector-algorithmic calculation of the power and electromagnetic moment of an asynchronous electric drive feeding from a single-phase network is developed, with vector-algorithmic commutation, and software for calculating parameters. The software part of the complex allows to regulate the speed of rotation of the motor by vector-algorithmic switching of transistors or, using pulse-width modulation (PWM), set any engine speed. Also sensors are connected to the hardware-software complex at the inlet and outlet of the cyclone. The developed cyclone with an inserted complex allows to receive high efficiency of product separation at various entrance speeds. At an inlet air speed of 18 m / s, the cyclone's maximum efficiency is achieved. For this, it is necessary to provide the rotational speed of an asynchronous electric drive with a frequency of 45 Hz.

1. Introduction

The ambient air surrounding the person is continuously exposed to contamination. The air of industrial premises is polluted by the emissions of process equipment or during technological processes without the localization of waste materials.

Air purification is one of the most important tasks at all plants where powdered raw materials are received, stored and transported by air. Therefore, the purification and separation of air mixtures is one of the most important and widespread processes. Currently, there are three ways to clean the process air: gravity cleaning, inertial and filtration [1].

The quality of air purification in dust collectors depends not only on such parameters as the flow velocity, the size of the particles being separated, but also on the ratio of the geometrical dimensions of the dust collector. The advantage of cyclones is the simplicity of the configuration. Also, the cyclones provide a degree of purification of up to 90% and are characterized by a low degree of trapping of particles of a size of 20 μm or less. One of the main parameters of cyclones is the input speed.

Frequency converters are used in electric drives to regulate the speed of the electric drive. The use of a regulated electric drive makes it possible to exclude the mechanical component of complex mechanisms and simplify the kinematics of machines, and the development of technological processes and the application of automated and software control systems require maintaining high accuracy and speed of the working mechanisms of the mechanisms.

Frequency converters are used to accurately regulate the speed of the drive and at the same time provide protection such as inrush current and power consumption of the device. With their help, it is possible to carry out remote monitoring and control of an asynchronous electric drive.

2. Relevance

The efficiency of the existing cyclones is 80-90%, with a potential of 99%. Cleaning can also take place through filter cloths that need to be changed during operation, as well as cyclones that are unable to capture dust with a small particle size and have little durability (especially when cleaning gases from dust with high abrasive properties). The main parameters that characterize the operation of the cyclone are hydraulic resistance and cleaning efficiency, which depend on the design features of the apparatus and the speed of the dust-gas flow. The technical requirements for the cyclone, as well as for any dust collector, consist in the greatest possible degree of purification at the lowest energy costs.

Therefore, the regulation of the speed of the electric motor in cyclone systems is actual.

3. Principle of operation

The functional diagram of the cyclone is shown in Figure 1 [2].

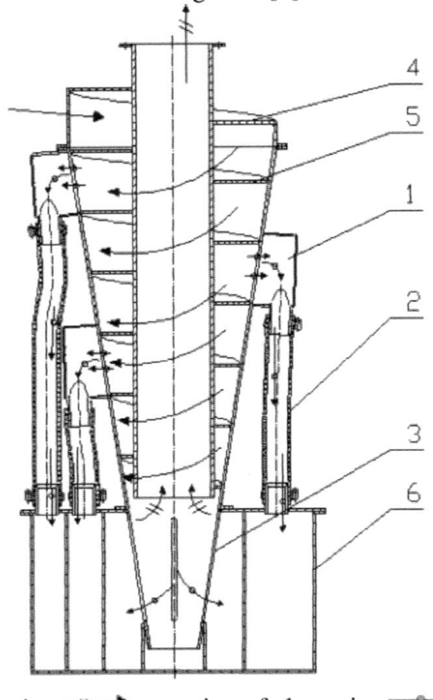

→ – motion of the raw air, → – motion of clean air, → – movement of dust.
1 – the box; 2 – flexible hose; 3 – attachment; 4 – snail; 5 – screw insert; 6 – container.

Figure 1. Functional diagram of the cyclone.

The air mixture on the material conduit enters the inlet snail 4 of the cyclone. The input snail serves to communicate the rotational axisymmetric motion to the product and thereby create conditions for the separation of the product from the stream. Exiting the swirler, the flow enters the conical part of the cyclone and acquires additional rotational screw motion due to the screw insert 5.

Large particles of the product rush to the walls of the cyclone due to the effect of the centrifugal force arising in the rotating flow, as a result of which their main mass enters the slit to catch large dust and precipitates in boxes located at different heights. The rest of the dust, along with the air, moves axially to the lower part of the cyclone, where, under the action of inertia, the dust enters the cone-cone 3 for additional cleaning of the air, and the purified air leaves the exhaust pipe at the lowest resistance.

Changing the gap between the cochlea to which the screw insert is fixed and the resistance of the cyclone changes with the outer cone. Within the permissible limits, raising the cochlea does not significantly change the efficiency of the cyclone.

To study the efficiency of the cyclone, it is necessary to use a frequency converter to regulate the speed of the air flow. A single-phase - three-phase frequency converter in a single-phase network can be used as a frequency converter. When this frequency converter is used, a certain sequence of transistors connected in accordance with the scheme [3] commences in both positive and negative half-periods of the supply voltage to create a magnetic field of the stator of the asynchronous electric drive in order to provide fixed magnetic flux positions. This method makes it possible to obtain the necessary current direction in the stator windings, thereby creating a rotating magnetic field, which provides vector-algorithmic switching of the windings of the asynchronous electric drive.

Control transistors are included in three zero-phase single-phase counter-parallel rectification circuits designed to enable the direction of rotation of the magnetic flux vector of the stator field to change. The corresponding middle point of the rectification circuit is connected to the input of the stator windings. The windings of the asynchronous electric drive are connected to a star. Control transistors need to be selected based on currents and voltages passing through the motor.

For the study of the modes of operation of the engine using vector-algorithmic commutation, current sensors (Figure 2*a*) and voltage are used (Figure 2*b*).

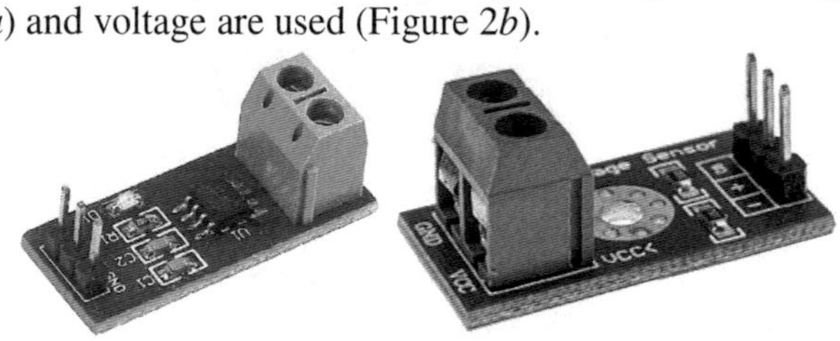

a) b)

Figure 2. Current and voltage sensors used to study the characteristics of an asynchronous motor.

The current sensor is implemented on the ACS712 module on the Hall effect. The current sensor can measure current up to 30A and is connected to the ADC of the controlling microcontroller. The voltage sensor is implemented on the standard module "voltage sensor for Arduino".

4. Software and hardware

To study the operation of an asynchronous electric drive using vector-algorithmic switching, software was developed that receives data from current and voltage sensors. The use of methods for determining the electromagnetic moment and the power of an asynchronous three-phase electric drive fed from a single-phase alternating current network is impossible with vector-algorithmic commutation, since using vector-algorithmic commutation there is no continuous sinusoidal voltage, since the control transistors in a certain sequence apply voltage to two or three windings of an asynchronous three-phase electric drive and, accordingly, a simultaneous inequality of these voltages in each moment of time on different windings of the electric drive. Therefore, a method was used in which the average value of flux linkage is found and a method for vector-algorithmic calculation of the power and electromagnetic moment of an asynchronous electric drive fed from a single-phase network is developed for vector-algorithmic commutation. Depending on the type of connection of the windings of the asynchronous three-phase electric drive, the calculation procedure will differ because in view of the different voltage applied to the windings of the electric drive, there will be a different way of adding these vectors. Then the calculation method will be the same and consists in relation to the power on the motor shaft when powered from a single-phase circuit and a three-phase circuit, respectively.

The oscillogram of the stator windings is shown in Figure 3.

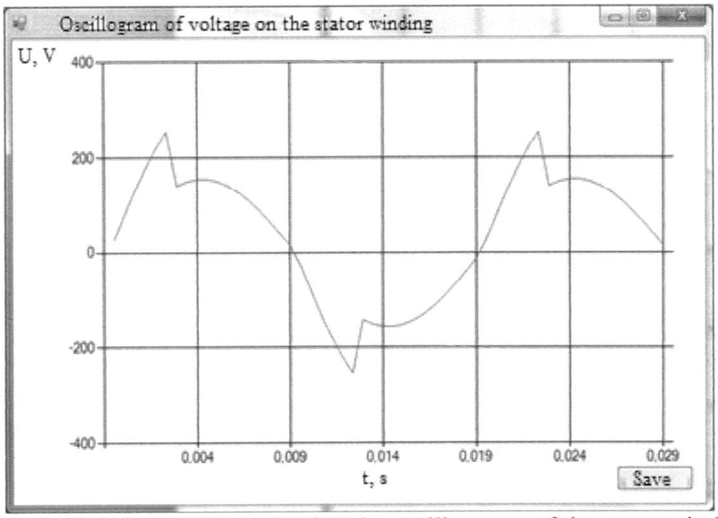

Figure 3. The window for outputting the oscillograms of the stator windings.

The software-hardware complex consists of several functional blocks connected to the microcontroller, presented in Figure 4.

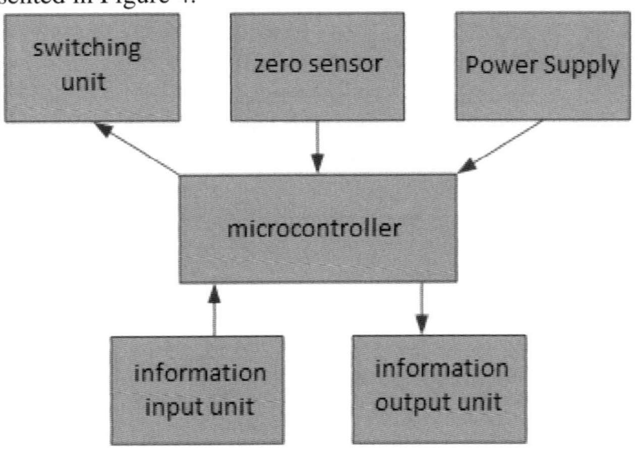

Figure 4. Block diagram of the control system.

The software part of the software-hardware complex [4] allows to regulate the speed of rotation of the motor by vector-algorithmic switching of transistors or, using pulse-width modulation (PWM), set any engine speed. Connected sensors of current and voltage allow in real time to receive data from the engine and, based on the mechanical characteristics of an asynchronous motor, to calculate the speed of rotation of the motor shaft, and also to study the operation modes of an asynchronous electric drive in vector-algorithmic commutation and fix emergency regimes. When using a cyclone battery, the developed software-hardware complex allows to record the increase or decrease of the load on the cyclones and, accordingly, to correct the engine speed. Also sensors complex at the inlet and outlet of the cyclone are connected to the software-hardware.

After the development and testing of the control system, experimental studies were carried out [5].

At the first stage of experimental research, the dependence of the air velocity on the radius of the cone was revealed (Figure 5).

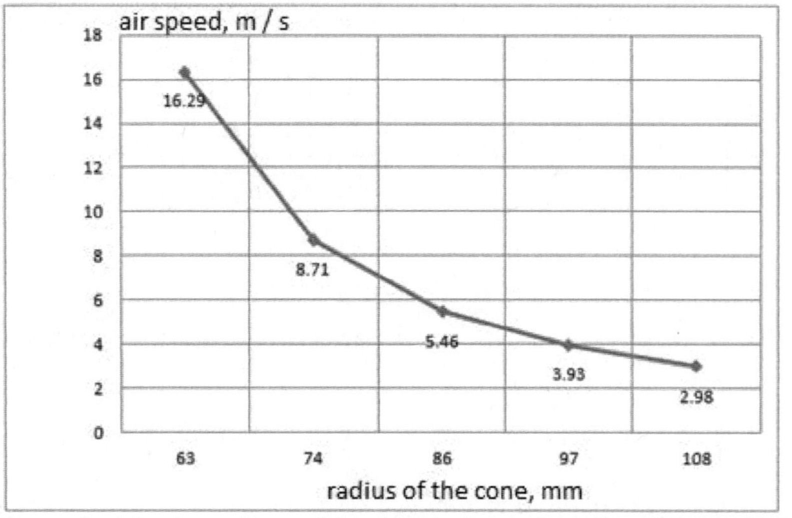

Figure 5. The graph of the dependence of the air velocity on the radius of the conical part of the cyclone.

As can be seen from Figure 5, with an increase in the radius of the cone, the airflow velocity decreases, which leads to a decrease in the efficiency of the cyclone operation.

The next part of the experimental studies consisted in determining the air velocity at the inlet to the cyclone, at which the maximum separation efficiency was achieved.

Experimental studies (Figures 6, 7) were carried out using a single-phase three-phase frequency converter to obtain different air velocities. The experiments were carried out on a conventional cyclone and a cyclone with a screw insert.

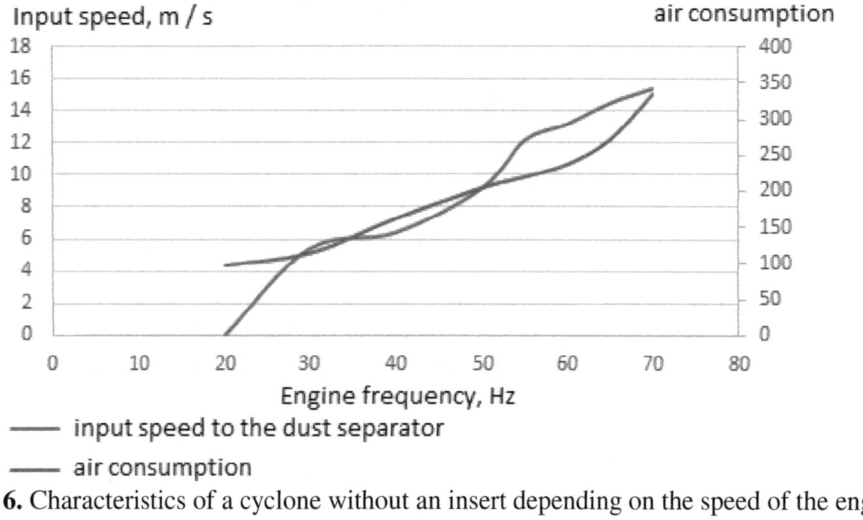

Figure 6. Characteristics of a cyclone without an insert depending on the speed of the engine.

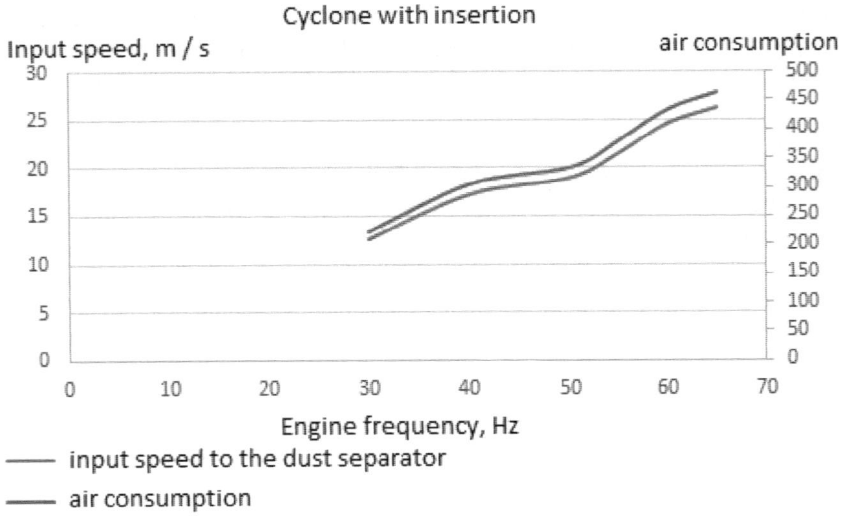

Figure 7. Characteristics of a cyclone with an insert depending on the speed of the engine.

For the experiment, weighed samples of the product weighing 100 gr. and poured into the receiving cyclone hopper.

It can be seen from Figures 6 and 7 that a cyclone with an insert develops a high input speed, which allows increasing the cleaning efficiency of a cyclone separator.

The results of the experiment carried out for the flour fed to the air flow are shown in the graph shown in Figure 8.

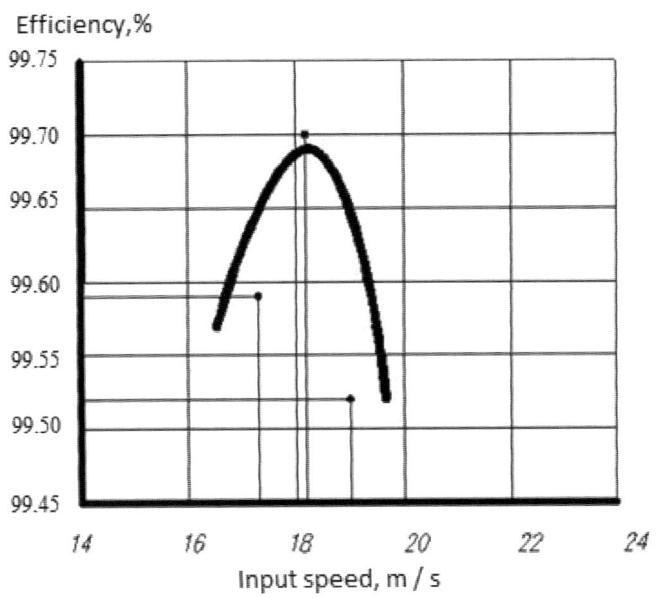

Figure 8. The plot of the efficiency of the separation of the product from the input air velocity.

It can be seen from Figure 8 that the developed cyclone allows obtaining high efficiency of product separation (more than 99%) at different input speeds. At an inlet air speed of 18 m / s, the cyclone's maximum efficiency is achieved. For this, it is necessary to provide the rotational speed of an asynchronous electric drive with a frequency of 45 Hz.

The most effective separation of the solid fraction of the flour from the aerodisperse flow is observed at the plant at an input speed of 18.2 m / s and a concentration of 43.2 g / m^3 (99.7%).

At a speed less than or higher than this value, a reduction in separation efficiency was observed.

5. Conclusion

The developed system allows to explore an asynchronous motor when it is connected to a single-phase alternating current network, and also to simulate the processes occurring in it. The developed software-hardware complex allows to regulate the speed of rotation of an induction motor connected to a single-phase alternating current network, which affects the efficiency of cyclone purification. The carried out experiment showed that the highest degree of purification is achieved at a frequency of rotation of the electric drive of 45 Hz.

Acknowledgement

The study was carried out with the financial support of the Russian Foundation for Basic Research in the framework of the scientific project No. 17-48-220056.

References

[1] Xu W W, Li Q, Zhao Y L, Wang J J and Jin Y H 2016 *IOP Conference Series: Materials Science and Engineering* **129** 012022 doi:10.1088/1757-899X/129/1/012022

[2] Zlochevskiy V L 2012 *Aerovintovoy tsiklon-separator [Aerovoltaic cyclone separator]* Patent RF № 2442662 (in Russian)

[3] Stal'naya M I, Eremochkin S Y, Borisov A P, Titova A A, Korolev D A 2016 *Shirokoreguliruemyy tranzistornyy preobrazovatel' chastoty dlya elektrodvigatelya peremennogo toka [Wide-adjustable transistor frequency converter for AC motor]* Patent RF № 2016108693/07 (in Russian)

[4] Nikolaeva V K, Borisov A P 2017 *Proceeding of Mezhdunarodnoe nauchnoe periodicheskoe izdanie po itogam Mezhdunarodnoy nauchno-prakticheskoy konferentsii* 191–193 (in Russian)

[5] Zlochevskiy V L, Borisov A P 2016 *Proceeding of Technologies for the food and processing industry of aic - healthy food* 75–83 (in Russian)

IOP Conf. Series: Materials Science and Engineering **289** (2017) 012040 doi:10.1088/1757-899X/289/1/012040

Reseach of electron irradiation effects upon plexiglas strain during bending test

N A Voronova[1], A I Kupchishin[2,3], M N Niyazov[4], B G Taipova[4], A T Abdukhairova[5]
[1]Senior researcher, Abay Kazakh National Pedagogical University, Almaty, Kazakhstan
[2]Head of laboratory, Abay Kazakh National Pedagogical University
[4]Researcher, Abay Kazakh National Pedagogical University
[5]Senior researcher, Abay Kazakh National Pedagogical University
E-mail: ankupchishin@mail.ru

Abstract. Experiments on dependence of strain (ε) vs stress (σ) during bending tests of non-irradiated and irradiated (by electrons) samples of ribbed plexiglas have been carried out. Obtained curves are described satisfactory within linear model, ε grows linearly as soon as σ increases.

1. Introduction

It is commonly known that various materials have got different mechanical and technological properties, which predetermine goods quality and their machinability. The properties are diagnosed by proper tests for strain, compression, flexing, hardness, etc. Bending test for example is carried out to check jump-joints [1 –3]. The test checks plasticity of sample and its joints and material ability to take up tailor-made flexing. It can be bending static and bending impact test (dynamic) [4]. Flat symmetrical bending test is a test when all external forces applied to the material act in one plane of symmetrization [5]. Herewith bending axis of the sample is in the same plane with external forces. Bending tests are carried out for plastic and fragile materials. Strain character of fragile and plastic samples differs. If plastic material is tested for bending, it will fail to break and determine bending strength due to large deformations. Other mechanical characteristics as for example proportionality limit, yield stress can be easily determined by tests [6 – 9]. Tests are critically important for strength and stiffness checking of fragile materials because usually they are destructed due to permanent strain [10, 11]. Maximum load and maximum deflection prior destruction point are calculated. Ultimate strength is calculated at maximum load. Bending tests can be carried out by two methods: to apply single force to center of a sample between supports or to apply two forces symmetrically between supports [12, 13]. Test results depend mostly on material type, load conditions, forms and sizes of samples. Modification of properties and structure of polymer and composite materials by irradiation with various particles including electrons are of special interest [14, 15]. Herewith various defects are formed (vacancies, interstitial atoms, various clusters, broken links, new compounds, etc. [16, 17]. The work is devoted to research of stress and electron irradiation effects upon strain of ribbed plexiglas during flat bending and applying single force. So far, such research has never been carried out before.

2. Experimental procedure

We developed and prepared an experimental facility to research dependence of strain ε vs stress σ. It consists of the following parts: strength sensor, indicator of clock type, tool for sample fixing, connecting wires and interface to see the data. It is installed at lower part of RU-50 rupture facility, which is a base for experiments. The facility also has a unit with loading nozzle and supports. Ribbed plexiglas of two shapes (pyramids and strips) has been taken as a test material. Thickness (*h*) of plexiglas samples is 3 mm in a convex part of ribbed material, width (*b*) is 5 mm and length is 40 mm. The smallest thickness in a concave part of ribbed material for pyramids is 1 mm, for strips is 2 mm. Research has been done at temperature 23 °C and relative humidity 50 %. Electron irradiation of samples has been carried out at linear accelerator ELA-6 with energy 2 MeV in air environment. Vacuum in the accelerator system has been maintained at level of 10^{-6} mm of mercury column. Samples for irradiation have been placed at distance of 40 cm from accelerator exit window. Beam current is 0.3 μA/cm^2, irradiation dose is – 100 kGy. Experiments have been carried out at a testing facility, which provided uniform speed 10 mm/min, relative movement of loading nozzle and supports as well as metered load with error ±1 % and bending with error ±2. An arithmetical average of all measurements considered to be an experimental result; During bending test destruction due to breaking, compressing and bending stress occurs. If various types of destruction happen during samples loading, it means that obtained values of bending stress are statically heterogeneous. In this case all results shall be analyzed and ongoing processes shall be researched.

3. Results and discussion

As the result of experimental research the dependences of strain ε vs stress σ for non-irradiated and irradiated samples of ribbed plexiglas have been obtained where flat bending with regard and without regard of horizontal component occur (Figures 1*a*, *b*). Figures show experimental results, lines show calculations. Physical and mechanical properties have been reviewed in coordinates ε, σ. Dependence σ vs ε is physically meaningless because σ is an argument and ε is a function, which shows in [18–21].

Dependence of ε vs σ based on experimental data have calculated as follows: bending stress is calculated using the equation [22]:

$$\sigma = \frac{M}{W}, \tag{1}$$

where *M* – bending moment; W – sample section modulus. Herewith bending moment (M) is found from expression:

$$M = \frac{FL}{4}, \tag{2}$$

where *F* – load; *L* – distance between supports. Sample section modulus (*W*) is calculated using:

$$W = \frac{bh^2}{6}. \tag{3}$$

Equations (1) – (3) result in:

$$\sigma = \frac{3FL}{2bh^2}. \tag{4}$$

Bending stress (σ_f) with regard of horizontal component of bending moment during bending *z* is calculated using the equation

$$\sigma_f = \frac{3FL}{2bh^2}\left(1+\frac{4z^2}{L^2}\right), \tag{5}$$

where z – bending in the center of sample between supports.

Extension elongation is calculated as bending – sample operating length ratio (ε = z/L). Expression (5) results in:

$$\varepsilon_f = \frac{1}{2}\sqrt{\frac{2\sigma_f bh^2}{3FL} - 1}. \tag{6}$$

Results show that extension elongation of ribbed plexiglas samples (non-irradiated and irradiated) grows linearly when stress is applied incrementally. Herewith strip ribbed plexiglas has higher strength and extension elongation than pyramid ribbed plexiglas. Plexiglas irradiated with 100 kGy samples lose plasticity by 20 and 40 % comparing to non-irradiated materials respectively. Material strength does not change after irradiation. Visual analysis shows that (transparence) decreases, samples are colored to brown. Counting of horizontal component does not affect significantly upon extension elongation. Experimental data are described linearly with good accuracy:

$$\varepsilon = \alpha\sigma, \tag{7}$$

where α – slope. α = 0.14 for non-irradiated strip ribbed plexiglas, α = 0.2 for pyramid ribbed plexiglas. α = 0.1 and 0.13 for irradiated materials respectively.

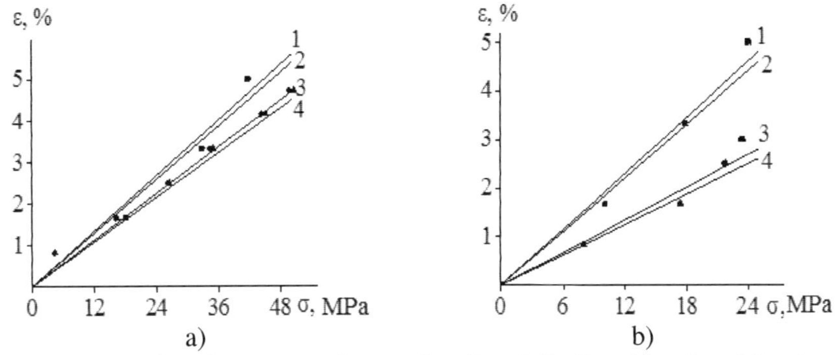

Figure 1. Dependence of strain vs stress for non-irradiated (1, 2) and irradiated by electrons (3, 4) strip (*a*) and pyramid (*b*) ribbed plexiglas samples during flat bending with regard of horizontal component (squares and stars) and without regard of horizontal component (dots and triangles) respectively.

4. Conclusions

1. Experiments in dependence of strain vs stress during flat bending tests of non-irradiated and irradiated ribbed plexiglas samples have been carried out. It is found out that irradiation leads to visible change of stress-strain behavior of material resulted in plasticity decrease. Extension elongation of irradiated strip plexiglas decreases by 20 % and of irradiated pyramid plexiglas by 40 %. Material strength does not change after irradiation.

The obtained curves of dependence ε vs σ are described satisfactory within linear model.

Acknowledgments

The study was supported by the Grant of the Ministry of Education and Science of the Republic of Kazakhstan.

References

[1] Belyaev N M 1976 *Resistance of materials* (Moscow: Nauka)

[2] Briskman B A, Tlebaev K B 2008 *High Performance Polymers* **20** 186 – 114

[3] Rubashkin A G, Chernilevskiy D V 1975 *Laboratory and practical works on engineering mechanics* (Moscow: Vysshaya Shkola)

[4] Jiazhen Sun, Yuefang Zhang, Xiaoguang Zhong 1994 *Polymer* **35** 2881 – 2883

[5] Tabata Y, Ikeda S, Tabata Y, Suzuki H, Miyoshi T, Katsumura Y 2008 *Rad. Phys. Chem.* **77** 401 – 408

[6] Oshima A, Ikeda S, Katoh E, Tabata Y 2001 *Rad. Phys.Chem.* **62** 39 – 45

[7] Katoh E, Sugisawa H, Oshima A, Tabata Y, Seguchi T, Yamazaki T 1999 *Rad. Phys. Chem.* **54** 165 – 171

[8] Lappan U, Geißler U, Lunkwitz K I 2000 *Rad. Phys. Chem.* **59** 317 – 322

[9] Rama K, Layek, Arun K Nandi 2013 A rewiewon synthesis and propertiesof polymer functionalized grephene 5087 – 5103

[10] Lappan U, Fuchs B, Geiβler U, Scheler U, Lunkwitz K 2002 *Polymer* **43** 4325 – 4330

[11] Lappan U, Fuchs B, Geiβler U, Scheler U, Lunkwitz K 2003 *Rad. Phys. Chem.* **67** 447 – 451

[12] Lunkwitz K, Lappan U, Fuchs B, Scheler U 2004 *J. Fluor. Chem.* **125** 863 – 873

[13] Komarov F F, Kupchishin A I, Pivovarov S P, Tlebaev K B, Kusainov A T, Rukhin A B, Pozdeeva T V 2012 *J. of engineering physics and Thermophysics.* **85** 455 – 458

[14] Luchnikov P A, Vetrova V V, Zubkov N P, Surzhikova O A, Chernova N I 2017 *IOP Conf. Series: Material Science and Engineering* **189** 012027 doi: 10.1088/1757-899X/189/1/012027

[15] Luchnikov P A, Surzhikova O A 2016 *IOP Conf. Series: Material Science and Engineering* **189** 012014 doi: 10.1088/1757-899X/189/1/012014

[16] Ivchenko V A 2016 *IOP Conf. Series: Materials Science and Engineering* **110** 012003 doi: 10.1088/1757-899X/110/1/012003

[17] Surzhikov A P, Galtseva O V, Vasendina E A, Vlasov V A, Nikolaev E V 2016 *IOP Conf. Series: Materials Science and Engineering* **110** 012002 doi: 10.1088/1757-899X/110/1/012002

[18] Kupchishin A I, Niyazov M N, Voronova N A, Kirdiashkin V I, Abdukhairova A T 2017 *IOP Conf. Series: Materials Science and Engineering* **168** 012017 doi: 10.1088/1757-899X/168/1/012017

[19] Kupchishin A I, Taipova B G, Voronova N A 2016 *IOP Conf. Series: Materials Science and Engineering* **168** 012015 doi: 10.1088/1757-899X/168/1/012015

[20] Kupchishin A I, Taipova B G, Kupchishin A A, Kozhamkulov B A 2015 *Mechanics of composite materials* **51** 115 – 118

[21] Kupchishin A I, Taipova B G, Kupchishin A A, Voronova N A, Kirdyashkin V I, Fursa T V 2016 *IOP Conf. Series: Material Science and Engineering* **110** 012037 doi: 10.1088/1757-899X/110/1/012037

[22] Afanasyev A M, Maryin V A 1975 *Laboratory workshop on resistance of materials* (Moscow: Nauka)

IOP Conf. Series: Materials Science and Engineering **289** (2017) 012041 doi:10.1088/1757-899X/289/1/012041

Properties of planar structures based on Policluster films of diamond and AlN

A F Belyanin[1], A P Luchnikov[2], S A Nalimov[3], A S Bagdasarian[4]
[1]Professor, Central Research Technological Institute "Technomash", Moscow, Russia
[2]Senior research scientist, Moscow Technological University (MIREA), Moscow, Russia
[3]Research scientist, Central Research Technological Institute "Technomash", Moscow, Russia
[4]Professor, Institute of Radio-engineering and Electronics of RAS, Moscow, Russia
E-mail: xamdex@gmail.com

Abstract. AlN films doped with zinc were grown on Si substrates by RF magnetron reactive sputtering of a compound target. Policluster films of diamond doped with boron were formed on layered Si/AlN substrates from the gas phase hydrogen and methane, activated arc discharge. By electron microscopy, X-ray diffraction and Raman spectroscopy the composition and structure of synthetic policluster films of diamond and AlN films were studied. Photovoltaic devices based on the AlN/PFD layered structure are presented.

1.Introduction

A layered structure on the basis Policluster films of diamond (PFD) and aluminum nitride (AlN) are promising for their use in devices of the microwave acoustoelectronics, in photodetectors of ultraviolet range and other devices of electronic equipment [1]. The layered structure of the PFD/AlN, doping metal impurities and nitrogen ions may alter the structural properties of the PFD to give them the inherent properties of semiconductors with high mobility of charge carriers [2, 3]. Such materials are needed to create functional radiation-resistant semiconductor devices. In such structures the AlN layer serves as a layer with n-type conductivity [4]. Diamond-like film coating formed in vacuum by sputtering of carbon target in high frequency (HF) magnetron and arc discharges [5, 6]. In [5] it is shown that during ion doping of carbon coatings is the formation of CNx compounds with a prevailing content of sp^2-phase with reduction of the cluster size. By doping with metal atoms and nitrogen it is possible to modify the structural and mechanical properties of diamond-like films. Technological impact of the components of the plasma treatment of the films leads to creation of additional defects and amorphization of the structure [7, 8].

The influence of fabrication regimes on the structure and composition Policluster films of diamond and AlN films that determine the performance characteristics of the photodetectors based on the structure of the PFD/AlN was studied.

2 Samples and research methodology

Samples in the form of a planar layered structure of Si/AlN/PFD on a substrate of single crystal silicon was formed by consistent growth of AlN films and the PFD on the substrate of Si{111}. First an AlN layer was formed by HF magnetron reactive sputtering of aluminum Al in the environment, gas composition of Ar+ (40...60) % vol. N_2 at a pressure of ~ 0,6...0,9 Pa. Power radio-frequency discharge at a frequency of high-frequency discharge of 13.56 MHz at a voltage of 200...350 V was ~ 1.5 to 3 kW. The magnitude of the electric displacement on the substrate when distance target–substrate of 60 mm was 30 V. The growth rate of the AlN film at substrate temperature 450...570 K was ~ 3 um/h. In the growth process, the AlN film was doping by zinc (the structure Si/AlN:Zn) to give it semiconductor properties. A composite target of Al and Zn plates were an area of 0.1...0.3 % of the area of the sputtering target.

On substrates of Si/AlN:Zn grown Policluster films of diamond from the gas phase H_2 and CH_4 in the activated arc discharge in the gas medium composition (1.5 to 4) % CH_4+H_2 at a pressure of > 10^4 Pa, the discharge power of 1200...1500 W. The growth rate Policluster films of diamond substrate at a temperature of 1100 to 1470 K was ~ 20 µm/h. the Layer of PFD in the process of growth was doping by evaporation of the bromine from the graphite container at a temperature of ~ 850 K.

On the surface structure of the AlN/PFD magnetron sputtering of Ti in Ar applied a translucent film of Ti with a thickness of ~ 40 nm at a pressure of 9.3 Pa. The substrate is pre-annealed luminous flux by the method of [9].

3.The structure of the films

The structure of the obtained films was investigated a method of raster microscopy using electron microscope (REM) Carl Zeiss Supra 40-30-87 and X-ray diffractometer Rigaku D/MAX-2500/PC (Cuk_α -radiation). Raman spectra films of AlN/PFD using a spectrophotometer type LabRAM HR 800 He-Ne laser emitting 632.8 nm were stadied. The concentration of dopants in the AlN films and the PFD controlled on the device CAMEBAX electron microprobe spectral analysis. The concentration of impurities in the AlN films was controlled separately as in X-ray amorphous and crystalline phases by Stripping voltammetry method and methods of selective dissolution of the phases of the film layer.

4. Results and their discussion

In Figure 1 shows X-ray diffraction pattern of the layered structure of Si/AlN/PFD when the thickness of the films: PFD – 12 µm; AlN – 3,6 µm.

The structure of the planar film of nanostructures Si/AlN/PFD are a mixture of amorphous and crystalline phases.

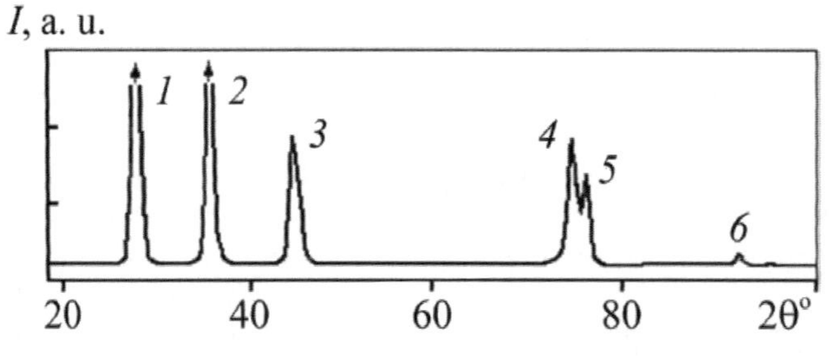

Figure 1. X-ray diffraction pattern from a layered structure of Si/AlN/PFD. The intensity peak of the scattered radiation corresponds to: 1 – 111 Si; 2 – 0002 AlN; 3 – 111 Diamond; 4 – 0004 AlN; 5 – 220 Diamond; 6 – 311 Diamond.

X-ray diffraction pattern obtained from structures with Si/AlN/PFD (Figure 1) showed the presence of crystalline phases of AlN spatial P63mc group and diamond group Fd3m. The crystallinity of the AlN films depends on technological modes of their formation and was 0...90 % vol. in the layer, and Policluster films of diamond were > 99 % vol.

During forming films of AlN layer is formed of polycrystalline phase fibrous (columnar) structure with orientation of the fibers in the direction of the texture axis <0001>. Axial columnar texture is formed depending on the sequence of the subsequent growth stages: emergence, coalescence of nuclei, geometric selection and reduction orientation between crystallites. Formation of AlN films in the direction of <0001> is determined by the atomic roughness of the growth surface. The growth layer of the AlN film proceeds in the direction <0001>, which coincides with the direction of the screw axis of the symmetry 6_3. This structure represents the composite reinforced thin needle-like fibers with strongly roughened side surface, the spaces between them completely filled with them coherent X-ray amorphous phase. According to REM needle fiber diameter is ~ 0.2 µm. Axis orientation <0001> texture relative to the surface of the substrate is determined by the flow direction of particles deposited particles in the layer. The fibers in the film consist of a group of coherent fibers. In terms of the broadening of diffraction peaks and calculated the crystallite size of crystalline phase of doped AlN films is 15...50 nm, and the undoped – is 35...65 nm when the axial orientation of textured crystallites with respect to the axis texture < 3°.

Studies of PFD films by REM showed that they consist of grains elongated in the growth direction with a transverse size of 2 to 8 µm on the surface of the layer. Film PFD possess two conical texture: <111> (with angle polarstar cone ~ equal to 8°) and <110> (~ 17.5°) with lamellar structure of the grains. The size of crystallites in doped and undoped boron films PFD depends on the temperature of their synthesis. So when the film thickness of 20 to 200 nm grains formed corresponding to the thickness of the layer.

In Figure 2 shows Raman spectra Policluster films of diamond (*a*) of thickness $h = 12$ µm AlN films (*b*) with a thickness of $h = 3.2$ µm. The prepared films were legionares Zn (curves 1 and 2). 1 Film contain additional phase Zn and ZnN_x (curves 1) with crystallinity $J < 10$ % vol.) and in the absence of additional phases (2) when $J \approx 40$ % vol. and undoped AlN film (3) has a crystallinity $J \approx 80$ % vol.

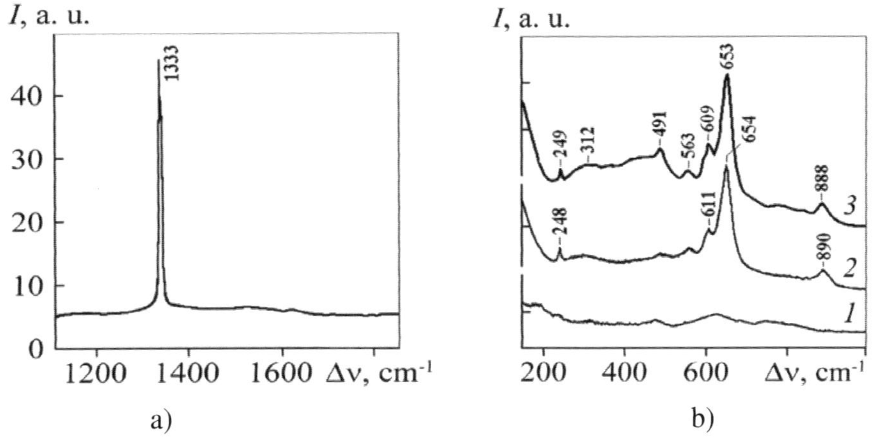

Figure 2. Raman spectra Policluster films of diamond (*a*) and AlN films (*b*).

In the Raman spectra (Figure 2*a*) doped films, the PFD has an intense diamond band at the KR Δv is equal to 1333 cm^{-1} with strip half-width $\Delta v_{1/2} = 3,4...4.5$ cm^{-1}. The intensity of the band diamond phase is much more intensity blurred bands of diamond-like carbon (~ 1550 cm^{-1}). View the Raman spectra of the AlN films depends on the structure of the structure (Figure 2*b*). Form, intensity and location of bands in the Raman spectra for AlN films obtained are substantially different under various technological regimes. The observed broadening of the bands ($\Delta v_{1/2}$> 100 cm^{-1}), characteristic for the crystal structure and the appearance of additional bands speaks about x-ray amorphous structure of the films. The periods of the lattice of undoped AlN films on 0.06…0.86 % higher than the theoretical value. In the adsorption layer of the formed film is observed in the interaction of ions N, Al and Zn, is formed in the high-frequency discharge plasma. The increase in the concentration of dopant in the films of AlN decreases the degree of crystallinity and at a concentration of ~ 2 at. % does not exceed ~ 15 % vol. Alloying impurity in the AlN film is in the form of individual atoms in crystalline and in amorphous phases. The X-ray amorphous inclusion Zn and Zn_3N_2 are located along the boundaries of fibers. The concentration of Zn in the AlN films depends on the ratio of the area of spray target areas of the plate Al and the impurity source of Zn, and the coefficients of their spraying.

By electron-beam probe spectral analysis to determine the content of Zn in the doped AlN films, which is > 3 wt %. According to the analysis of X-ray amorphous phase of AlN, the content of Zn is 0.72 wt %. The main share falls on the impurity amorphous Zn and ZnN_x. The periods of the lattice of the crystalline phase of AlN films: Zn in the degree of crystallinity of ~ 12 % vol. and concentration of Zn to ~ 2 wt %. less than the theoretical. Clearly defining the conductivity type are point defects interstitial impurity Zn. in the crystalline phase AlN.

The resulting films of PFD had p-type conductivity with the maximum concentration of boron atoms (up to ~ $4 \cdot 10^{19}$ cm^{-3}) with the activation energy of electrical conductivity of ~ 0.1 eV.

5. Film photodetector

In Figure 3 shows graphs of the voltage on the pads of layered nanostructures AlN:Zn/PFD:B upon irradiation with light in the UV range. Policluster films of diamond is obtained at temperatures: 1 – 1250 K; 2 – 1400 K.

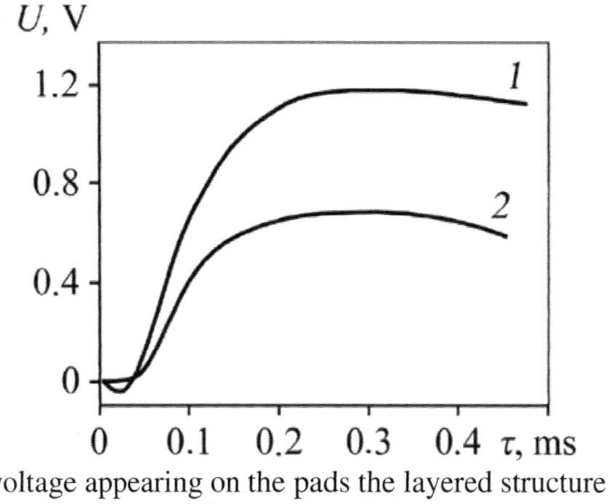

Figure 3. The voltage appearing on the pads the layered structure AlN:Zn/PFD:B during UV-irradiation.

During exposed to UV irradiation on the pads (layer semitransparent Ti) layered structure AlN:Zn/PFD:B, giving the voltage to 1.2 V (Figure 3). In the studied heterostructures, the AlN layer contained the total Zn concentration of ~ 0.16 at. % (the ratio of concentration of Zn in the amorphous and crystalline phase in the AlN film was of ~ 2.5 when the degree of crystallinity of about 40 % vol.). The concentration of boron in the PFD was estimated to be ~ 1019 cm^{-3}. AlN nanostructures:Zn/PFD:B in which the formation of the PFD layer was carried out at temperatures > 1300 K have a lower sensitivity when exposed to UV radiation.

6. Conclusions

The obtained layered planar structures Si/AlN/PFD on silicon substrates by sequential deposition in vacuum of layers of films of AlN and PFD by methods of HF-magnetron reactive sputtering and arc discharge. During the deposition of AlN films and the PFD have been alloying zinc and boron, respectively. Film structure is AlN/PFD during irradiating with light of the UV range generates an electrical signal of amplitude 1.2 V. The combination of Policluster films of diamond with films of other wide bandgap semiconductor materials in planar layered structures allows you to create new passive and active nanostructures for solid-state microelectronics.

Acknowledgments

This work was supported by the Ministry of Education and Science of the Russian Federation (№3.8074.2017/8.9 basic part of the State task).

References

[1] Belyanin A F, Bagdasaryan A S 2017 *Successes of modern radio electronics* 30–38 (in Russian)
[2] Sang L, Liao M, Sumiya M A 2013 *Sensors* **13** 10482–10518 doi:10.3390/s130810482
[3] BenMoussa A et al 2009 *Diamond and Related Materials* **18** 860–864 doi:10.1016/j.diamond.2008.11.013
[4] Teremetskaya I G et al 1995. *Applications of Diamond Films and Related Materials. Third International Conference. Gaithersburg* 469–472
[5] Piliptsou D G et al 2017 *IOP Conf. Ser.: Mater. Sci. Eng.* **168** 012103 DOI:10.1088/1757-899X/168/1/012103
[6] Borisov V V et al *2017 Nanomaterials and Nanostructures – XXI Century* **8** 18–25 (in Russian)
[7] Luchnikov P A et al 2017 *IOP Conf. Ser.: Mater. Sci. Eng.* **168** 012092 doi:10.1088/1757-899X/168/1/012092
[8] Luchnikov P A et al 2017 *IOP Conf. Ser.: Mater. Sci. Eng.* **189** 012014 doi:10.1088/1757-899X/189/1/012014
[9] Aleutdinov A D et al 2017 *IOP Conf. Ser.: Mater. Sci. Eng.* **81** 012069 doi: 10.1088/1757-899X/81/1/012069

IOP Conf. Series: Materials Science and Engineering **289** (2017) 012042 doi:10.1088/1757-899X/289/1/012042

The study of initial permeability temperature dependences for LiTiZn ferrite ceramics

A B Petrova[1], A L Astafyev[2], A V Ershov[3]
[1]Junior Researcher, Tomsk Polytechnic University, Tomsk, Russia
[2]Engineer, Tomsk Polytechnic University, Tomsk, Russia
[3]Undergraduate, Tomsk Polytechnic University, Tomsk, Russia
E-mail: abk9@tpu.ru

Abstract. Results of obtaining and analyzing the temperature dependences of initial permeability of ferrite ceramics are presented in the paper. It was shown that the level of the defective state of ferrite ceramics can be obtained from the value of two parameters α and β of the phenomenological expression describing the experimental dependences. The results showed that the main criterion of the defect state is the parameter β/α, which is related to the elastic stresses in the material. An indicator of the structure perfection is also the value of the maximum of the initial permeability near the Curie temperature.

1. Introduction

Ferrite ceramics of the lithium group are the materials with well-established magnetic properties [1–3]. At present, in the manufacture of ferrite products, there are various defectiveness controlling methods, based on measuring those properties of the material that appreciably react to changes in chemical composition. These are methods for measuring the true heat capacity, saturation magnetization, and magnetic permeability in the temperature interval including the Curie point, the initial paraprocess susceptibility. Such methods are laborious, selective to certain defects types and are low-sensitivity.

Nowadays microwave ferrites are widely used as magnetic materials for phase shifters, circulators, phased array elements and other microwave technology elements. In the process of their manufacture according to the classical ceramic technology, the probability of a residue of unreacted oxides and phases of the synthesis reaction intermediate products is high. Such defects, along with the material porosity, cause the deterioration of its magnetic characteristics [2].

Therefore, for ferrite ceramic products it is required to develop a modern highly sensitive magnetic method for controlling the structural perfection (defective state) of the resulting material. The base of this method can serve as a measure of the temperature dependence of the ferrite ceramics initial permeability. Such dependences are one of the most structurally sensitive and serve to determine the phase homogeneity of ferrites and to determine their Curie point. To optimize new technological processes, an operative control of the whole set of defects that affect the magnetic parameters of ferrite ceramics is necessary.

The use of structurally sensitive characteristics for studying the chemical inhomogeneity of ferrite materials has shown that quantitative comparisons of homogeneity and defect state of materials are possible only in the presence of approximating expressions for the controlled parameters temperature dependences. In work [4], a phenomenological expression for the temperature dependence of initial permeability $\mu_i(T)$ was obtained. In ferrite physics, the most of papers related to temperature studies of the initial permeability and susceptibility were devoted to the study of the impact of additives or regimes of sintering ceramics on the Curie point. In modern works devoted to the study and analysis of the temperature dependences of the ferrites magnetic properties, the impact of homogeneity on the drop sharpness in the temperature dependence of initial permeability near the Curie point was pointed out [5, 6].

The maximum value of the initial permeability – μ_{max}, was determined from the maximum of the temperature dependence μ_i.

Initial permeability has been found to be related with magnetization and ionic structure, thereby the thermal spectra of permeability can be also considered as a test of the formation and homogeneity of the ionic structure of the samples. Thepresentworkreportstheresultsof approbation of the method for estimating the defective state of ferrite ceramics on the basis of mathematical processing of the temperature dependence of initial permeability.

2. Experimental techniques

The sintering process and specimen preparation method were the same as reported in [7–10]. The specimens were manufactured using a standard ceramic production process from an industrially synthesized mixture (Russian designation of the grade “3ZSC-18”) by thermal sintering of the compacts in air in a laboratory resistance furnace at the temperature 1423 K for one hour. The mode of sintering was selected to follow that used in the industrial production of the 3ZSC-18 ferrite.

The compaction of toroidal shape samples was carried out using one-sided cold pressing with a hydraulic press PGr-10. The pressure was 200 MPa. Pressure impact time interval was three minutes.

The samples were sintered in laboratory conditions in air in a resistance furnace with a built-in thermostat. For approbation of the method for estimating the defective state of ferrite ceramics, samples were sintered in laboratory conditions from charge synthesized by ceramic technology. The sintering regime parameters was T_S = 1150 °C, t_s = 1 h. After sintering, the toroidal shape samples possessed the following dimensions: 18.5×14×2 mm.

In present work ceramic samples density and porosity were measured by hydrostatic weighing with high-precision analytical balances Shimadzu AUW 220D.

Electron micrographs were obtained using scanning electron microscope Hitachi TM-3000. The average grain size was calculated by the intercept method.

The saturation magnetization (M_s) was measured at room temperature with the vibrating sample magnetometer with the maximum field of 10 kOe.

The phase composition of the samples was controlled using an ARL X'TRA X-ray diffractometer (Switzerland) with a Peltier Si(Li) semiconductor detector and Cu$k_\alpha$radiation. XRD patterns were measured for 2θ = (10 – 80)° with a scan rate of 0.02 °/s. The phase compositionsof the examined samples were determined using the PDF-4 powder data base of the International Centre for Diffraction Data (ICDD). The XRD patterns were processed by the full-profile Rietveld analysis using the Powder Cell 2.5 software.

The reference material of the ferrite ceramics of grade 3CH-18 (SL-187) had the toroidal shape (outer diameter D = 20.5 mm, internal diameter d = 10.9 mm, height h = 3.5 mm). They were purchased at the Ferrit-Domen Research Institute for comparative studies of the temperature dependences of initial permeability. For reference material the following sintering regime was used: T_S=960 °C, t_s=8 hours.

For measuring the temperature dependences of initial permeability, an installation on the basis of an automatic bridge LCR-819 (digital meter L, C, R) and a special measuring cell with a built-in heater was constructed. The frequency of the magnetizing field was 10 kHz. The level of the test signal was 0.05 V.

The assembled device used a heating cell (Figure 1) of the original design. The measuring cell represents a stainless steel cylinder 1. At the top, the cylinder closes with a fluoroplastic lid 2 with a sample fixing device 3 and a thermocouple 4. The cover also had apertures for the output of the conductors 5 from the winding of the measured sample 7. In the lower part of the cylinder, to equalize the temperature in the sample area, was fixed a copper cup 11, at the bottom of which there is a heater

8. For increasing the heating uniformity, the upper part of the copper cup was closed with a aluminum heat shield. All the elements were located in an aluminum case 10, on which there are connectors for connecting the ground wire and the wires supplying the heating element. The case was thermally insulated from the cell heating parts by a powder of light-weight fireclay bricks 9.

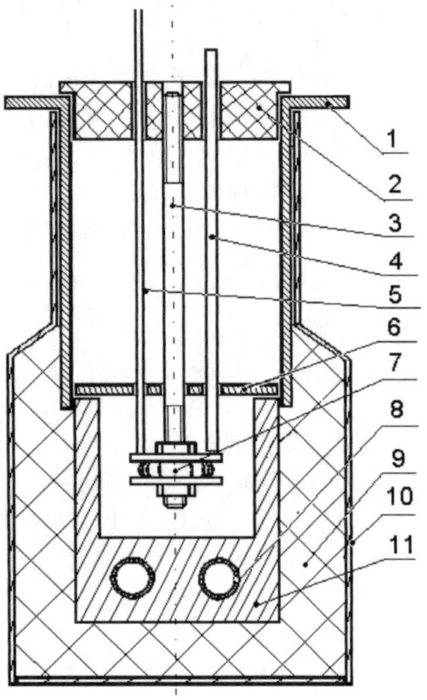

Figure 1. Measuring cell: stainless steel cylinder (1), fluoroplastic lid (2), lower electrode with sample fixing device (3), thermocouple (4), conductors (5), thermal screen (6), sample (7), heater (8), powder of light-weight fireclay bricks (9), aluminium case (10), copper cup (11).

A single-layer winding was wound on the sample from 46 turns of the MG TF-0.12 wire with fluoroplastic insulation, which retains its properties up to the upper limit of the measurements. The sample with the winding was fixed inside the measuring cell. The ends of the winding through the holes in the cover were taken out to the connection device of the automatic bridge (LCR-819). The heating of the measuring cell was made by a built-in heater. The current in measuring cell in was set by a thermocontroller. The thermostat was powered from the transformer. The temperature was controlled by a thermocouple of the chromel-alumel group. Thermocouple junction was pressed against the sample.

Toroid samples were used as the cores of the inductance coil. The inductance L was measured by a bridge method for an inductor with a core of a ferrite toroid. The measurement of L was carried out on a low-frequency meter LCR-819 in an electromagnetic field with a frequency of 1 kHz with an intensity of 0.1 Oe. A ferrite toroid with a winding from one layer of a thin copper wire with fluoroplastic insulation was placed in a special measuring cell with built-in heaters. The measurements were carried out with a slow cooling of the sample from a temperature 50 degrees higher than the Curie point, in order to exclude the influence of the prehistory on the initial permeability. The temperature was controlled by a thermocouple of the chromel-alumel group (type *K*). Thermocouple's measuring junction was pressed against the sample. Installation on the basis of an automatic bridge LCR-819 and a voltmeter V7-78/1 was constructed to measure the temperature dependence of initial permeability. The installation provided automated measurements of the inductance and temperature of the sample with a given polling rate. Thus, it allowed specifying a different number of measured points and thereby setting the measurement resolution. The latter condition was required to refine the complicated dependences μ_i (T) for samples with two or more magnetic phases, and for samples with Hopkinson peak. The data from the LCR-819 (L) meter and V7-7/1 (T) voltmeter were received as a text file on a PC using a specially written module of the LABView software package. The calculation of μi *(T)* and their mathematical processing was carried out using the Origin 9.

The initial permeability (μi) of the toroidal ferrite samples was calculated by using the Equation

$$\mu_i = \frac{L \cdot 10^7}{2 \cdot h \cdot N^2 \cdot \ln(D/d)} \qquad (1)$$

where L – inductance coil (H); N – number of turns in a coil;h – toroid height (m); D, d – outer and inner diameterof toroid (m).

The Curie point T_c was determined from the temperature dependence of μ_i by making a tangent to the curve of the rapid drop section of μ_i. The intersection of this tangent with the temperature axis determined the value of T_c. Also Tc was determined by mathematical processing $\mu_i(T)$.

3. Result and discussion

Figure 2 presents a microstructure of the sintered ferrite ceramics surface.

Figure 2. SEM micrograph of ferrite ceramics laboratory sample.

As can be seen from the Figure 2 ferrite is characterized by a typical polycrystalline structure with grains and grain boundaries. As can be seen from the Figure 2, the average grain size is in the range of 30–150 μm.

Samples sintered under laboratory conditions and those manufactured by Ferrit-Domen Co were used to test the method for estimating the defective state of ferrite ceramics. Their parameters are shown in Table 1.

Table 1. Ferrite samples parameters.

Sample	Density, g/cm³	Porosity, %	Grain size, μm	T_c, °C	M_s, emu/g
3CH-18, T_s=1150 C, t_s=1 h	4.29	4.5	20	277	36.1
SL-187, T_s=960 C, t_s=8 h	4.3	4.4	47	277	34.1

The phenomenological expression (2) was used to a mathematical analysis of the experimental temperature dependences.

$$\mu_i^* = \frac{1+\mathrm{x}}{1+N\cdot\mathrm{x}} \qquad \mathrm{x} = \left[\frac{\left(1-\frac{T}{T_c}\right)^{\delta}}{\alpha\left(1-\frac{T}{T_c}\right)^{\gamma}+\beta}\right]^{g} \qquad (2)$$

In the expression (2) the coefficients α, β, γ and δare determined by the relations: $\alpha = \frac{K_1(0)}{M_s^r(0)}$, $\beta = \frac{\lambda_s(0)}{M_s^r(0)} \cdot \sigma$, $\delta=(r-n)f$, $\gamma=(m-n)f$. In these relations σ is average level of elastic stresses.

With positive values of γ and δ, function (2) describes an asymmetric peak shape, distinctive for typical dependences $\mu_i(T)$.

Indicators g and r are set according to the selected magnetization mechanism. For example, for the model of rotation of magnetic moments (the case of single-domain particles) $g = 1$, $r = 2$. For the spherical bending model of 180 °domain walls fixed at grain boundaries and other structural defects $g = 2$, $r = 2$ (Smith and Wijn model), for cylindrical bending of domain walls (Kersten model) $g = 0.5$, $r = 3$. The Smith and Wijn model is most frequently used for polycrystalline ferrites [11].

X-ray diffraction data confirmed the formation of well-defined homogeneous single phase cubic spinel structure for all samples. Samples have similar values of T_c and M_s, density and porosity, but the average grain size for them is significantly different. The least-squares method (the Levenberg-Marquardt algorithm) was used to process the experimental temperature dependences of initial permeability in software package Origin 9.

Table 2 shows the results of mathematical processing for the temperature dependences of initial permeability of samples sintered both in laboratory conditions at T_s= 1150 ° C for 1 h (Figure 3) and in the factory conditions at T_s = 960 °C for 8 h (Figure 4). Also Table 2 presents the values of the maxima of the temperature dependences of initial permeability near the Curie point, μ_{max}.

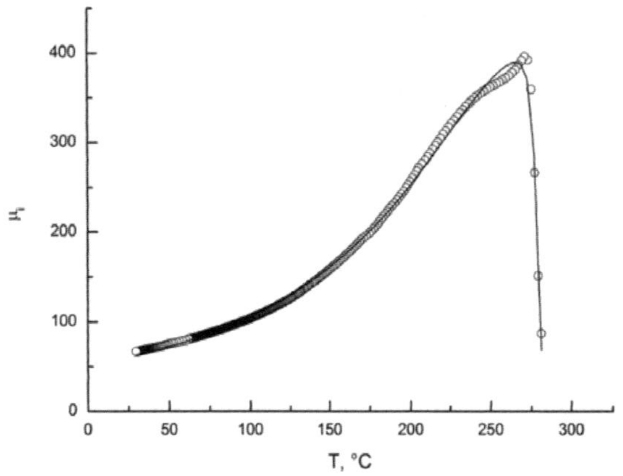

Figure 3. Temperature dependence of initial permeability for laboratory ferrite ceramics (symbols). The solid line is the calculated curve in expression 2 (g=2).

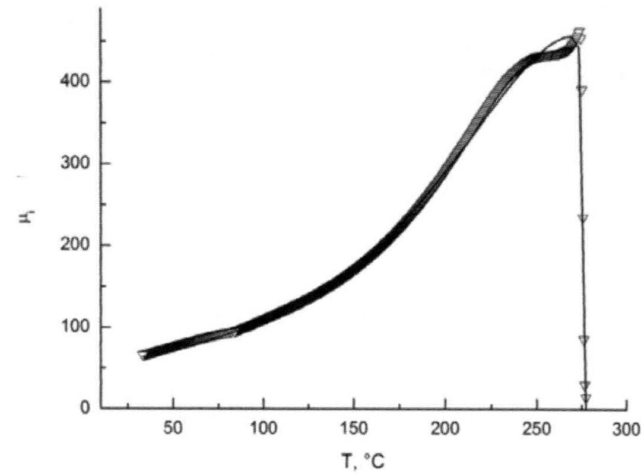

Figure 4. Temperature dependence of initial permeability for reference material SL-187.

Table 2. The fitting results.

Sintering regimes	α	β	N	T_c, °C	μ_{max}	β/α
T_s=1150 °C t=1 h	0.1368±5.45455E-4	1.96921E-6±1.19331E-6	0.00275±1.77083E-5	283±0.28558	395	2.3×10^{-6}
T_s=960 °C t=8 h	0.13425±0.0004	2.278E-6±9.0966E-7	0.00216±4.4101E-6	277.26748±0.0249	464	2×10^{-5}

Figure 3 shows the temperature dependences of initial permeability for laboratory ferrite ceramics and Figure 4 for reference material. The solid lines in these Figures are the calculated curve in expression (2). As can be seen from these Figures, the calculated curve describes rather well the experimental points, except for the Hopkinson peak.

The Hopkinson peak on the temperature dependence of initial permeability obtained for samples of both types (Figure 3 and Figure 4) reveals the effect of an increase in the initial permeability in the vicinity of the Curie point due to a sharp drop in the crystallographic magnetic anisotropy. Such behavior is due to the transition of magnetic particles (domains) to the superparamagnetic state, which is inherent in the single-domain state in the case when the domain size and the average grain size are the same, for example, in the case of nanoparticles [12–22].

It can be seen from Table 1 and Table 2 that the maximum temperature dependence of initial permeability (μ_{max}) is observed for the reference sample, in spite of the considerable porosity.Consequently, the most perfect structure is characteristic for reference samples. Perhaps this behavior is due to the large grain size of reference samples of ferrite ceramics. Herewith a larger number of magnetic domains being formed in the bulk of the grains, compared to laboratory samples.

4. Conclusions

A method for estimating the defective state of ferrite ceramics was developed and tested on the basis of measurements of the temperature dependence of initial permeability.

It is shown that the main criterion of the defect state is the β/α parameter of the phenomenological expression associated with the elastic stresses in the material.

The result of mathematical processing showed that the level of the defective state of the sample sintered in the laboratory is lower in comparison with the reference samples. An indicator of the perfection of the structure is also the value of the maximum of the initial permeability near the Curie point.

Acknowledgements

The research is funded from The Ministry of Education and Science of the Russian Federation in part of the "Science" program.

References

[1] Zheludev I S 1968 *Physics of Crystalline Dielectrics* (Moscow: Nauka) p 464 (in Russian)
[2] Rankis G Zh 1981 *The dynamics of the magnetization in polycrystalline ferrites* (Riga: Zinatne) p 187 (in Russian)
[3] Surzhikov A P et al 2013 *Russ. Phys. J.* **56** 681–685 doi:10.1007/s11182-013-0085-6
[4] Malyshev A V et al 2016 *Ceramics International* **42** 16180–16183 doi:10.1016/j.ceramint.2016.07.137
[5] Ghodake J S et al 2015 *J. Magn. Magn.Mater.* **378** 436–439 doi:10.1016/j.jmmm.2014.11.041
[6] Parvin R, Momin A A, Hossain A K M 2016 *J. Magn. Magn.Mater.* **401** 760–769 doi: 10.1016/j.jmmm.2015.10.130
[7] Lysenko E N et al 2015 *Journal of Thermal Analysis and Calorimetry* **122** 1347–1353 doi:10.1007/s10973-015-4849-9
[8] Surzhikov A P et al 2014 *Russ. Phys. J.* **57** 621–626 doi: 10.1007/s11182-014-0284-9
[9] Frangulyan T S et al 2006 *Russ. Phys. J.* **49** 506–510 doi: 10.1007/s11182-006-0133-6
[10] Surzhikov A P et al 2012 *Journal of Thermal Analysis and Calorimetry* **108** 1207–1212 doi:10.1007/s10973-011-1734-z
[11] Neel L 1963 *C. R. Acad. Sci.* **257** 2917–2932
[12] Slama J et al 2011 *Journal of electrical engineering* **62** 239–243 doi: 10.2478/v10187-011-0038-7
[13] Manikandan M et al 2015 *Phys. Status Solidi A* **212** 2179–2185 doi: 10.1002/pssa.201532229
[14] Chinnasamy C N et al 2002 *J. Magn. Magn.Mater.***238** 281–287 doi: 10.1016/S0304-8853(01)00928-3
[15] Hoque S M et al 2011 *Materials Sciences and Applications* **2** 1564–1571 doi: 0.4236/msa.2011.211209
[16] Hu J et al 2007 *Physica B* **400** 119–123 doi: 10.1016/j.physb.2007.06.026
[17] Ghodake J S et al 2015 *J. Magn. Magn.Mater.* **378** 436–439 doi: 10.1016/j.jmmm.2014.11.041
[18] Momin A A et al2017 *J. Magn. Magn.Mater.* **423** 124–132 doi: 10.1016/j.jmmm.2016.09.078
[19] Hoekstra B et al 1978 *Journal of Applied Physics* **49** 4902–4907 doi: 10.1063/1.325524
[20] Alten W J et al 1967 *J. Appl. Phys.* **38** 3023–3024
[21] Xu Z C 2003 *J. Appl. Phys.* **93** 4746–4749 doi: 10.1063/1.1562745
[22] Slama J et al 2011 *Journal of electrical engineering* **62** 239–243 doi: 10.2478/v10187-011-0038-7

IOP Conf. Series: Materials Science and Engineering **289** (2017) 012043 doi:10.1088/1757-899X/289/1/012043

Mechanochemical synthesis and physico–chemical investigations of new materials for gas sensors

E G Shubenkova
Associate Professor, Omsk State Technical University, Omsk, Russia
E-mail: shubenkova-e@mail.ru

Abstract. Solid solutions of the InSb-ZnTe semiconductor system containing up to 20 mol.% of ZnTe were synthesized for the first time. The role of mechanochemical treatment in the process of obtaining solid solutions of this system is shown. Solid solutions in the InSb-ZnTe system have been identified by Raman spectroscopy, and the optical properties of its components have been studied. On the basis of an analysis of the anti-stokes spectral radiation distribution the solid solutions formation was identified both on the dependence of the spectral distribution maximum's shift on the composition of the $InSb_{1-x}$-$ZnTe_x$ system, and by estimating the radiation intensity of the initial binary semiconductors at frequencies corresponding to the LO- and TO- vibrations of the binary compounds crystal lattice. The values of the band gap for InSb, $(InSb)_{0.95}(ZnTe)_{0.05}$ and $(InSb)_{0.9}(ZnTe)_{0.1}$ were calculated, their values were 0.22 eV, 0.30 eV and 0.38 eV, respectively.

1. Introduction

Interest in solid solutions based on indium antimonide and zinc telluride is primarily associated with their unique electron-optical properties, which makes it possible to obtain promising materials with controlled electronic and other physicochemical properties.

Using the A^3B^5, A^2B^6 type compounds and solid solutions based on them as components of the sensor element provides a sensitivity to a broad range of toxic and chemically-active gases in the environment and gas emissions [1–6]. The basis of semiconductor sensors are thin layers (250 nanometers) InSb, ZnTe and their solid solutions deposited on appropriate substrates. Selection of sensitive material for semiconductor sensors represents obtain samples of different composition, and studying the sensitivity of their surface in adsorption different gases [4]. The main difficulty in obtaining and investigating solid solutions is the duration and laboriousness of the isothermal diffusion method used in their synthesis. Therefore, the development of new approaches to the complex multicomponent solid solutions's synthesis is a needed task.

In work solid solutions were synthesized from mixtures of initial components (InSb and ZnTe powders), taken in certain molar ratios. Preliminary mechanochemical treatment of the binary semiconductors starting mixtures preceded the process of their alloy. Mechanochemical treatment contributes to the emergence of new active sites on the freshly formed solids surface and to acceleration of diffusion processes. As a result, it allows to increase the reactivity of substances and to intensify the course of solid-phase reactions [7].

The optical properties of the synthesized samples were studied whith using the Raman spectroscopy method (Raman scattering method). Based on the frequencies's values of the crystal lattice vibration, both the binary components of the four-component semiconductor system and the obtained solid solutions were identified. In work the band gap was calculated both for the initial

IOP Conf. Series: Materials Science and Engineering **289** (2017) 012043 doi:10.1088/1757-899X/289/1/012043

compounds and for solid solutions. In addition, Raman spectroscopy method provides the possibility of obtaining information not only about the states within the forbidden band, which is especially important in the case of doping, but also the types of atoms and their coordination environment, which determines their reactivity [8, 9].

2. Experimental

Optical studies of the initial binary compounds and synthesized samples of solid solutions were carried out by Raman spectroscopy method on the RFS-100 Fourier spectrometer, with a resolution of 1 cm^{-1}. The samples were excited by laser radiation at room temperature (yttrium-aluminum garnet with neodymium YAG:Nd^{3+} ($Y_3Al_5O_{12}$:Nd^{3+}), λ = 1064 nm). The obtained data were analyzed in the anti-stokes scattering region in the observation range 0 – 4000 cm^{-1} and in a narrow range of crystal lattice frequencies (0 – 400 cm^{-1}).

On the basis of an analysis of the anti-stokes spectral radiation distribution the solid solutions formation was identified both on the dependence of the spectral distribution maximum's shift on the composition of the $InSb_{1-x}$-$ZnTe_x$ system, and by estimating the initial binary semiconductors radiation intensity at frequencies corresponding to the LO- and TO- vibrations of the binary compounds crystal lattice.

The experimental data, their reproducibility and accuracy was monitored by parallel measurements, and the use of computational algorithms of mathematical statistics.

3. Results and considerations

The results of the spectral distribution of the anti-stokes and stokes radiation of the four-component solid solutions system $InSb_{1-x}$-$ZnTe_x$, exposed in air, are shown in Figures 1 and 2.

The obtained dependences show that in the spectra of the starting compounds in the anti-stokes region there are narrow peaks corresponding to the LO and TO frequencies of their oscillations crystal lattice: 180 cm^{-1} and 191 cm^{-1} for indium antimonide; 206 cm^{-1} (LO- component) and 177 cm^{-1} (TO-component) for zinc telluride.

Analysis of the spectra presented above showed, that the peak, defined as the peak of the interband recombination (corresponding to the transition "conduction band - valence band"), is present both in the spectrum of indium antimonide and in the spectra of synthesized solid solutions, and is fixed in the frequency range 2500 – 3300 cm^{-1} [9]. The difference between solid solutions and starting compounds is manifested in its intensity as well as the position of the maximum, which shifts to higher frequencies with increasing zinc telluride content.

Strong exciton absorption is characteristic of A^2B^6 compounds [8], therefore the certain amount of ZnTe contained in a InSb at the formation of solid solutions causes the observed luminescence effect in the anti-stokes region. As mentioned above, it is due to, firstly, to the appearance of impurity levels in the band gap of indium antimonide by reason of the doping, and, secondly, due to interband transitions.

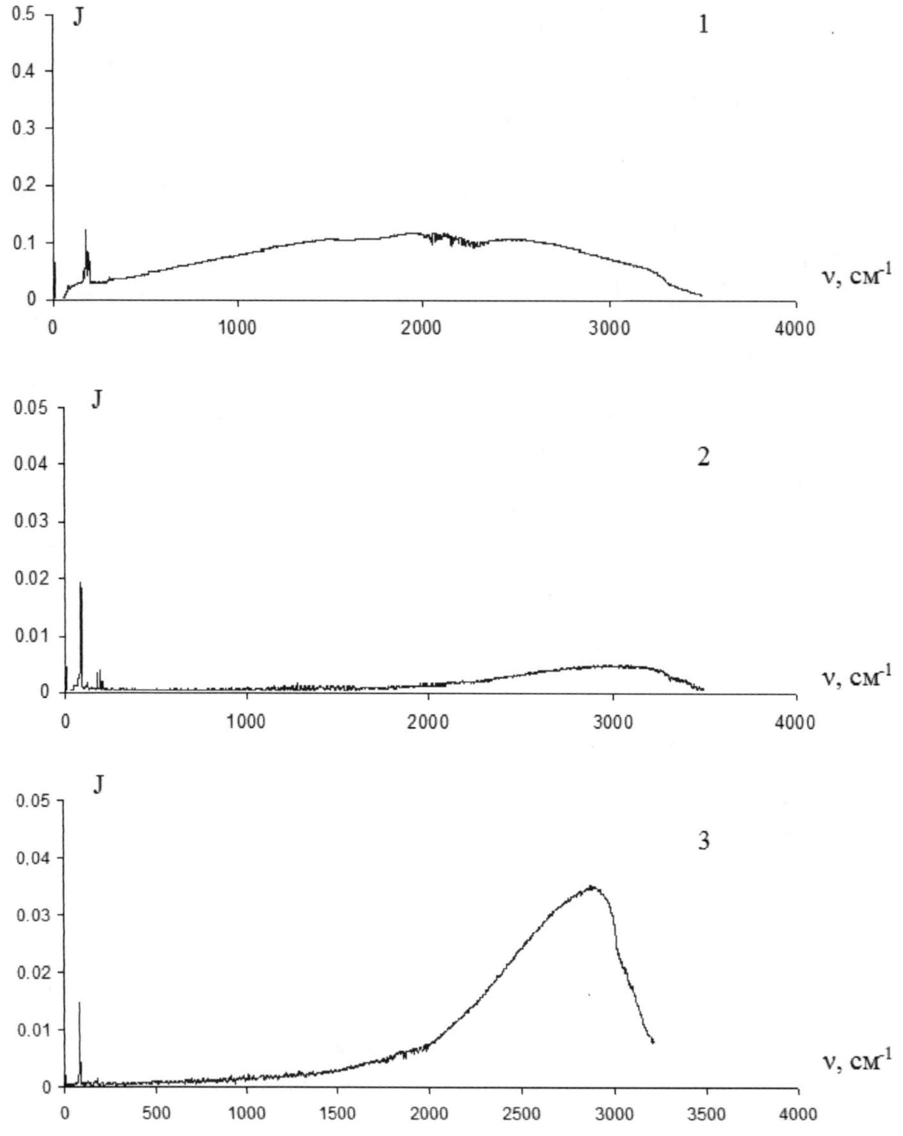

Figure 1. Raman spectroscopy: 1 – InSb; 2 – $InSb_{1-x}$-$ZnTe_x$ - mechanical mixture; 3 – $InSb_{1-x}$-$ZnTe_x$ – solid solution, where $x = 0.05$ (5 mol.%).

In addition, analysis of Raman spectra (Figure 1) made it possible to establish that at the stage of mechanochemical treatment of the initial mixture of binary components InSb and ZnTe, partial formation of solid solutions is occurred. This is confirmed by the optical vibrations of initial components lattices are presented not only in the spectrum of the mechanical mixture, in addition, in the spectrum of the mechanoactivated mixture $InSb_{1-x}$-$ZnTe_x$ (where $x = 0.05$), as well as in the spectrum of a solid solution with the same concentration of ZnTe. The presence in spectrum of mechanoactivated mixture $InSb_{1-x}$-$ZnTe_x$ (where $x = 0.05$) in the range 2500 – 3500 cm^{-1} a broad band of low intensity indicates that the synthesis of solid solutions at the stage of mechanochemical treatment occurred in part.

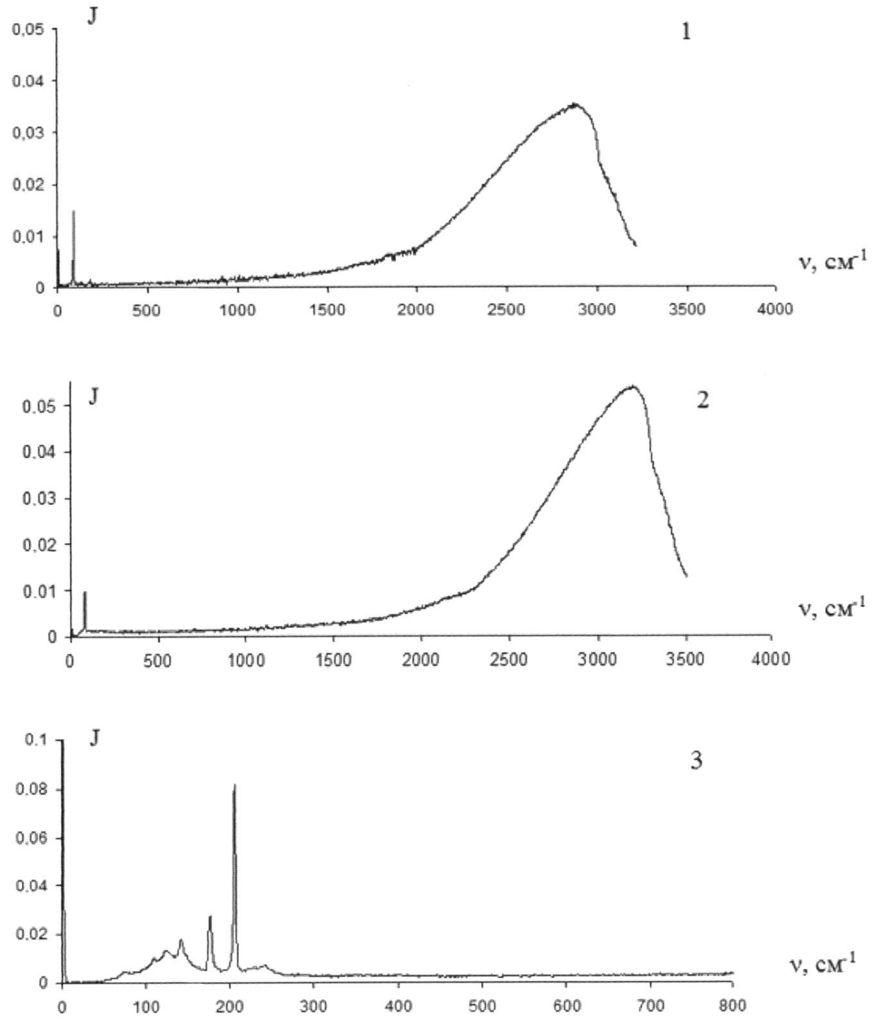

Figure 2. Raman spectroscopy: 1 – $InSb_{1-x}$-$ZnTe_x$ – solid solution, where $x = 0.05$ (5 mol.%); 2 – $InSb_{1-x}$-$ZnTe_x$ – solid solution, where x = 0.10 (10 mol.%); 3 – ZnTe.

The mean energies corresponding to the interband transitions were calculated on the basis of Raman spectra. The calculated values correspond to the band gap width and make up 0.22 eV, 0.30 eV and 0.38 eV for InSb, $(InSb)_{0.95}(ZnTe)_{0.05}$ and $(InSb)_{0.9}(ZnTe)_{0.1}$ respectively and vary linearly depending on the solid solutions composition. When forming unbroken row of substitutional solid solutions, the most of the parameters are linear function of composition, and that fact confirms the formation in $InSb_{1-x}$-$ZnTe_x$ solid solutions of cation-anion complexes distributed uniformly [1, 2, 8, 9].

4. Summary

Solid solutions of the InSb-ZnTe semiconductor system containing up to 20 mol.% of ZnTe were synthesized for the first time. The role of mechanochemical treatment in the process of obtaining solid solutions of this system is shown. Solid solutions in the InSb-ZnTe system have been identified by Raman spectroscopy (Raman scattering method), and the optical properties of its components have been studied.

Further study of the semiconductor multicomponent solid solutions surface properties in real conditions will allow us to identify samples that have the greatest sensitivity to toxic microimpurities and to develop recommendations for their use as sensitive elements of gas sensors and sensory arrays.

References

[1] Kirovskaya I A, Shubenkova E G 2009 *Russian Journal of Physical Chemistry A* **83** 2322–2330 doi: 10.1134/S003602440913024X
[2] Kirovskaya I A et al 2010 *Russian Journal of Physical Chemistry A* **84** 661–667 doi: 10.1134/S0036024410040242
[3] Cattrall R W 1997 *Chemical Sensors* (New York: Oxford University Press Inc.)
[4] Shubenkova E G 2017 *IOP Conference Series: Materials Science and Engineering* **189** 012009 doi: 10.1088/1757-899X/189/1/012009
[5] Yamazoe N et al 1996 *Materials Science and Engineering: B* **41** 178–181 doi: 10.1016/s0921-5107(96)01648-0
[6] Dai Z R et al 2003 *Advanced Functional Materials* 9–24
[7] Boldyrev V V 1996 *In Reaktivity of solids. Past, present and future* (Oxford: Blakwell)
[8] Yu Peter Y Cardona Manuel 2010 *Fundamentals of Semiconductors: Physics and Materials Properties* (Berlin Heidelberg: Springer-Verlag)
[9] Yeliseyev A A Lukashin A V 2010 *Functional nanomaterials* (Moscow: FIZMATLIT)

IOP Conf. Series: Materials Science and Engineering **289** (2017) 012044 doi:10.1088/1757-899X/289/1/012044

New microfocus bremsstrahlung source based on betatron B-18 for high-resolution radiography and tomography

M M Rychkov[1], V V Kaplin[2], E L Malikov[3], V A Smolyanskiy[4], I B Stepanov[5], A S Lutsenko[4], V Gentsel'man[4], I K Vas'kovskiy[4]

[1]Head of Laboratory, National Research Tomsk Polytechnic University, Tomsk, Russia

[2]Senior Research Fellow, National Research Tomsk Polytechnic University, Tomsk, Russia

[3]Researcher, National Research Tomsk Polytechnic University, Tomsk, Russia

[4]Engineer, National Research Tomsk Polytechnic University, Tomsk, Russia

[5]Deputy Director for Development, National Research Tomsk Polytechnic University, Tomsk, Russia

E-mail: rychkov@tpu.ru

Abstract. New microfocus source of hard bremsstrahlung (photon energy > 1 MeV), based on the betatron B-18 with a narrow Ta target inside, for high-resolution radiography and tomography is presented. The first studies of the source demonstrate its possibilities for practical applications to detect the microdefects in products made from heavy materials and to control gaps in joints of parts of composite structures of engineering facilities. The radiography method was used to investigate a compound object consisting of four vertically arranged steel bars between which surfaces were exposed gaps of 10 μm in width. The radiographic image of the object, obtained with a magnification of 2.4, illustrates the good sensitivity of detecting the gaps between adjacent bars, due to the small width of the linear focus of the bremsstrahlung source.

1. Introduction

The production of a microfocus radiation source based on relativistic electron beams is important for high resolution radiography and tomography. Ordinary betatrons generating secondary hard radiation caused by interaction of the internal electron beam with the target (typically a thick target), that is larger in its area than the cross section of the millimeter-size beam, are used for obtaining the images of a number of objects. But, in [1, 2] the idea was proposed to use internal target much smaller than diameter of the electron beam of the cyclic accelerator to reduce the focal spot of the generated Bremstrahlung. Here, if the beam will circulate for a sufficiently long time on the radius of the micro-target location, then, due to betatron oscillations, electrons will fall on such a target with a sufficiently high efficiency.

This paper presents the results obtained for generation of linear-microfocus Bremsstrahlung under interaction of 18 MeV electrons with thin target which was oriented along the direction of the internal beam of the B–18 betatron in order that the electrons can interact with the narrow front face of the target. Magnified images of the compound steel object have been obtained using the radiations generated in the narrow internal Ta targets, the width of which is approximately 100 times smaller than the diameter of the electron beam. The formation of the object structure image with participation of the absorption and phase contrast effects is shown.

2. Experimental setup

The experimental setup is shown in Figure 1. A thin Ta target was placed on a thin vertical goniometer holder inside the equilibrium orbit of the accelerated electrons, Figure 1*a*. Additional magnetic field generated by the dump coil within 30 μs reduced the orbit radius and the electrons fell onto the internal target. The generated radiation escaped through the 50 μm Mylar window of the experimental chamber, passed through an investigating object and fell on the Agfa D-4 X-ray film placed at a distances of 114 cm from the target, Figure 1*b*.

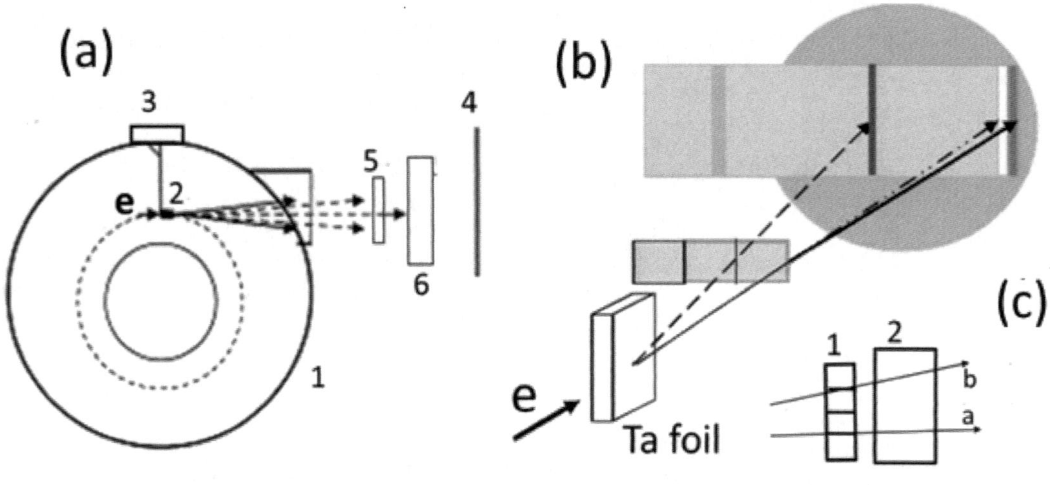

Figure 1. The experimental scheme: *a)* 1 – betatron chamber, 2 – narrow target, 3 – internal goniometer, 4 – X-ray film, 5 – object of investigation, 6 – steel-plate X-rays absorber; *b)* Locations of narrow target on the electron beam and object on radiation beam, and also the scheme of object radiogram; *c)* 1 – steel object (view from above), 2 – steel-plate X-rays absorber.

The 13-μm-thick Ta target with the vertical $L = 5$ mm and horizontal $T = 4$ mm (along the direction of the electron beam) sizes were used in the experiments. The object of investigation was installed on an external goniometer at a distance of 48 cm from the target. The steel plate could be placed at a distance of 24 cm from the object in order to absorb the X-rays (photon energy $E_\gamma < 1$ MeV). The object photographs obtained using hard bremsstrahlung ($E_\gamma > 1$ MeV gamma-rays), were processed with the scanner to obtain negative and positive images of the object structure for subsequent analysis.

3. Experimental results

Figure 2 shows the photograph of angular distribution of the radiation generated by 18 MeV electrons in the 13 μm Ta target oriented at angle $\theta_o = 0.3°$ with respect to the electron beam. The angular distribution of radiation generated in the Ta target differs significantly from the distribution obtained using the 50 μm Si target [3]. In [3] the radiation beam photographs were formed mainly by 5 – 50 keV X-rays. Therefore, in that case, the investigation of absorption contrast and size of the radiation source focus was carried out using the *Duplex IQI* microstructure (thin wires) [4].

In our case of Ta target, the radiation photograph is formed by hard radiation, since the soft part of the generated spectrum is suppressed by absorption in the heavy target material. This is evidenced by the presence of the image of the flange of the exit window of betatron chamber, which were not detected using the softer photons generated in the Si target, because of their less penetrating power. The *Duplex IQI* can be used effectively with radiation energies up to 400 keV. When using energies in the MeV range the results may not be completely satisfactory. Therefore, in order to investigate the

IOP Conf. Series: Materials Science and Engineering **289** (2017) 012044 doi:10.1088/1757-899X/289/1/012044

formation of the image contrast by hard bremsstrahlung we used a compound steel object consisted of four bars, Figure 1*c* (1), with dimensions of 9×9×35 mm^3 with the 10-μm-width gaps between their polished surfaces. The gaps were as a model of inaccuracy of the assembly of the object, which should be detected by radiography using the created source.

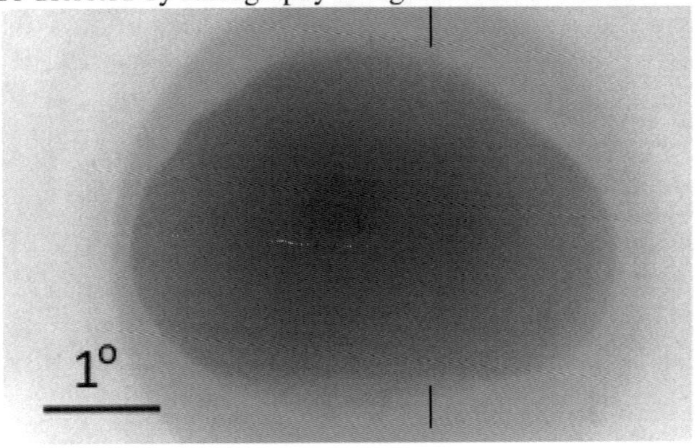

Figure 2. Photographs of angular distribution of X-rays at 13 μm Ta target orientation θ_o = 0.3° with respect to the direction of electron beam. Vertical marks show the projection of target plane on X-ray film.

The image of the object with three 10-μm gaps between the bars was obtained with a magnification of 2.4 using hard bremsstrahlung (Eγ > 1 MeV) generated in the Ta target. There exists also the image of the 0.5 mm thick steel plate attached to the object. The negative (a) and positive (b) images of the object presented in Figure 3*a*, *b* show good enough resolution of the gaps. It can be also seen that the gap image width increases from the right gap to the left one. This increase of the gap image is due to the different orientation of the bar surfaces forming the gaps, relative to the axis of radiation cone.

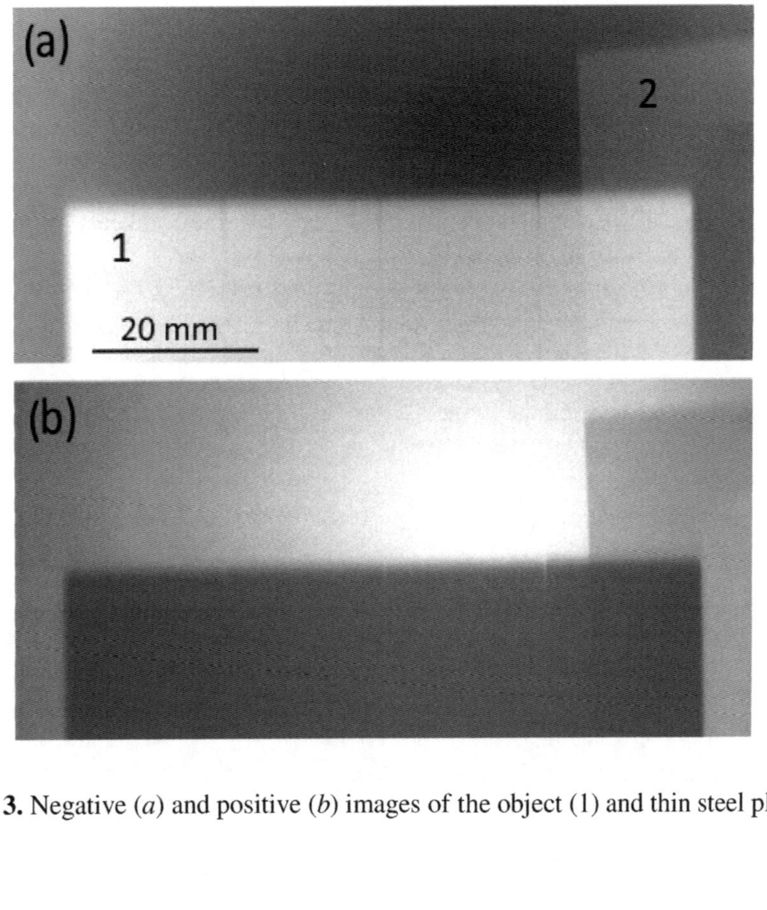

Figure 3. Negative (*a*) and positive (*b*) images of the object (1) and thin steel plate (2).

The object was oriented so that the surfaces of the steel bars, which formed the right gap was parallel to the part of gamma-rays that passed along these surfaces, Figure 1*c* (arrow a). The surfaces of the middle and left gaps have the angles of inclination (for example, arrow b in Figure 1*c*), relative to the directions of the gamma-rays $\theta_z = 1°$ and 2°, respectively. Therefore, one obtains the images of the projections of these gaps that formed by gamma rays passing at the angles θ_z to their surfaces (for example, arrow b in Figure 1*c*). This effect is clearly seen in Figure 4 where the gap images additionally increased are shown.

Figure 4. The increased images of left (*a*), middle (*b*) and right (*c*) gaps.

The width S_z of gap image can be estimated as $S_z = L_2\,((t_z + l\bullet\theta_H) + S_H)\,/\,L_1 - S_H$, where t_z and l are the width and length of the gap, L_1 and L_2 are the distances from the target to the sample and up to the X-ray film, θ_H is the horizontal angle of emission of radiation from the target in the direction of the gap and $S_H = (t + T\bullet\theta_H)$ is effective horizontal size of the source; t and T – target thickness and length along electron beam. It is seen that S_z increases with increasing the angle θ_H that is equal, in our case, to the value of the angle θ_z.

In addition, the images in Figure 3 show an enhanced contrast of the images of the right edge of the object and the left edge of thin steel plate. This is due to the phase contrast effect [5], since the source size in this case is close to 13 µm. The image of the left edge of the object is blurred because the image is formed by radiation emitted from the target at an angle of about $\theta_H = 3°$ to the surface of the target. Probably, in this case the phase contrast effect is not realized because of the large effective size of the S_H radiation source, which is defined above and equal, in this case, $S_H = 222$ µm. In order to achieve best resolution of the defects or media interfaces in the images formed by an absorption or phase contrast, it is necessary to orient them along the direction of the plane of the internal target. The dependence of effective size S_H of a source extended along the electron beam on the position of the object in the radiation cone was investigated in [3].

Vertical effective size S_v of the source is equal to the diameter D_e of the electron beam because is practically independent on the vertical emission angle θ_V, since the angular width of the emission cone is several degrees. In our case, $D_e = 1.48$ mm, therefore the images of the object and thin steel plat upper edges are blurred due to a large vertical size of the radiation source.

Figure 5 shows the images additionally increased of middle and right gaps without (*a*) and with the additional steel plate, (2) in Figure 1*c*, with the thickness of 20 mm (*b*) or 40 mm (*c*).

Figure 5. Images of middle and right gaps without (*a*) and with 20 (*b*) and 40 mm (*c*) steel plate.

The steel plates were additionally used as X-ray absorbers, which increased the fraction of radiation in the gamma-ray region of the spectrum of bremsstrahlung. This also simulated a situation where the gaps are in the steel bulk. It is seen that at increasing the thickness of additional steel plate up to 40 mm one can also detect the 10-μm-width gaps between the steel bars of the compound object.

4. Conclusion

The study has shown the possibility to successfully generate hard radiation in a narrow target which width is about one hundred times smaller than the diameter of the betatron electron beam, and to use this radiation for obtaining the magnified high resolution images of micro-defects into products made of heavy materials with participation both the absorption and phase contrast effects in formation of the images.

In our case, the radiation spectrum of the betatron generated in narrow internal targets extends from several keV to 18 MeV. For light targets, the images of non-thick objects are formed by a soft part of the radiation spectrum [3]. In our case of heavy target, the radiation spectrum is dominated by hard radiation due to strong absorption of radiation of the soft part of the bremsstrahlung spectrum in the target. The radiation generated in such target is applicable for obtaining images of thick objects made from heavy materials.

The images of a compound object consisted of four steel bars demonstrated the high resolution of a series of 10 μm gaps between adjacent bars due to the small horizontal size of the focal spot of the linear microfocus bremsstrahlung source. The results also demonstrate the edge phase contrast due to the high degree of spatial coherence of the radiation.

The obtained results attest to the high quality of the radiation beam generated by new microfocus source based on compact betatron that can also be used in a laboratory physical experiment, for example, in materials science to study internal interfaces of media, microdefects and microinclusions in the heavy composite materials. In our case, the 18-MeV betatron-based linear microfocus source generates bremsstrahlung with a spectrum up to the electron energy, while the microfocus X-ray tubes widely used for various purposes have so far reached the photon energy of 750 keV [6].

Acknowledgements
This work is supported by the Russian Science Foundation, project № 17-19–01217.

References

[1] Pushin V S, Chakhlov V L 1997 A method for producing bremsstrahlung focal spot of small dimensions in a cyclic accelerator of charged particles *Patent RU* 2072643 (in Russian)
[2] Kasyanov V A, Mikhalchuk A A, Pushin V S, Romanov V V, Safronov A S, Chakhlov V L, Matte M M 1998 *Devices and experimental procedure* **1** 41
[3] Rychkov M M, Kaplin V V, Sukharnikov K, Vaskovskii I K 2016 *JETP Lett* **103** 723–727 doi: 10.1134/S0021364016110114
[4] The Computerised Information Technology Ltd [Electronic resource] URL: http://www.cituk-online.com/acatalog/
[5] Wilkins S V, Gureyev T E, Gao D, Pogany A, Stevenson A W 1996 *Nature* **384** 335–338
[6] The Nikon Metrology X-ray Products [Electronic resource] URL: http://x-sight.co.za/products/54/750-kv-microfocus-x-ray-source/